REVIEWS IN
ECONOMIC GEOLOGY

(ISSN 0741–0123) Volume 1

FLUID–MINERAL EQUILIBRIA IN HYDROTHERMAL SYSTEMS

ISBN 0-9613074-0-4

The Authors:

R. W. HENLEY
Chemistry Division
D.S.I.R.
Geothermal Research Centre
Private Bag
Taupo, New Zealand

A. H. TRUESDELL
M.S. 910
U.S. Geological Survey
345 Middlefield Road
Menlo Park, CA 94025

P. B. BARTON, JR.
M.S. 959
U.S. Geological Survey
Reston, VA 22092

J. A. WHITNEY
Department of Geology
University of Georgia
Athens, GA 30602

Series Editor: JAMES M. ROBERTSON
New Mexico Bureau of Mines & Mineral Resources
Campus Station
Socorro, NM 87801

SOCIETY OF ECONOMIC GEOLOGISTS

Frontispiece The Great Geyser and Strokkur erupting simultaneously

FLUID-MINERAL EQUILIBRIA IN HYDROTHERMAL SYSTEMS

Table of Contents

Chapter 3

CHEMICAL GEOTHERMOMETERS FOR GEOTHERMAL EXPLORATION
(AHT)

Chapter 4

GASEOUS COMPONENTS IN GEOTHERMAL PROCESSES
(RWH)

Chapter 5

MORE MILEAGE FROM YOUR GAS ANALYSES: THE GAS GEOTHERMOMETERS
(AHT)

Chapter 6

HYDROLYSIS REACTIONS IN HYDROTHERMAL FLUIDS
(RWH)

Chapter 7

pH CALCULATIONS FOR HYDROTHERMAL FLUIDS
(RWH)

Chapter 8

REDOX REACTIONS IN HYDROTHERMAL FLUIDS
(PBB)

Chapter 9

METALS IN HYDROTHERMAL FLUIDS
(RWH)

Chapter 10

STABLE ISOTOPES IN HYDROTHERMAL SYSTEMS
(AHT)

Chapter 11

AQUIFER BOILING AND EXCESS ENTHALPY WELLS
(RWH)

Chapter 12

VOLATILES IN MAGMATIC SYSTEMS
(JAW)

Chapter 13

HIGH TEMPERATURE CALCULATIONS IN GEOTHERMAL DEVELOPMENT
(RWH)

Chapter 14

HIGH TEMPERATURE CALCULATIONS APPLIED TO ORE DEPOSITS
(PBB)

FOREWORD

In June 1982, the Society of Economic Geologists (SEG) Council appointed an Ad Hoc Committee on SEG Short Courses to develop a set of recommendations for initiating and operating a new short course series. That committee's report, representing a major and sustained effort on the part of a number of individuals, but especially Bill Kelly and Phil Bethke, was formally presented in February 1983 in Atlanta. Included in the report was the suggestion that a printed set of "short course notes" be produced as part of each course offering.

In November 1983, the Society of Economic Geologists sponsored its first official short course -- on Fluid-Mineral Equilibria in Hydrothermal Systems -- held prior to the annual meetings of the Geological Society of America and Associated Societies in Indianapolis, Indiana. Primary course organizers and lecturers were R. W. Henley, A. H. Truesdell, P. B. Barton, Jr., and J. A. Whitney. These individuals produced a text for the course that has become the first volume of Reviews in Economic Geology, the newest publishing venture of the SEG.

Reviews in Economic Geology is designed to accompany the Society's short course series. Present plans call for a volume to be produced annually in conjunction with each new short course. A volume will first serve as a textbook for the short course, and subsequently will be available to SEG members and others at a modest cost.

Normally, a volume will appear in its final, published form in time for its related course. This first time, however, the Indianapolis short course was supplemented by a "pre-production" model of Volume 1. What follows here is a substantially revised and, we hope, improved version of that book; it incorporates course participants' suggestions and a modicum of organizational changes and standardization on the part of the Series Editor.

In addition to the authors, significant contributors to the production of this volume include: Dan Hayba of the U.S. Geological Survey - Reston, who convinced his word processor to speak clearly to mine, a feat of no small proportions; Jim Cheek of the University of New Mexico Printing Plant, whose artistic sense and layout experience were instrumental in creating the cover design; Lynne McNeil of the New Mexico Bureau of Mines and Mineral Resources, whose professionalism with a word processor, patience with the Series Editor, and continued good humor made the final editing and formating much less of a chore than it might have been. Lastly, I gratefully acknowledge the support and encouragement of Frank Kottlowski, Director of the New Mexico Bureau of Mines and Mineral Resources, whose perception of his staffs' professional obligations and personal opportunities gives each of us just enough rope.

James M. Robertson
Series Editor
Socorro, NM

PREFACE

The approach to hydrothermal ore deposits through geothermal chemistry benefits from the detailed physical, chemical and isotopic data now available from the exploration and exploitation of a large number of active geothermal systems. These provide an understanding of the processes common to both geothermal and hydrothermal ore deposits.

The present text has arisen from a number of seminars on geothermal chemistry presented in recent years. These include the Geothermal Institute (University of Auckland, New Zealand), Geothermal Resources Council (San Francisco, 1982), the Society of Economic Geologists Short Course (Indianapolis, 1983) and shorter courses in Mexico and Switzerland. These courses focused on applications in the geothermal industry but encompassed applications to epithermal ore deposits. A chapter contributed by Jim Whitney deals with the chemistry of magmatic gases in order to extend the quantitative approach of this text to a wider range of hydrothermal environments.

The present text is very much a combined effort, but for the convenience of the reader we have indicated by initials, in the table of contents, the principal author of each chapter.

Objectives

The text is designed as a practical but informal guide to the more frequently encountered hydrothermal calculations in common use today and not as a thermodynamic text. We hope that it may help to dispel some of the mystery surrounding such calculation procedures and make them available to a larger number of practicing geothermal scientists, economic geologists and geochemists. We also hope that it may help to bridge a growing information gap between applied geothermal science and research into the origin of hydrothermal ore deposits.

Format

The text is designed primarily for the practising geothermal scientist or economic geologist as a self-help guide but the format is such that it may in total or in part form the basis of university undergraduate or graduate courses. Only an elementary knowledge of thermodynamics and physical chemistry is assumed with references to the standard literature given for background reading. The text also provides a literature resource and compilation of commonly used data and equations. Of necessity much of the descriptive matter is highly condensed and is not intended as a substitute for reading the available literature on geothermal systems and hydrothermal ore deposits.

Since many readers will be unfamiliar with geothermal phenomena, we have included a few illustrative plates, in many cases with a historical perspective provided in the caption.

Each chapter has been designed as a separate entity cross-referenced to others to maintain continuity. If used in graduate or undergraduate courses, a basic geochemistry and thermodynamics course will be a prerequisite. Each chapter of this text will require some 3-4 hours teaching laboratory time <u>and</u> time for assignments based on the text. The material in Chapter 2 is unfamiliar to most earth scientists and may require an introductory lecture; particular hurdles are often encountered in the use of Steam Tables and the concept of the steam or water fraction.

The calculations contained in this book are not difficult but can be arduous without calculator assistance. In particular iterative calculations are frequently used and many of the calculations are required routinely. For these applications a programmable calculator is highly desireable and in order to present programs for the more involved calculations we have standardized on the Hewlett Packard 41C series. Other programmable calculators and of course (at higher cost) computers would serve as well.

Some illustrative programs for the HP41C are given in Appendix I. Readers will develop individual programing styles and applications during the course of study and are encouraged to develop their own versions of these programs. One of the most useful programs which may be applied to sets of field or experimental data is the curve-fitting program provided in the HP41C Standard Applications Manual or Applications Modules. It is reproduced in Appendix I by courtesy of Hewlett-Packard.

Students should be encouraged to write their own calculator programs but class use of the illustrative programs provided in Appendix I allows rapid progression to the essential conclusions without diversion by programming problems. These programs may be rapidly provided to a class using an HP41C light wand and bar codes.

Team problem solving is a useful method of rapidly achieving good class results.

Answers to some of the more intricate problems and/or comments on their solution are provided in Appendix II. Such problems are identified in the text by the symbol $ in the adjacent left margin.

Figure and plate credits are given in the back of the book.

Acknowledgements

In completing this guide to hydrothermal chemistry a number of acknowledgements are due. The authors' approaches to the chemistry of geothermal systems have been greatly influenced by Jim Ellis, Don White, and Bob Garrels, and this work bears some of their imprint. More recent influences have been our colleagues at Chemistry Division, DSIR (New Zealand) and at the U.S. Geological Survey, as well as the many students participating in earlier courses. The perceptive reader may recognize these workers in the Frontispiece.

One of us (RWH) wishes to thank DSIR for overseas study leave and the Fulbright Foundation for a 1983 Travel Award, both of which provided the opportunity to complete this text. Also a special word of thanks to Meg and the boys for their patience during the mammoth task of preparing this publication.

Review comments from Sue Kieffer and Rosemary Vidale were greatly appreciated and have contributed substantially to the text. Thanks are also due to Phil Bethke for his enthusiasm for the whole project, and to Pan Eimon for providing her sketch of the Creede district. Our thanks go to all those who worked so hard to prepare the manuscript, especially to Corrine Weaver, Irene Harrell, Pat Dick, and Sharon Thorne; to Dan Hayba for drafting most of the diagrams; to Paul Delaney for programming the steam table; and to the drafting and photographic personnel at the U.S. Geological Survey.

Dick Henley, Al Truesdell, Paul Barton

BIOGRAPHIES

RICHARD W. HENLEY received a BSc. in Geology in 1968 from the University of London and a PhD. in geochemistry from the University of Manchester in 1971 following experimental studies of gold transport in hydrothermal solutions and the genesis of some Precambrian gold deposits. He was Lecturer in Economic Geology at the University of Otago, New Zealand from 1971 to 1975, and at Memorial University, Newfoundland until 1977. Research interests have focussed on the mode of origin of a number of different types of ore deposits including post-metamorphic gold-tungsten veins, porphyry copper, massive sulphide, and placer gold deposits. He is currently Head of the Geothermal Chemistry Section of the Department of Scientific and Industrial Research at Wairakei, New Zealand, and a visiting lecturer at the Auckland Geothermal Institute. His present research includes a number of isotope and chemical studies relating to the exploration and development of geothermal systems and geothermal implications for the origin of ore deposits.

ALFRED H. TRUESDELL received a B.A. in chemistry and geology from Oberlin College in 1956 and, after a few years working on uranium mineralogy at the U. S. Geological Survey, received a M.A. and Ph.D. in Geochemistry, Petrology, Mineralogy and Physical Chemistry from Harvard in 1962. Since then he has done experimental and modeling studies on low and high temperature mineral-solution equilibria, specializing, since 1967, in geothermal systems. He has worked on exploration for geothermal systems in the U. S., Mexico, Indonesia, Taiwan, China, India, and the Azores. At present his main interest is in the effects of development on reservoir processes and chemistry. His main field areas are the exploited fields of Cerro Prieto (Mexico), Larderello (Italy) and The Geysers. He is a member of several societies and a past editor of Geochimica et Cosmochimica Acta.

PAUL B. BARTON, JR. received a B.S. in Geology and Mineralogy from Pennsylvania State University in 1952 and a PhD in Geology (specialization in Economic Geology) from Columbia University in 1955. Since then he has been a research geologist with the U. S. Geological Survey in the Washington area, although he did spend three years as Deputy Chief for Scientific Program in the Survey's Office of Mineral Resources. His research has dealt with the thermodynamic properties of minerals (particularly those in the system Fe-S Zn, Cu, As, Sb, Bi), and he has been active in the application of such results to mineral deposits, particularly to epithermal base and precious metal ores, to massive sulfide ores, and to Mississippi Valley ores. He is a member of several professional societies (including being Past President of The Society of Economic Geologists) and the National Academy of Sciences. He has served on several National Research Council committees, including chairing a panel on non-fuel minerals.

JIM WHITNEY received B.S. and M.S. degrees in Geology and Geophysics, and Earth and Planetary Sciences from Massachusetts Institute of Technology in 1969, and the PhD in Geology from Stanford University in 1972. Since that time, he has taught and conducted research at the University of Georgia where he is currently Professor and Head of the Geology Department. His research has included experimental and field studies of granitic systems, solubility of metals in chloride solutions at high temperatures, activities of gaseous species in silicic magmas, and the possible contribution of magmatic fluids to ore deposits.

Chapter 1
INTRODUCTION TO CHEMICAL CALCULATIONS

This text is designed to introduce you to the practical concepts and calculations involved in interpreting the chemistry of high-temperature fluids in geothermal systems and hydrothermal ore-forming environments. It is intended that the energetic reader will learn to understand chemical principles, handle routine calculations and follow specialized chemical studies involved in geothermal exploration and exploitation and in ore genesis.

Although the emphasis of the text is on the interpretation of the chemistry of active geothermal systems, the principles involved are equally relevant to the interpretation of fossil hydrothermal ore-forming environments. Many gold-silver ore deposits, for example, have been shown to have formed in the near-surface region of hydrothermal systems similar in fluid chemistry and setting to those active today (White, 1981; Henley and Ellis, 1983). Combination of a knowledge of the principle processes within the active geothermal systems, the thermodynamics of complex ion formation, mineral-fluid equilibria and stable isotope systematics provide a framework which may assist in reconstruction of the hydrological regime within a fossil hydrothermal system where ore deposition occurred. This in turn may become useful in ore search. A chapter dealing with the hydrothermal chemistry of magmatic systems is included later in order to encompass a wider range of ore depositing environments and perhaps the root zones of the active geothermal systems.

After a short introduction to the types of geothermal fluids and chemical calculations, successive chapters will address the interpretation of water and gas analyses from geothermal wells. When we understand the reservoir compositions of some geothermal fluids and their relations to rock chemistry and temperature, we will consider the chemical and isotopic changes that occur in the natural transport of this fluid to the surface, derive and use chemical geothermometers and mixing relations, and map the surface chemistry of a hot spring system. After these studies of natural fluids at depth and at the surface, we will study chemical changes that occur during the exploitation of geothermal fluids and how to anticipate and avoid some of the problems of scaling and corrosion. We cannot cover all aspects of geothermal chemistry in the space available but we do hope to dispel at least some of the mystery surrounding chemical calculations and allow you to use chemical data with confidence. In many of the procedures discussed here a range of chemical relationships and concepts are brought together into what at times may appear a confusing muddle. At times of stress therefore the reader may find it relaxing to turn to a similar eclectic path of the type described by Adams (1979).

The recurring problem throughout the text is how to relate measured concentrations in geothermal water and steam to equilibria in the underlying reservoir. The following paragraphs provide a brief summary of the basic chemical relationships used in hydrothermal calculations. For fuller discussion of these topics, the reader should turn to the texts recommended at the end of the chapter. We start with an outline of concentration units commonly employed to report analytical data and then summarize some of the thermodynamic tools at our disposal.

> Boxes like this are scattered throughout this book. They are used to set off useful digressions and special problems. Other problems are given in continuity in the text.

CONCENTRATION UNITS

Chemical analyses of water samples are usually reported in mg/kg solution (which equals parts per million (ppm) by weight) or in mg/l (of solution), while chemical calculations are generally in molal units (moles/kg of solvent, in most cases here water or steam condensate) abbreviated m.

Exact conversion of mg/kg analytical units to molal(m) units requires the equation,

$$\text{molal conc. of i} = \frac{\text{mg(i)/kg}}{1000 \times \text{GFW(i)}} \times \frac{1000}{1000 - \Sigma\,(\text{mg/kg})/1000}$$

where GFW(i) is the gram formula weight of i and Σ (mg/kg) is the sum of the concentrations of all dissolved constituents.

For concentration data in mg/l, a similar expression is used with the addition of the specific volume of the solution, V_1, in the second term

$$\text{molality (i)} = \frac{\text{mg(i)/l}}{1000 \times \text{GFW(i)}} \times \frac{1000\ V_1}{1000 - \Sigma\,(\text{mg/l})\ V_1/1000}$$

For dilute solutions (salinity < seawater) the second term in these equations is close to one and may be omitted.

Another useful concentration unit particularly for dissolved gases is the mole fraction given by,

$$X(i) = \frac{\text{moles(i)}}{\Sigma\ \text{moles}}$$

where Σmoles includes all substances in any volume or mass of a mixture. For other concentration units see Garrels and Christ (1965).

MASS ACTION CONSTANTS

The most fundamental chemical equation is the reaction. For example, if two bicarbonate ions react to form carbon dioxide(gas), carbonate ions and water, we write

$$2HCO_3^- = CO_2(g) + CO_3^= + H_2O \qquad (1)$$

Basic thermodynamics tells us that the change in the Gibbs free energy for a reaction at equilibrium is zero so that

$$\Delta G_{CO_2} + \Delta G_{CO_3^=} + \Delta G_{H_2O} - 2\,\Delta G_{HCO_3^-} = 0 \qquad (2)$$

We can write equations that relate a thermodynamic concentration of each reacting species to its free energy. In general the thermodynamic concentration unit for solutes in aqueous solutions and for components of solid solutions is the activity and the thermodynamic concentration unit for gas components in either vapor or liquid phases is the fugacity. For substances with ideal solution behavior the molality or mole fraction may be substituted for the activity and the pressure for the fugacity. In most of the book we do not use gas activities but retain fugacities or, where permissible, pressures because these quantities are similar to those measured.

The equation that relates the fugacity of CO_2 to its free energy is

$$\Delta G_{CO_2} = \Delta G^\circ_{CO_2} + RT \ln f_{CO_2} \qquad (3)$$

where ΔG° is the free energy when CO_2 is at unit fugacity (called the standard free energy, see box below), R the gas constant, T the temperature in degrees Kelvin and f is the fugacity of the gas. At low pressures (as found in most geothermal systems) fugacity coefficients approach unity, and partial pressure may be used in place of fugacity in equation (3).

This equation can be derived by assuming that CO_2 behaves as an ideal gas for which $(\partial G/\partial P) = \underline{V}$ and $P\underline{V} = RT$ (both for one mole at constant temperature). Then $(\partial G/\partial P) = RT/P$, $\int dG = RT\int dP/P$ and $\Delta G = RT \ln P + c$, where c is a constant. If P = 1 bar in the standard state then $c = \Delta G°$ and $\Delta G = \Delta G° + RT \ln P$. For non-ideal solid or aqueous solutions, the thermodynamically effective concentration is the activity; for gases, the fugacity.

It is important to realize the connection between the standard state and the concen-centration units (activity, fugacity, pressure, molality, mole fraction, etc.) used in equation (4). The standard free energy of formation is the free energy in the standard state which is defined by the concentration being equal to one. In our simple derivation this is evident because the term RT ln P becomes zero when P = 1 and the constant of integration, c becomes the standard free energy of formation, $\Delta G°$. So long as the thermodynamic data are in the matching standard state (unit activity, unit fugacity, etc.) it is quite acceptable to mix units in an equilibrium constant. The choice of concentration units depends also on the ideality or non-ideality of mixing. The activity coefficients of uncharged species in dilute solutions are so near one that molality may be substituted for activity. For further definition and discussion of standard states see the recommended reading.

We can combine equation (2) with a set of equations like equation (3) to obtain the mass action constant and its relation to the change in standard free energy of the reaction (at equilibrium $\Delta G = 0$).

$$0 = \Delta G°_R + RT \ln \frac{P_{CO_2}\, a_{CO_3^=}\, a_{H_2O}}{a^2_{HCO_3^-}} \qquad (4)$$

or

$$\Delta G°_R = -RT \ln K \qquad (5)$$

where

$$G°_R = G°_{\Sigma products} - \Delta G°_{\Sigma reactants} \qquad (6)$$

and K is the equilibrium (or mass action) constant. In calculations of isotopic equilibria, we will use fractionation factors which are essentially the same as equilibrium constants.

To use these equations we must relate activities to analytical concentrations, and we must obtain values of equilibrium constants or free energies. Free energies and calculations of equilibrium constants are discussed as they are used.

ACTIVITY - CONCENTRATION RELATIONS

Generally, under most geothermal or epithermal conditions, we can calculate activities from concentrations using a few simple rules:

1. All simple solids and water may be assumed to have activities equal to 1. This is not true for components of solid solutions and is decreasingly valid for water as salinity increases above that of seawater.

2. At most hydrothermal pressures, gas activities (or fugacities) may be assumed to be equal to partial pressures. This is true whether or not a vapor phase is present. We will discuss later the relation of pressures to gas concentrations in liquid.

3. Activities of dissolved species (ions and molecular or uncharged species) may be calculated from some version of the extended Debye-Hückel equation. This equation calculates the activity coefficient (γ_i) which when multiplied by the concentration (m_i) gives the activity ($a_i = \gamma_i m_i$). The equation has the general form, for a species, i,

$$-\log \gamma_i = \frac{A z_i^2 I^{1/2}}{1 + \mathring{a}_i B I^{1/2}} + bI \qquad (7)$$

in which z is the ionic charge, I the ionic strength and A, B, $\mathring{a}_i$ and b are constants. The ionic strength is a measure of the average electrical field surrounding an ion in solution. It is defined as $I = 1/2\ \Sigma m_i z_i^2$. For most geothermal solutions I is equal to the sum of m_{Na^+} + m_{K^+}. The constants A and B are related to the properties of the solvent and are listed in Table 1.2. $\mathring{a}_i$ represents the distance of closest approach between specified ions but is neither the sum of the crystallographic radii nor the ionic hydration radii of the individual ions. Helgeson et al. (1981) discuss rigorous procedures for the calculation of $\mathring{a}_i$ and values commonly employed in geothermal fluid-mineral calculations are listed in Table 1.1. b is called the deviation function and is the principal extension term in the Debye-Hückel equation. Up to 250°C, b has values in the range 0.03 to 0.05 when concentrations are up to 3 molal (Helgeson, 1969). The computation of the extended term is discussed in full by Helgeson et al. (1981).

The limit to which the extended Debye-Hückel equation can be successfully employed is somewhat controversial. Helgeson et al. (1981) suggest that where the i, I and b terms are rigorously derived it may be employed up to ionic strengths of 9 molal. Harvie and Weare (1980), in reviewing the application of the Debye-Hückel equation, suggest that at low temperatures good agreement occurs between observed mineral solubilities and those calculated via equation (7) up to about 1 molal above which a virial expansion should be used. Such an expansion has been developed by Pitzer (1973, 1981). Fortunately most of the geothermal fluids explored for power generation have ionic strengths much less than 2 molal allowing confident use of equation (7). The majority of epithermal ore depositing fluids appear to have had similar low ionic strengths.

> It is useful to program equation (7) and solve for various temperatures and ionic species to produce graphs of activity coefficients as a function of ionic strength. Values of I = 0.03 and 0.3 molal will be useful for later problems. Neutral species (z = 0) have activity coefficients close to 1 in dilute solutions. Appendix I includes a simple program (ION) for the calculation of activity coefficients using equation (7).
>
> Plot graphs of $\gamma_{HCO_3^-}$ and $\gamma_{CO_3^=}$ vs ionic strength (0.01 to 3m) at 25° and 250°C.

4. In isotopic studies, activities of isotopic species ($H_2{}^{16}O$ or DHO for example) are assumed equal to their concentrations. The concentration scale used for isotopes is discussed later.

MASS AND HEAT BALANCES

Any conservative quantity (total mass of any element or isotope, any extensive thermodynamic quantity) may be used to write a material or energy balance equation. Thus, the total amount of carbon in our example (equation 1) remains constant and, using molal concentration units (moles(i)/kg solvent), we can write a <u>mass balance equation</u>. For example,

$$m_{C,total} = m_{CO_2,aq} + m_{HCO_3^-} + m_{CO_3^=} \quad (8)$$

We commonly use an <u>enthalpy or heat balance</u> for a geothermal reservoir fluid (of enthalpy = $H_{reservoir}$) that is separated into steam (enthalpy = H_v) and liquid water (enthalpy = H_l), where y is the mass fraction of steam that is formed; i.e. the <u>steam fraction</u>.

$$H_{reservoir} = y\ H_v + (1-y)\ H_l \quad (9)$$

Table 1.1 -- Values of ionic charge, z and ion size parameter, å , for the common ionic species in geothermal fluids (after Truesdell and Jones, 1974). å is given in units of 10^{-8} cm.

H^+	Na^+	HCO_3^-	HS^-	$H_3SiO_4^-$	$H_2BO_3^-$	F^-	$SO_4^=$	NH_4^+	HSO_4^-
1	1	1	1	1	1	1	2	1	1
9.0	4.0	4.5	4.0	4.0	4.0	3.5	4.0	2.5	4.0

OH^-	$CO_3^=$	Cl^-	Li^+	K^+	Ca^{++}	Mg^{++}
1	2	1	1	1	2	2
4.0	4.5	3.5	6.0	3.0	6.0	8.0

Table 1.2 -- Values of the Debye-Hückel Coefficients A and B along the vapor pressure pressure curve of water (from Helgeson et al., 1981). Units for A are $kg^{1/2}mole^{-1/2}$ and for B are $kg^{1/2}mole^{-1}cm^{-1} \times 10^8$ (the product å x B cancels the 10^8 factor).

t (°C)	A	B	t	A	B
0	0.4913	0.3247	180	0.7575	0.3605
10	0.4976	0.3261	190	0.7829	0.3629
20	0.5050	0.3276	200	0.8099	0.3655
30	0.5135	0.3291	210	0.8387	0.3681
40	0.5231	0.3307	220	0.8697	0.3707
50	0.5336	0.3325	230	0.9030	0.3734
60	0.5450	0.3343	240	0.9391	0.3762
70	0.5573	0.3362	250	0.9785	0.3792
80	0.5706	0.3381	260	1.0218	0.3822
90	0.5848	0.3401	270	1.0699	0.3855
100	0.5998	0.3422	280	1.1238	0.3889
110	0.6158	0.3443	290	1.1850	0.3926
120	0.6328	0.3465	300	1.2555	0.3965
130	0.6507	0.3487	310	1.3381	0.4009
140	0.6697	0.3510	320	1.4369	0.4058
150	0.6898	0.3533	330	1.5584	0.4114
160	0.7111	0.3556	340	1.7138	0.4178
170	0.7336	0.3580	350	1.9252	0.4256

The solution of this equation and its application will become almost second nature to you. Similar equations for dissolved constituents, gases and isotopes will also be used frequently. For example,

$$C_{Cl,reservoir} = y\, C_{Cl,vapor} + (1-y)\, C_{Cl,liquid} \qquad (10)$$

$$C_{CO_2,reservoir} = y\, C_{CO_2,vapor} + (1-y)\, C_{CO_2,liquid} \qquad (11)$$

$$\delta D_{reservoir} = y\, \delta D_v + (1-y)\, \delta D_l \qquad (12)$$

where concentrations, C, are in molal or related units and δD is the concentration of deuterium (this unit is explained in Chapter 10).

The routine use of equations (10) to (12) is discussed more fully in Chapter 2.

We should note at this point that in dealing with reactions involving dissolved components, a wide range of dissolved species - or complexes - may be present. For lead, for example, $PbCl^+$ and $PbCl_2$ are some of the complexes in solution. Mass and charge balance equations, as well as equilibrium constants used in solubility calculations should include all the complexes that are present. In some cases where we know (or we think we know) that a particular complex dominates the others, we can carry out first order calculations using only that principal complex.

Carbon dioxide in solution occurs as the two complex uncharged species formally written as CO_2,aq and H_2CO_3 (carbonic acid). Because the equilibrium constant for the reaction between these two species is poorly known, it has become customary to use a pseudo-species H_2CO_3 or CO_2,aq to stand for total unionized CO_2 in solution depending upon the context of the calculation.

CHARGE BALANCE

In solving many systems of chemical equations, an additional relation is often required and this may be provided by the charge balance. Since charged solutions are abhorred by nature, the sum of positive charges must equal that of negative charges.

$$\Sigma m_i z_i = 0 \tag{13}$$

For example, if only H_2O, CO_2, HCO_3^-, $CO_3^=$, H^+, Na^+ and OH^- were present in the solution

$$m_{Na^+} + m_{H^+} = m_{HCO_3^-} + 2m_{CO_3^=} + m_{OH^-} \tag{14}$$

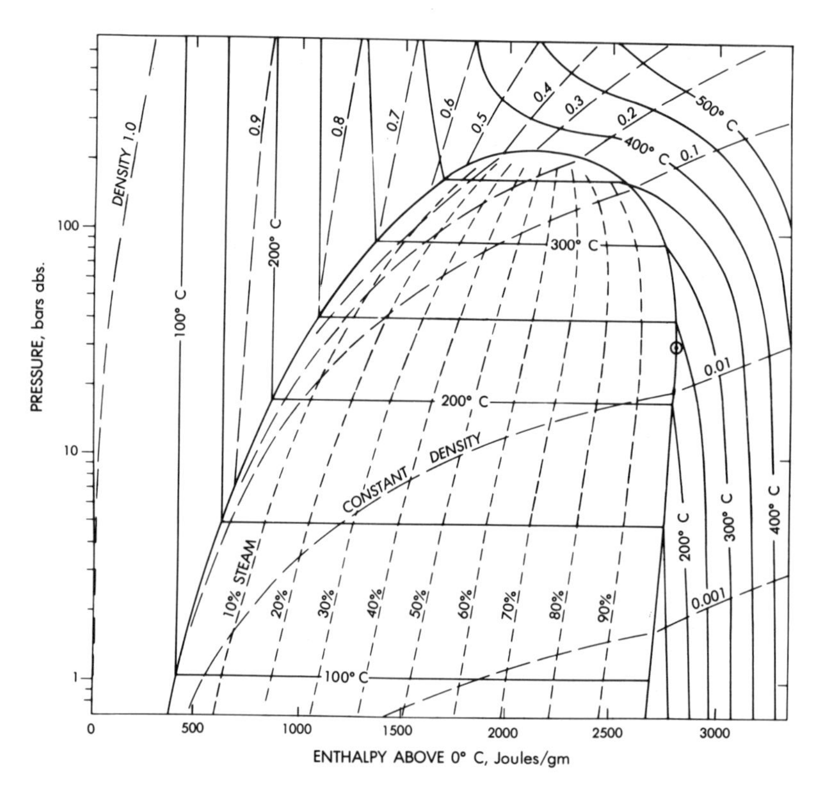

Figure 1.1. Enthalpy-temperature-pressure-density relations for water to 500°C and 600 bars. The diagram summarizes the data presented in Steam Tables such as Appendix III (modified from White et al., 1971).

THERMODYNAMICS OF WATER

In high temperature geothermal and hydrothermal systems, boiling is a process of first rank importance. Boiling* occurs naturally during reduction of pressure by upflow, by explosive removal of rock and water, by producing and sampling geothermal wells, or during heating by igneous intrusion. Thermodynamic data for water and steam are required to calculate mass and enthalpy balances for boiling and condensation processes. The data used in this book are contained in the abridged steam tables provided in Appendix III. The pressure-temperature-enthalpy-density relations are also shown in Figure 1.1, redrawn from White et al. (1971).

Because of the variety of thermodynamic data compilations currently in common use, we have chosen units for convenience of reference rather than consistently using SI or c.g.s. units. Thermodynamic data for water are given in Joules per gram (Keenan et al., 1969) with fluid specific volume in cm^3/g, and pressures in bars. Units of concentration are given in mg/kg (for analytical data) and in molal or millimolal units. Tables of conversion factors and symbols used in the text and a listing of atomic weights are given on the end papers.

---

* The term "boil" simply implies transition from a liquid to a vapor state, regardless of composition or number of components. Boiling is synonomous with effervescence and is the reverse of condensation. In many geothermal applications processes take place on the boiling point curve for the system fluid (i.e. pure H_2O, H_2+CO_2, H_2O+NaCl$\pm$ CO_2) implying coexistence of two phases, liquid and vapor.

SUMMARY OF CHEMICAL COMPUTATIONS

In general, two types of equations - products and sums - are used in our computations. Equilibrium expressions are products of concentrations, activities, fugacities, and pressures (or their reciprocals) that equal constants (e.g. mass action, distribution, solubility, or fractionation). Mass, energy, or charge balances are sums of concentrations (usually moles or isotope units), thermodynamic quantities (usually free energies or enthalpies), or charges (charge per ion times ion concentration) that equal a constant or zero. The balances apply exactly to the analyzed or calculated concentrations (of mass, energy or charge) but the equilibria usually apply to activities (thermodynamically effective concentrations) or fugacities. For dissolved species, the connection between concentrations and activities is the activity coefficient which can be approximated reasonably well using the Debye-Huckel equation and its variations. Fugacities are related to pressures through experimentally derived fugacity coefficients.

These equations are sufficient to solve any chemical problem although the approximations made (especially with regard to activity coefficients) may make the result unreliable. In addition, the thermodynamic data may be inadequate, or worse, the reaction studied may not be the right one!

For a number of reactions, the assumption of equilibrium is justified in geothermal fluids which have been held at moderately high temperatures (>100 C?) for relatively long time periods (years). In any case the equilibrium state may be the only one calculable and serves as a reference for disequilibrium states.

RECOMMENDED READING

Adams, D., 1979, Hitchhikers Guide to the Galaxy: Crown Publications, 215 p. Recommended in times of stress.

Darken, L. S., and Gurry, R. W., 1953, Physical Chemistry of Metals: McGraw-Hill, New York, 535 p. Chapter 9 (p. 206-234) recommended.

Denbigh, Kenneth, 1957, The Principles of Chemical Equilibrium: Cambridge Univ. Press, Cambridge, 491 p. A difficult but comprehensive treatment. Chapters 2-4 and 10 are highly recommended.

Garrels, R. M., and Christ, C. L., 1965, Solutions, Minerals, and Equilibria: Harper and Rowe, New York, 450 p. Chapters 1 and 2 (p. 1-73) recommended.

Krauskopf, K. B., 1969, Thermodynamics used in Geochemistry; Chapter 3 in Handbook of Geochemistry: Wedepohl, K. H., exec. editor, Springer Verlag, New York, p. 37-77. Condensed but highly recommended.

REFERENCES

Harvie, C. E., and Weare, J. H., 1980, The prediction of mineral solubilities in natural waters: the Na-K-Mg-Ca-Cl-SO_4-H_2O system from zero to high concentration at 25°C: Geochimica et Cosmochimica Acta, v. 44, p. 981-997.

Helegeson, H. C., 1969, Thermodynamics of hydrothermal systems at elevated temperatures and pressures: American Journal of Science, 267, p. 729-804.

Helgeson, H.C., Kirkham, D. H., and Flowers, G. C., 1981, Theoretical prediction of the behavior of aqueous electrolytes at high pressures and temperatures: Calculation of activity coefficients, osmotic coefficients, and apparent molal and standard and relative partial molal properties to 600°C and 5 kb: American Journal of Science, 281, p. 1249-1516.

Keenan, J. H., Keyes, F. G., Hill, P. G., and Moore, J. G., 1969, Steam Tables - Thermodynamic properties of water including vapor, liquid and solid phases (International Edition - metric units): Wiley, New York, 162 p.

Pitzer, K. S., 1973, Thermodynamics of electrolytes I: Theoretical basis and general equations: Journal of Physical Chemistry, v. 77, p. 268-277.

Pitzer, K. S., 1981, Characteristics of very concentrated solutions; in Rickard, D.T. and Wickman, F.E., Chemistry and Geochemistry of Solutions at High Temperatures and Pressures: Pergamon Press, Oxford, p. 249-264.

Truesdell, A.H., and Jones, B.F., 1974, WATEQ, a computer program for calculating chemical equilibria of natural waters: Journal Research, U.S. Geol. Survey, v. 2, p. 233-248.

White, D.E., Muffler, L.J.P., and Truesdell, A.H. 1971, Vapor-dominated hydrothermal systems compared with hot water systems: Economic Geology, v. 66, p. 75-97.

Chapter 2
CHEMICAL STRUCTURE OF GEOTHERMAL SYSTEMS

In this chapter we shall examine the different types of water which may occur in geothermal systems and relate this range of water types to the basic processes which dominate their chemistry. We shall learn how, from the chemistry of water discharged from wells, we can obtain specific information about the deep fluids in a geothermal system and how they relate to natural discharges at the surface. Skills developed in this way may then be used in exploration to obtain deep system information from analyses of natural discharges. In later chapters we shall learn additional techniques based on chemical data to obtain essential information about reservoir behavior before and during exploitation. This chapter is concerned largely with non-volatile components (NaCl, SiO_2, etc); gases are briefly mentioned but discussed in much more detail later. The chemical and physical processes discussed here and in later chapters apply equally to hydrothermal ore deposits but in their case chemical data must be estimated from fluid inclusion, stable isotope, mineral paragenesis, and stability data.

It would be impractical to discuss all types of geothermal systems in this text; they occur in a range of tectonic settings (Fig. 2.1) and here we shall focus on the higher temperature systems which occur in areas of active volcanism. The principal features of these systems are outlined below and for more detail the reader is referred to reviews by Ellis and Mahon (1977), Elder (1981), Rybach and Muffler (1981) and Henley and Ellis (1983). Derivation of the evidence for the hydrologic and chemical structure of geothermal systems forms the basis of the later chapters.

STRUCTURE OF ACTIVE GEOTHERMAL SYSTEMS

Geothermal systems arise where input of heat (usually magmatic) at depths of a few kilometres, sets deep groundwaters in motion. These groundwaters are usually meteoric in origin but in some systems deep fossil marine (connate) or other saline waters may be present (e.g., Salton Sea, California). Systems near the coast may be fed by seawater or both meteoric water and sea water (e.g., Svartsengi, Iceland). It is possible that the magmatic heat source may add some water and dissolved constituents like HCl, CO_2, and SO_2 at temperatures over 500° C but because of dilution and reaction during convective upflow, this possibility is very difficult to prove except by analogy with deeply dissected "fossil" geothermal systems such as those now mined as porphyry copper and molybdenum deposits. In these deposits isotopic data indicate the early introduction of an aqueous phase evolved within the magma body itself.

As hot waters rise convectively, they react with their host rocks, dissolving some constituents such as silica and altering primary minerals to develop a new mineral suite, the constituents of which reflect the reaction temperature and chemistry. Depending on their source, the <u>chloride</u> waters which have evolved in the deep system commonly have chloride contents ranging up to ten thousand parts per million (mg/kg) (Table 2.2) but in some systems (e.g., Salton Sea, California) chloride may be as high as 155,000 mg/kg. The locations of these and other active geothermal systems mentioned in the text are shown in Figure 2.1.

In general, temperatures in the upper 1/2-2 km of the deep systems follow a <u>boiling point to depth</u> relationship. The boiling that occurs in this region results in the transfer of gases (CO_2, H_2S, CH_4, etc.) into a vapor phase. This phase then migrates independently to the surface to form fumaroles. It may encounter near surface cold groundwater into which it will condense to form <u>steam-heated</u> waters; oxidation of H_2S in this environment produces <u>acid-sulphate</u> waters (with low chloride contents and pH in the range of 0-3.0) which react rapidly with host rocks to give <u>advanced argillic</u> alteration assemblages (kaolinite, alunite, etc). Since the heat content of steam is high relative to liquid water, these steam-heated waters may themselves boil at the water table giving rise to 'steaming ground', a common feature in geothermal areas. The deep chloride waters may also flow directly to the surface to form <u>boiling</u>, near neutral to alkaline pH, <u>high-chloride</u> springs, or may become diluted to give relatively <u>dilute chloride waters</u>. Depending on the hydrology of the surface system each of these water types may interact with each other to give hybrid water types (e.g., acid-sulphate chloride, etc.). Bicarbonate-rich waters occur where groundwaters dissolve CO_2 arising from deep gas exsolution. In this case the acidity due to the dissociation of dissolved CO_2 (carbonic acid) leads to rock alteration and the generation of sodium and bicarbonate as the dominant ions. Such waters are characteristic of the 'condensate zones' of vapor-dominant systems such as Kawah Kamojang, Indonesia, as well as on the margin of the more common liquid-dominant systems such as Broadlands, New Zealand (Mahon et al., 1980).

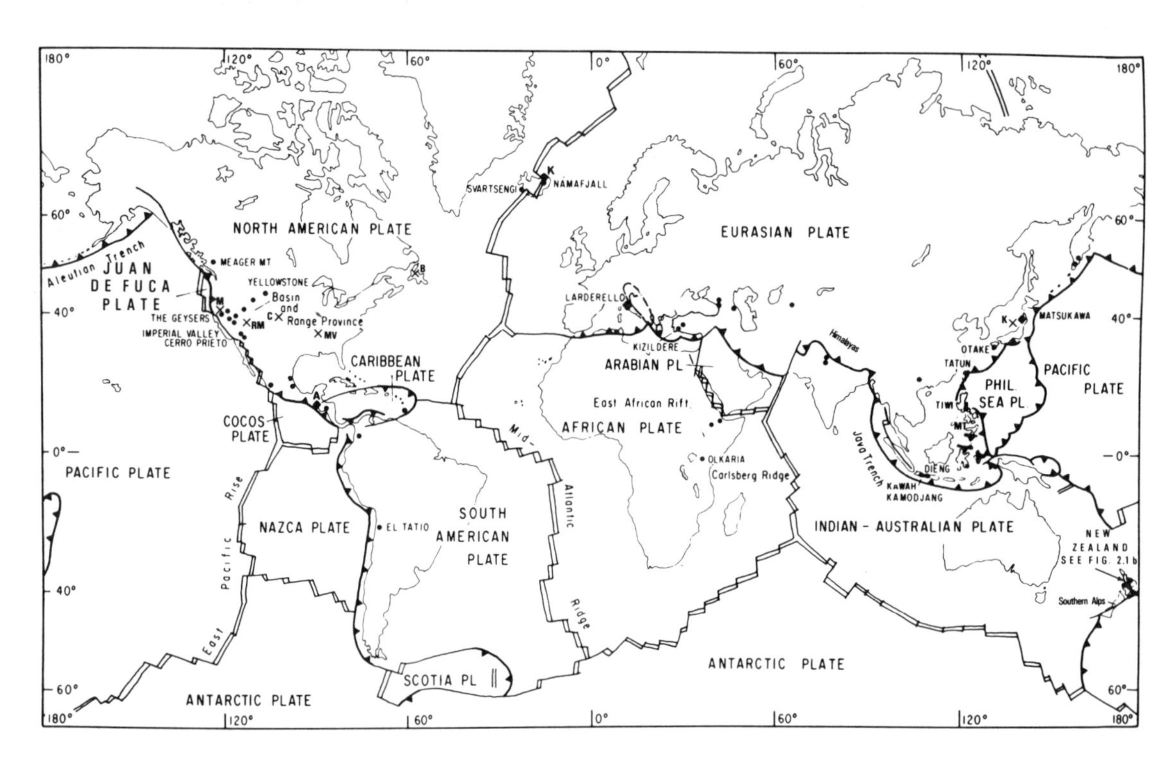

Figure 2.1a. Lithospheric plate boundaries provide the framework for the global distribution of major geothermal systems. Symbols: ● geothermal systems; A (Ahuachapan/El Salvador), K (Krafla/Iceland), T (Mahio-Tongonan, Phillipines), x mineral deposit localities; M (McLaughlin, California), RM (Round Mountain, Nevada), C (Creede, Colorado,) MV (Mississippi Valley), K (Kuroko District/Japan). (Adapted with permission from Rybach and Muffler, 1981).

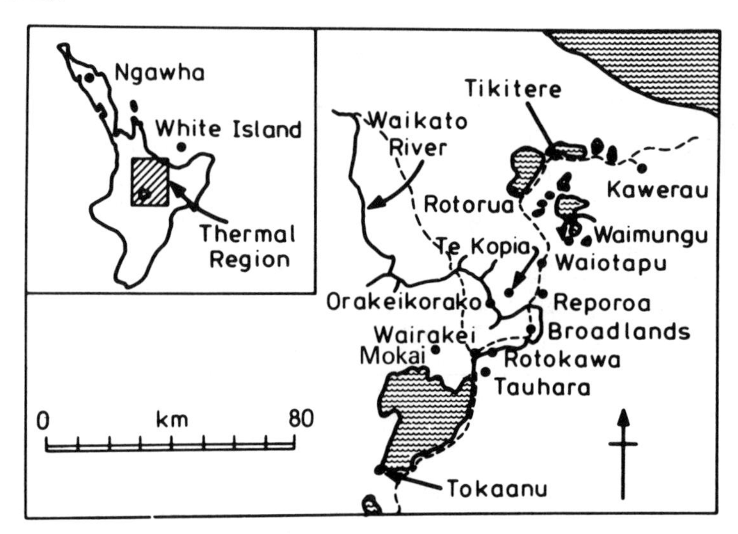

Figure 2.1b. Location of geothermal systems in the North Island, New Zealand. (Modified from Ellis and Mahon, 1977)

Table 2.1 -- Summary of water types in geothermal systems

	Approximate pH range	Principal anions
Groundwater	6-7.5	trace HCO_3^-
Chloride water	4-9	Cl, lesser HCO_3^-
Chloride-bicarbonate	7-8.5	Cl, HCO_3^-
Steam heated waters	4.5-7	$SO_4^=$, HCO_3^-, trace Cl^-
Acid-sulphate	1-3	$SO_4^=$, trace Cl^-
Acid-sulfate-chloride	1-5	Cl, $SO_4^=$
Bicarbonate	5-7	HCO_3^-
Dilute chloride	6.5-7.5	Cl, lesser HCO_3^-

BOILING POINT - DEPTH RELATIONS

As hot water rises toward the surface the pressure imposed on it by overlying fluid decreases. Eventually it reaches a level at which a vapor phase separates and migrates to the surface independently - i.e., boiling occurs. Above this level temperatures and pressures in the fluid are constrained on the liquid + vapor phase boundary shown in Figure 2.2a. The same phase boundary is conveniently displayed as a boiling point-depth curve (Fig. 2.2b) where pressure, P, has been translated into depth through the relation

$$dP/dh = \rho_w g$$

where g is the acceleration due to gravity and ρ_w the density of water at the boiling point, and h the height of the overlying column of water.

The boiling point - depth curve is a limiting condition giving the maximum temperature that a fluid may attain at a given depth or pressure provided that the liquid phase alone controls pressure. At Wairakei, for example, the deep fluid at depths greater than 1 km appears to have an enthalpy equivalent to that of steam-saturated water at 260 C. On Figure 2.2b draw the temperature - depth path of this fluid as it rises toward the surface. At what depth does the fluid start to boil?

For a saline fluid the vapor pressure and specific volume of the solution is related to the total salt concentration (Haas, 1971). In gassy systems the dissolved gas pressure leads to an increase in the "boiling point" depth for a given temperature so that relative to gas-poor systems like Wairakei, two phase conditions occur to much greater depths (see Chapter 4).

Convective or buoyancy pressures coupled with permeability restrictions may lead to excess pressures holding the fluid below its boiling point, e.g., shallow wells at Yellowstone (White et al., 1975). Grant et al. (1983) suggest that, in general, flow through geothermal systems requires that pressure gradients exceed hydrostatic by up to 10%, allowing for slightly higher temperatures at given depth than would be anticipated from the hydrostatic boiling point-depth relation discussed above.

Boiling is one of the three principal processes leading to temperature decrease in a rising fluid, the others are dilution and to a lesser extent, conduction. Dilution by cool groundwaters may prevent the deep fluid reaching the boiling point and this is readily recognized from chemical data.

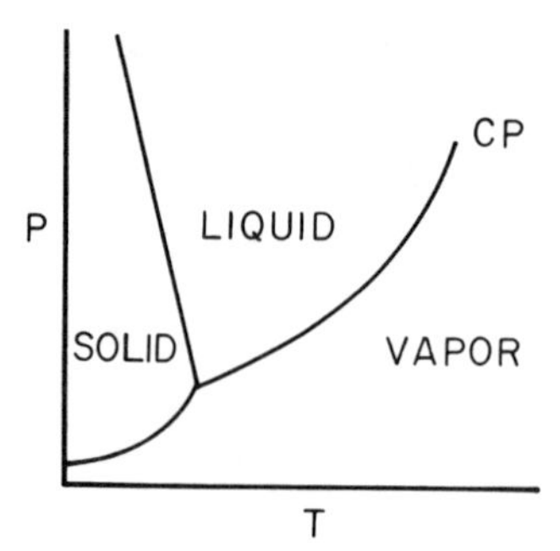

Figure 2.2a. Schematic P vs T diagram for water. C.P. = critical point.

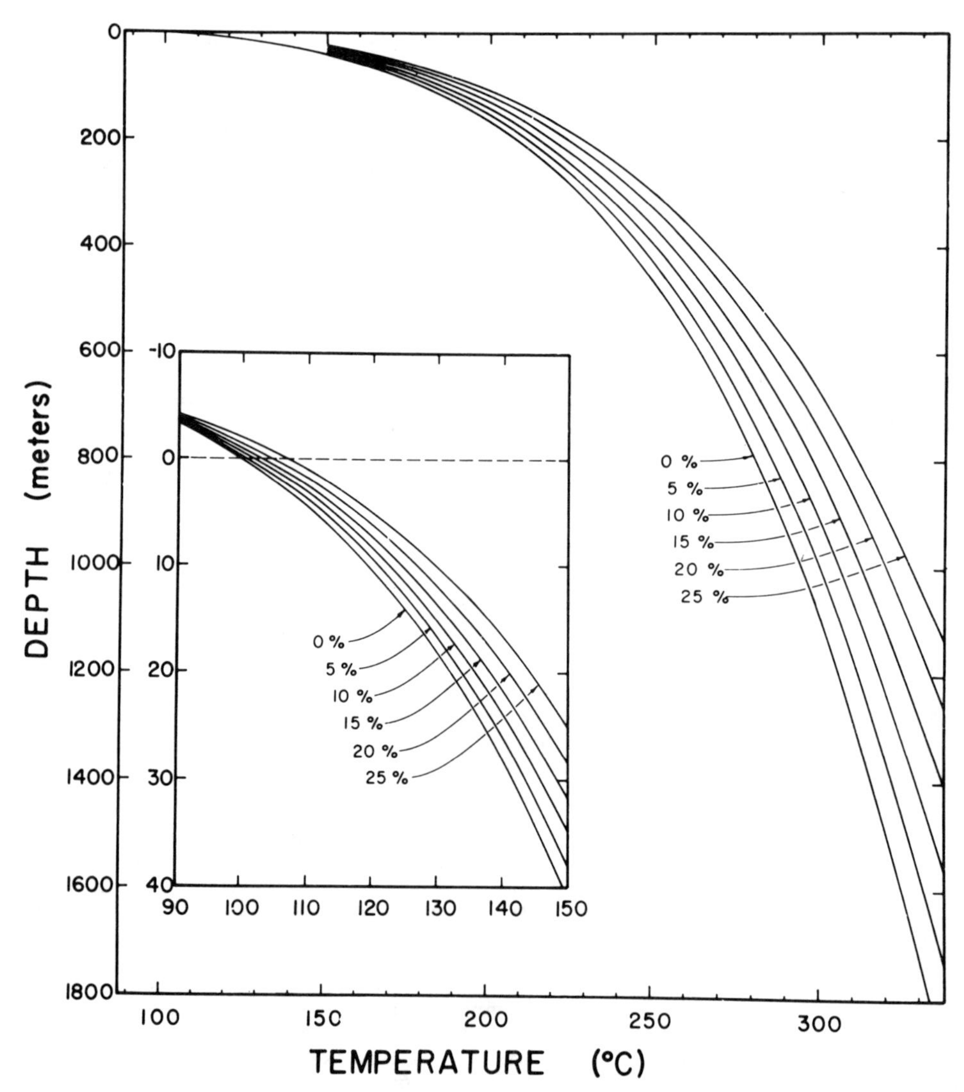

Figure 2.2b. Boiling-point curves for H_2O liquid (0 wt percent) and for brine of constant composition given in wt percent NaCl. The insert expands the relations between 100° and 150°C. The temperature at 0 meters for each curve is the boiling point for the liquid at 1.013 bars (1.0 atm) load pressure which is equivalent to the atmospheric pressure at sea level. The uncertainty is contained within the width of the lines. (Reproduced with permission from Haas, 1971)

Problem: Construct an approximate boiling point-depth curve from 0 to 200 m depth using 20 m intervals. Assume that the specific volume of water in each interval is the same as at the top of the interval. (Specific volume data for water are given in Appendix III.)

Figure 2.3 shows the structure of a typical geothermal system in silicic volcanic terrane like that in the Taupo Volcanic Zone, New Zealand. Notice the dynamic features of the system: recharge by meteoric waters, heat input at depth, convective upflow, deep mixing with meteoric water, transfer of steam to the surface and its interaction with groundwater, and the flow of deep fluid direct to the surface or its dilution and outflow to some hydraulic base level like a river or lake.

The hydrologic structure of systems in andesite strato volcanoes (Ahuachapan, El Salvador) is sketched in Figure 2.4. In these systems the same basic processes occur, but notice that chloride water springs occur several kilometers from the hot upflow part of the system

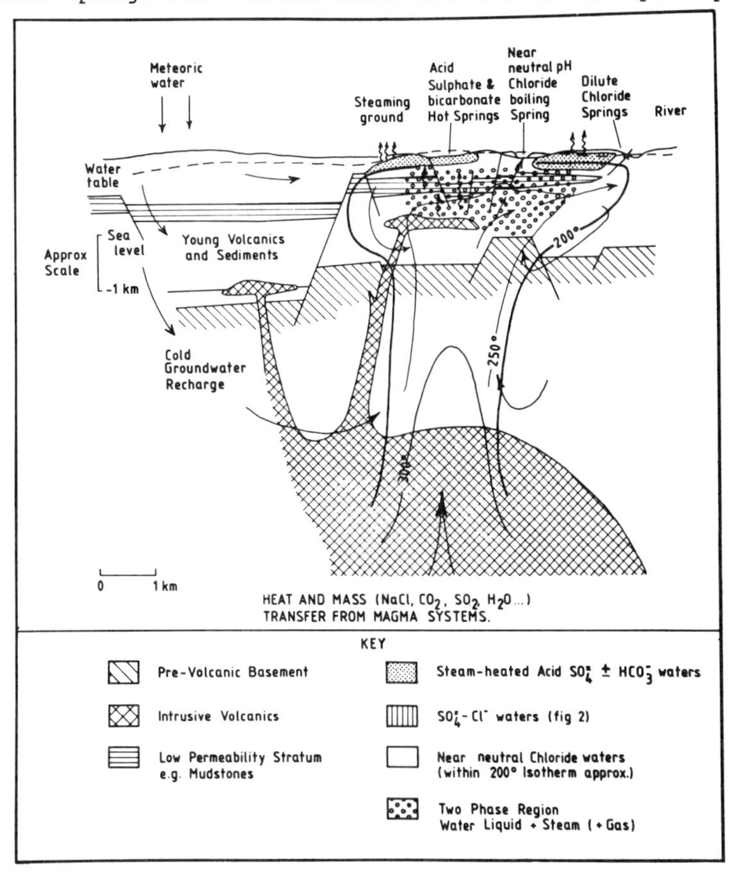

Figure 2.3. Schema of the main features of a geothermal system typical of those in silicic volcanic terranes. The system is supplied by ground-water in this case derived from meteoric water. Heat, together with some gases, chloride, water and some other solutes, is assumed to be supplied by a deeply buried magmatic system and results in a convecting column of near-neutral pH chloride water with two phase conditions in the upper part of the system. Steam-separation processes give rise to fumaroles and steam absorption by groundwater, with oxidation of H_2S at the water table, gives rise to isotopically enriched steam-heated acid sulfate and bicarbonate waters. Mixing may occur between the deeper chloride waters, steam-heated waters and fresh groundwater to give a range of hybrid waters. Outflows from the deep chloride system occur either as boiling alkaline springs often associated with silica terrances, or after mixing with cold groundwaters, as near-neutral pH relatively dilute chloride springs. (Reproduced from Henley and Ellis, 1983, with permission.)

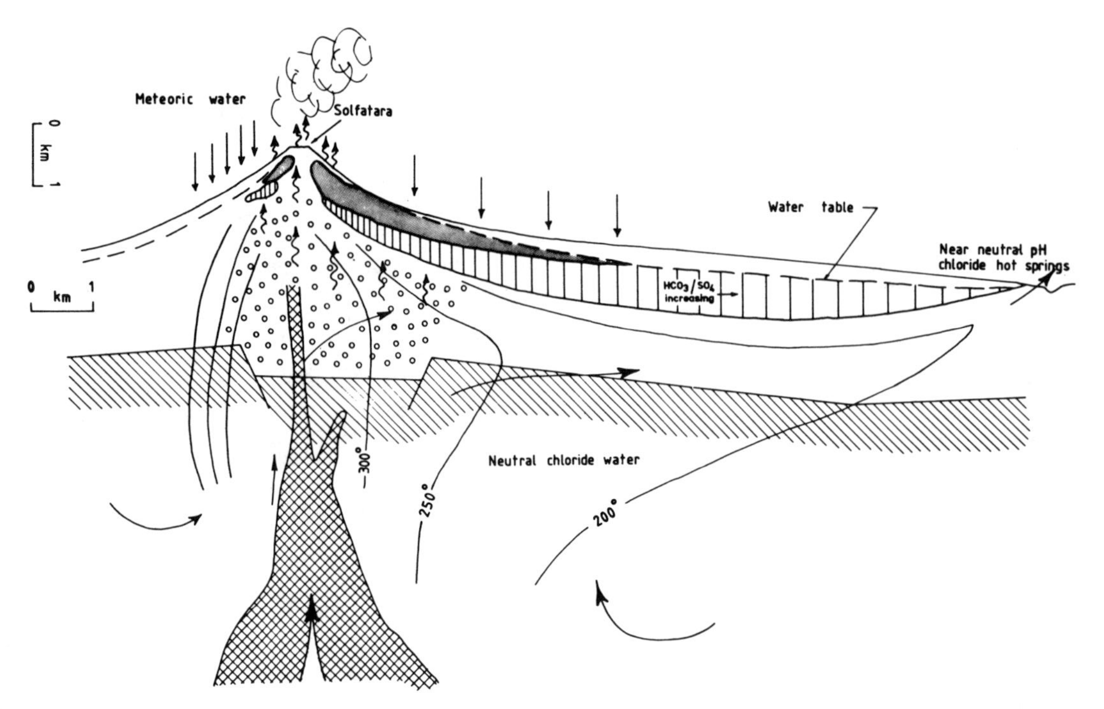

Figure 2.4. Schema of a geothermal system typical of active island-arc andesite volcanoes. Lower permanent water tables in tropical regions and the high relief of the volcanic structures result in a scarcity of chloride water discharges except at some distance from the upflow center. The latter may be revealed by fumaroles, intense rock alteration and steam-heated, often perched, aquifers. Near-surface condensation of volcanic gases and oxidation result in acid sulfate waters in the core of the volcano and an acid crater lake may also form (see Plate 4). (Symbols as for Figure 2.3.) (Reproduced with permission from Henley and Ellis, 1983.)

and the related steam flows to the surface. Hydrothermal systems also develop in tectonic basins in response to magmatic heating; examples are Cerro Prieto, Salton Sea, and the Great Basin, Nevada.

In some cases steam is the dominant mobile phase in an active system and an unusually thick "steam" zone is formed (Lardarello, Italy; the Geysers, California). (Such a zone may also develop as a consequence of extensive exploitation as has occurred at Wairakei.) Vapor-dominated geothermal systems were defined by White et al. (1971) as those systems in which vapor (steam) constitutes the continuous fluid phase and determines the change in pressure with depth. The role of mobile steam and relatively immobile liquid water in causing superheated or saturated steam without liquid water to be produced from these systems was later emphasized by Truesdell and White (1973) and Grant et al. (1980). Although vapor-dominated systems are valuable economically (the two largest producing fields in the world - The Geysers, California and Larderello, Italy - are vapor-dominated), and generally interesting scientifically, they are disappointingly uninformative from a geochemical viewpoint. Because no liquid water is produced the water-rock interactions that undoubtedly proceed in these systems cannot be studied directly. Even the gases in produced steam are not directly informative about subsurface conditions and reactions because a large and variable amount of liquid water is vaporized and mixed with reservoir steam as a consequence of production. Methods have been devised by Giggenbach (1980) and D'Amore and Celati (1983) to calculate the amount of vaporized water. These methods are discussed in Chapter 11 in context with other systems in which boiling occurs in the reservoir.

Examine Table 2.2 and Figures 2.3 and 2.4 and try to account in a simple manner for the origin of the different spring waters using these four categories:

1. Near neutral pH boiling chloride waters

2. Near neutral pH dilute chloride waters

3. Steam heated waters

4. Hybrid waters.

Table 2.2-- Major element analyses of spring waters from a number of geothermal areas (in mg/kg)

Location	Spring Temp (°C)	$pH_{20°}$	Na	K	Ca	Mg	Cl	B	SO_4	total HCO_3^-	SiO_2
Wairakei (NZ)											
Champagne Pool	99	8.0	1070	102	26	0.4	1770	21.9	26	76	294
Tauhara (NZ)											
A.C. Spring	70	5.9	56	14	14	6	8	0.1	105	375	230
Terraces; Iron Spring	76	7.3	403	46	-14-		537	9.8	105	276	250
Spa; Fissure Spring	95	8.0	820	59	-24-		1342	24	62	44	169
Spa Fumarole Condensate	98	2.5	30	2	16	>3	<7	-	-	-	-
Broadlands (NZ)											
Ohaaki Pool	95	7.1	860	82	2.5	0.1	1060	32	100	679	338
Waimangu (NZ)											
Frying Pan Lake	67	3.8	545	49	10.5	-	762	6.5	320	193	380
Ngawha (NZ)											
Jubilee Bath	50	6.5	870	79	8	2.5	1336	1020	500	333	186
Cerro Prieto (MEX)											
Spring N29	89	7.6	5120	664	357	4.6	8790	-	31	65	73
Tongonan (PHIL)											
Banati Springs	98	8.25	1990	211	86	0.4	3397	34.5	74	7	278

RECALCULATION OF GEOTHERMAL WELL DATA TO AQUIFER CONDITIONS

Now let us start to use chemical data from geothermal wells and springs to develop some quantitative understanding of these geothermal systems. To do so we first need to learn how to recalculate chemical data obtained from samples taken at the surface from geothermal wells or hot springs so that we may quantitatively allow for the effects of steam separation.

The Steam Fraction

When a geothermal fluid rises rapidly to the surface through a well or fissure it boils to produce steam and liquid water. Non-volatile components of the fluid, such as silica, remain in the liquid phase, whereas the volatile components such as CO_2 mostly enter the steam phase. For example, consider an aquifer fluid at 265°C rising to the surface and "flashing" to 1 bar (abs.) (100°C).

Since flow is extremely rapid, the amount of heat lost by the fluid to its surroundings may be considered negligible--and the overall process therefore 'adiabatic'. If the process occurs reversibly, the expansion may be considered isentropic, but this is unrealistic since we know that the aquifer fluid expands irreversibly as it exits the aquifer into the lower pressure environment of the discharging well. For an irreversible expansion, with $\delta q = 0$, the process is isenthalpic*, so that the steam fraction may be calculated as follows.

* Provided that changes in potential and kinetic energy are negligible

From your steam tables (Appendix III) obtain the following information:

$H_{l,\ 265°C}$ = J/gm $H_{l,\ 100°C}$ = J/gm

$H_{v,\ 100°C}$ = "

(H = specific enthalpy, subscripts l,t and v,t refer to liquid and vapor phases at the specified temperature, t.)**

** In this text t is used for temperatures in °C and T for those in °K.

Provided no heat is gained or lost by the fluid as it moves to the surface

$$H_{l,\ 265°C} = x\ H_{l,\ 100°C} + y\ H_{v,\ 100°C}$$

where x is the mass fraction of the initial fluid which remains as liquid, and y is the mass fraction which turns to steam at 100°C.

$$x + y = 1$$

$$\therefore H_{l,\ 265°C} = (1 - y)\ H_{l,\ 100°C} + y\ H_{v,\ 100°C} \qquad (1)$$

Rearrange this equation to obtain the steam fraction, y, and solve the equation for the example given above:

$$y =$$

In many steam tables, values of the heat of evaporation (L) are given. Since $L = H_v - H_l$, equation (1) may be simplified for heat balance as,

$$H_{l,\ 265°C} = H_l + y\ L$$

If the process were indeed reversible, the steam fraction calculated assuming constant entropy, would be significantly smaller (y = 0.27). In very high speed expansion processes an isentropic or quasi-isentropic assumption may be more appropriate (see, for example, Kieffer, 1982).

Concentration changes during phase separation

If the aquifer fluid involved in the above flashing process to 100°C had a chloride concentration of 1145 mg/kg, what would be its chloride concentration after steam loss?

For our example calculate the new concentration of chloride in the flashed water. Write a Cl mass balance (chapter 1) and solve for C_{Cl}, the concentration of chloride, in the separated water. If you assume that Cl is not soluble in steam below about 370°C, the equation you obtain is

$$C_{Cl,\ 100°} = C_{Cl,\ 265°}/(1-y) = 1145/(1-y) = \quad mg/kg \qquad (2)$$

Use the value of y which you just calculated to solve the equation.

Multiple Flash Steam Separation

In many fields steam is separated from well discharges at two consecutive pressures prior to rejection of the residual liquid into a weirbox where additional steam loss occurs at the local boiling point. For the deep fluid considered above calculate the chloride concentration of weirbox liquid if steam separation occurred in the following 3 stages; 10 b.g., 5 b.g., 0 b.g.*

$ How do you account for the difference in composition with respect to single stage of steam separation at the weirbox?

If you measure the flow of steam from a single stage separator at 6 b.a. as 26 tonnes/hour and of water leaving the weirbox (silencer) as 100 tonnes/hour how would you calculate the total discharge enthalpy? What is it?

$ *Warning! Watch out for gauge pressures. The pressure actually read from a gauge (b.g. = bars gauge) is equal to the absolute pressure (b.a. = bars abs.) minus the ambient pressure of the atmosphere - usually taken as 1 b.a.

The same procedure may be followed to obtain the relative concentrations of the more insoluble gases in the initial fluid and subsequently separated vapor phase.

Suppose the original reservoir fluid contained 0.2 mg/kg of hydrogen. Hydrogen is very insoluble in liquid water at low pressures so that during boiling it moves (partitions) into the steam phase. So the concentration of H_2 in the liquid $\simeq 0$, but H_2 in the steam is given by

$$H_{2_{v,\ 100°}} = 0.2/y = \qquad (3)$$

Components like CO_2, H_2S and NH_3 are slightly soluble in the low temperature flashed water so that the concentrations in each phase are related by

$$H_2S_{reservoir} = (1 - y)\ H_2S_{l,\ 100°} + y\ H_2S_{v,\ 100°} \qquad (4)$$

Equation (3) is a first order approximation useful for rapidly calculating concentrations of the more insoluble gases. A rigorous equation is developed in Chapter 4 but requires some knowledge of the gas distribution coefficients.

To summarize, the concentration, C, of any component, including heat, and isotopes, may be written,

$$C_{reservoir} = (1 - y)\ C_l + y\ C_v \qquad (5)$$

here l and v refer to the flashed water and steam phases respectively.

Equation (5) is the heat and mass balance equation. You will find that its use becomes second nature to you.

COMPARATIVE CHEMISTRY OF GEOTHERMAL WELL DISCHARGES

In order to compare analyses from different wells or fields, or the same well at different times, we need to recalculate them to a common basis - often this is the composition obtained by flashing down to 0 b.g. This condition corresponds to that of water in the "weirbox" of the discharging well from which the majority of water samples are obtained. The separation of water and steam at a geothermal wellhead are outlined in Figures 2.5a and b. The compositions of waters sampled at higher pressure may be readily recalculated to this reference condition using the heat and mass balance equation as we see in the following exercises.

1. Using the heat and mass balance equation, recalculate the Cerro Prieto (CPM 19A) analysis (given in Table 2.3) to obtain the composition the water separated from this discharge if flashed to 0 b.g. Start by finding the steam fraction, y, which would result if the water at the sampling pressure (6.55 b.g.) was allowed to flash to 100°C. Then

$$C_{weirbox} = C_{sample}/(1-y)$$

The answer is given in Table 2.4.

2. Similarly recalculate the Reykjanes and Ngawha analyses to weirbox conditions.

3. Complete Table 2.4 by using your steam tables and equation (5) to calculate the concentrations of silica and chloride in the aquifers feeding each of the wells in Table 2.3.

Table 2.3 -- Concentration of major components in waters discharged from geothermal wells

	Enthalpy	Separation Pressure	$pH_{20°}$	Na	K	Ca	Mg	Cl	B	SO_4	HCO_3	SiO_2
	J/gm	b.g.*		mg/kg at separation pressure								
Wairakei W24 New Zealand	1085	0	8.3**	1250	210	12	0.04	2210	28.8	28	23	670
Tauhara TH1 New Zealand	1120	0	8.0	1275	223	14	n.a.	2222	38	30	19	726
Broadlands BR22 New Zealand	1160	0	8.4	1035	224	1.43	0.10	1705	51	2	233	848
Ngawha N4 New Zealand	1023	5.4	7.6	1025	90	2.9	0.11	1475	1080	27	298	464
Cerro Prieto CPM19A Mexico	1203	6.55	7.27	7370	1660	438	0.35	13800	14.4	18	52	808
Mahio-Tongonan 103 Philippines	1580	0	6.97	7155	2184	255	0.41	13550	260	32	24	1010
Reykjanes 8 Iceland	1153	19.0	6.4	11150	1720	1705	1.44	22835	8.8	28	87	631
Salton Sea IID1 California	1279	0.0	5.2	62000	21600	35500	1690	191000	481.2	6	220	1150

N.B. b.g. - bars gauge. Separation pressures are usually given as pressure gauge readings so add 1 bar to obtain the pressure in bars (absolute) before using your steam tables.

N.B. The pH given in this table is that of the water sample as measured in the laboratory and NOT that of the aquifer fluid from which it was derived. The two pH values always differ significantly because of the transfer of acidic gases to the separated steam as well as because of the temperature difference. We shall develop methods for the calculation of pH_T, the aquifer fluid's pH in Chapters 6 and 7.

Table 2.4

	Enthalpy - temperature* °C	Weirbox Cl (mg/kg)	Weirbox SiO_2 (mg/kg)	Downhole Cl (mg/kg)	Downhole SiO_2 (mg/kg)
W24					
TH1					
BR22					
N4					
CPM 19A	274		928	10488	614
MAHIO 103					
REYKJANES 8					
SALTON SEA IID1					

* The enthalpy-temperature is the temperature of steam-saturated water which has the same enthalpy as that measured for the fluid discharged from the well. For most of the wells in Table 2.3, measured enthalpies are close to those of steam-saturated water at measured "downhole" aquifer temperatures. For steam-saturated water the enthalpy-temperature relation is available from the Steam Tables (Appendix III).

Sampling and Analysis of Geothermal Fluids

Since the exploration and development of the Wairakei geothermal system in the mid-1950's, standard sampling and analysis techniques have been established for natural and well discharges. These are summarized by Ellis and Mahon (1977), Giggenbach (1975), Klyen (1975), Thompson (1975) and Truesdell and Nehring (1978).

Plate 1 shows a typical well discharging at Cerro Prieto, Mexico and Figure 2.5a the location of suitable points for the separation of water and steam samples from a cyclone separator of the type shown. The enthalpy of the discharge in this case may be calculated from the relative flows (kg/sec) of the separated water and steam (Grant et al., 1983).

Sampling from wells at the exploration stage may be more difficult and requires installation of correctly located sample points in the pipeline through which the two phase mixture from the well is discharged (Mahon, 1961) (Fig. 2.5b). In this case a mini-separator may be used for the collection of a steam sample from the two phase discharge and a water sample may be taken from the weirbox.

The enthalpy of the discharge is commonly obtained using an empirical calculation based on the measured pressure at the end of the two phase discharge line with suitable correction for the gas content of the discharge (James, 1962, Grant et al., 1983). Another method based on gas analyses is given in Chapter 4.

Plate 1. A typical well at Cerro Prieto discharging to a cyclone separator (left) with waste water rejected to a silencer and weirbox (background) or through pipes to evaporation ponds.

In the next exercise we start to compare the concentrations of components in the geothermal fluids represented in the above tables and look for patterns which might indicate that controlling reactions may occur in the reservoirs regardless of geography or water source.

First plot a graph of SiO_2 vs "enthalpy temperature" using (a) the data directly from Table 2.3 for weirbox samples (0 b.g.) and then plot a separate graph using (b) the "enthalpy temperature" vs downhole silica concentration, the data you have calculated in Table 2.4.

If you can explain the relationships shown, you will have discovered a sensitive chemical geothermometer, i.e. you will have discovered that a chemical quantity measured at surface allows you to infer temperatures at depth.

Apart from the Philippines sample, your curve (b) of silica concentration vs temperature is close to that determined in the laboratory for the solubility of quartz*. Because rapid conductive cooling of a solution initially saturated with quartz involves no concentration change, curve (b) is often referred to as the quartz (conductive) geothermometer*. Curve (a) is related to it but avoids the need to calculate y-values. It is called the quartz (adiabatic) geothermometer. Other geothermometers will be discussed later in the course.

---

* Later in the course you will learn how to express the curve (i.e., the solubility of quartz) you have obtained as an equation such as

$$t^{\circ}_{\text{quartz, no steam loss}} = \frac{1309}{5.19 - \log C_{SiO_2}} - 273.15$$

where C_{SiO_2} is the SiO_2 concentration in mg/kg.

(This equation applies through the range 0-250°C.)

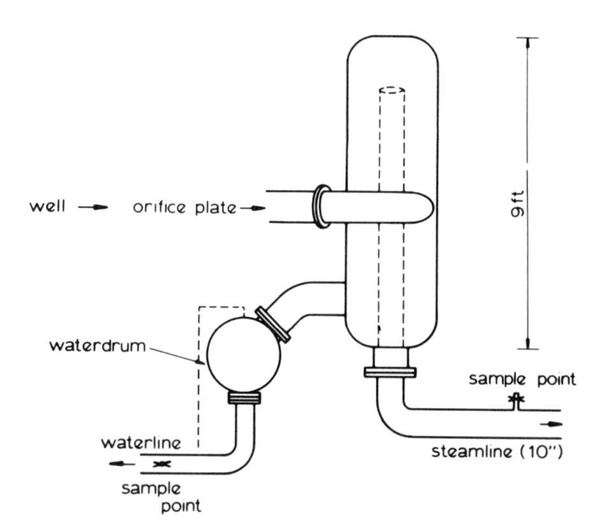

Figure 2.5a. Schematic section of a cyclone separator typical of those installed in geothermal fields. Water discharged from the separator may subsequently be further flashed at a lower pressure cyclone or in a weirbox. Samples from either are suitable for analysis of the liquid phase provided the sequence of separation pressures are known, but gas samples should be obtained from the vapor separated at the first stage separator. Why? (Reproduced with permission from Henley and Singers, 1982)

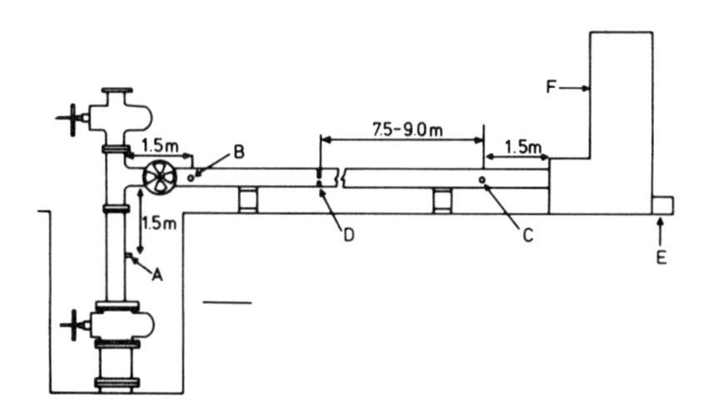

Figure 2.5b. Recommended position of sampling points on the surface piping of a geothermal well from which water and steam samples may be collected; A, B, and C are sample points; D, a constriction such as a back-pressure plate, valve or bend; f, silencer, E, weirbox. (From Ellis and Mahon, 1977).

$ 1. Try plotting aquifer chloride concentrations and Na/K ratios (from Table 2) versus enthalpy. Account for your results.

$ 2. How do you account for the anomaly shown by the Philippine well? (n.b. downhole temperature in this well is ≃ 305°C).

Total Discharge Compositions

In many wells as discussed in Chapter 11 the fluid discharged at the wellhead may not be directly related to the original aquifer fluid. This problem arises where steam from adjacent wells is drawn into the discharge of the sampled well or where additional heat is adsorbed from wallrocks. The Philippines example is such an anomalous - or excess enthalpy - well. Rather than assume without further qualification that concentrations obtained through equations (1) to (5) are representative of aquifer chemistry, the term " total discharge composition " should be used to describe the recalculated analysis based on discharge enthalpy.

The total discharge (often abbreviated T.D.) composition may also be used for comparing the chemistry of wells instead of the weirbox condition which we adopted above. Some problems may occur however if the well is of 'excess enthalpy' type - these are discussed in Chapter 11 but already we have one example, the Mahio-Tongonan well in the Phillipines.

MIXING OF GEOTHERMAL FLUIDS

As described earlier in this chapter, hot springs may arise where chloride water from depth reaches the surface directly, but in many systems underground mixing of fluids occurs to give hybrid fluids. In order to obtain hydrologic information from hot springs and wells we need to be able to recognize these processes and at least semi-quantitatively determine the relative proportions of each component of the mixture.

The heat and mass balance equations apply to any situation where phase separation or phase mixing occurs. If you mix 100 tonnes of 10°C water, whose enthalpy is 42 J/gm with 200 tonnes of 265°C water (1160 J/gm), the fully stirred mixture (300 tonnes) will have a total enthalpy of 2.36×10^{11} J. and temperature of 186°C since

$$(100/300 \times 42) + (200/300 \times 1160) = 791 \text{ J/gm}$$

and the hot and cold water fractions are .33 and .67 respectively.

This balance equation may be solved or expressed graphically (Fig. 2.6).

Now plot the variation of chloride concentration due to mixing on the same diagram and find the concentration of the 2:1 mixture considered above. Can you plot an equivalent diagram for the mixing of cold seawater (C_{Cl} = 19000 mg/kg) with the same hot chloride water considered above? Where would you expect to see such a mixing process occurring?

The same information can also be plotted on a single graph (Fig. 2.7) - often called a mixing diagram. In this example, the chloride concentration of the high temperature component is 1145 mg/kg and that of the cold component is 0.

Boiling of aquifer fluids may also be accommodated in 'mixing-diagrams' like Figure 2.7. The heat and mass balance equation used earlier in recalculating water sample analyses is the basis for the 'steam-loss' line shown in the figure.

> Take some time to consider these graphs in relation to equations (1), and (2). If the enthalpies and chloride concentrations of a series of deep wells fall on a steam-loss line in Figure 2.7, would you expect them to fall on or off a boiling-point depth curve such as is shown in Figure 2.2b ?
>
> Sketch a path in Figure 2.7 for (1) a deep fluid which boils before mixing, and (2) one that mixes before boiling.

GASES IN GEOTHERMAL FLUIDS

A number of gases occur dissolved in geothermal fluids. Table 2.5 gives illustrative analyses. Noble gases and some hydrocarbons are also present in somewhat lower concentrations. Carbon dioxide dominates the gas chemistry and as discussed later (Chapter 4) may play

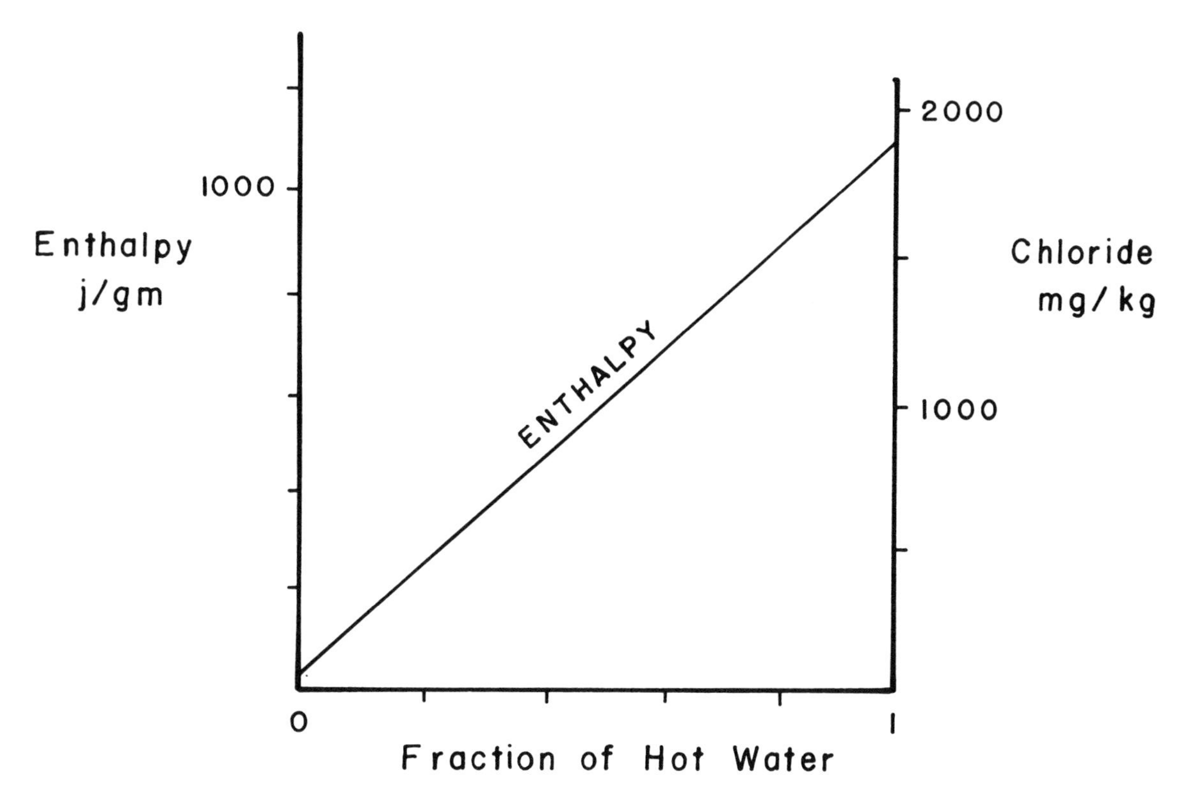

Figure 2.6. Mixing relations between hot water (265°C, Cl = 1145 mg/kg) and cold fresh water. The enthalpy of the mixed fluid is shown as a function of the fraction of hot water in the mixture. Plot the equivalent relation between hot water fraction and chloride concentration.

an important role in the physical state of the system. Reactions between gases such as CO_2 + $4H_2$ = CH_4 + $2H_2O$ are the basis of useful geothermometers. Much of the discussion of gas chemistry is left to individual chapters - partitioning during boiling (Chapter 4), gas geothermometers (Chapter 5).

As explained in Chapter 4, gas concentrations may be expressed in a variety of different units. In Table 2.5 the total gas concentration in the steam separated from an average Wairakei well is 0.2 millimoles gas/mole steam.

$$\text{i.e. } C_{\text{total gas}} = 0.2 = C_{CO_2} + C_{H_2S} + C_{H_2} + \ldots$$

The individual gases have been expressed as millimoles per mole (1000 millimoles) of gas, so that for the Wairakei example

$$C_{CO_2} = \frac{917}{1000} \times 100 = 91.7 \text{ mole\% of the combined gas.}$$

∴ In the steam

$$C_{CO_2} = 0.2 \times \frac{91.7}{100} = 0.18 \text{ millimoles/mole steam}$$

or

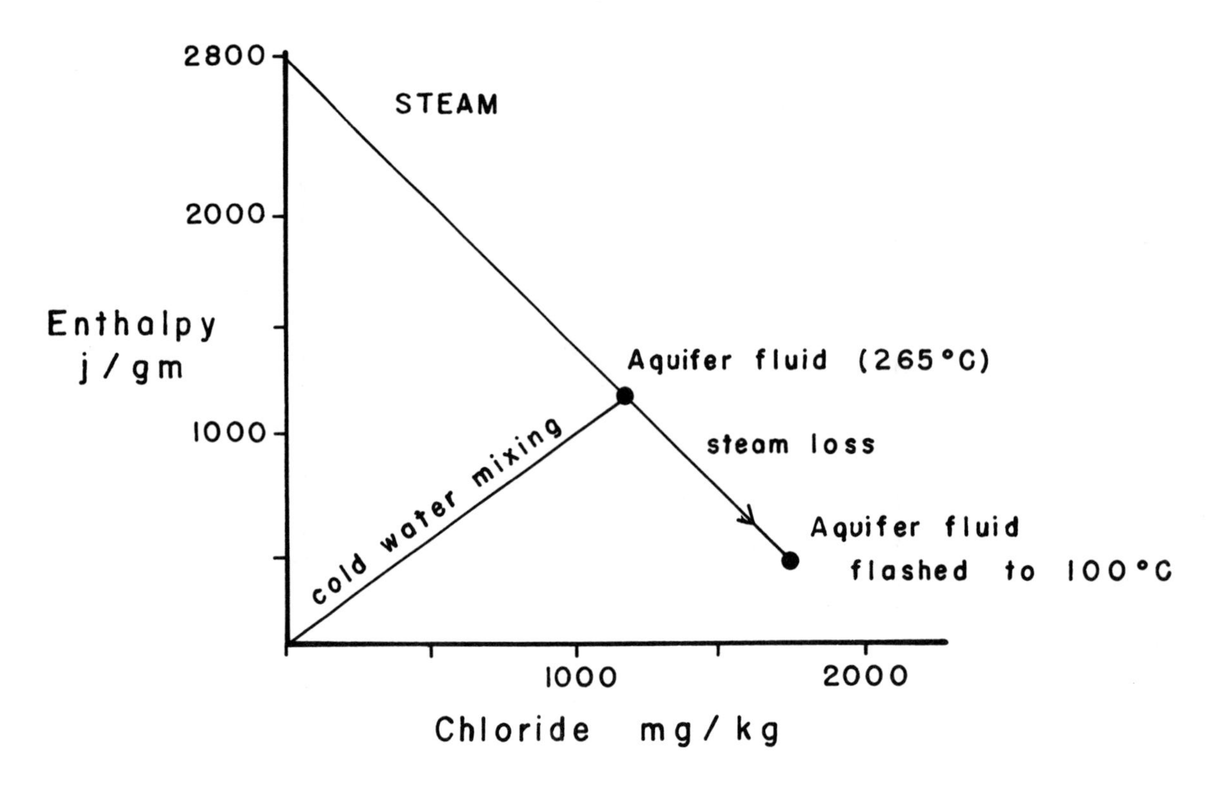

Figure 2.7. A typical enthalpy-chloride mixing diagram showing the effects of boiling and dilution on the 265°C aquifer fluid discussed in the text.

$$0.2 \times \frac{91.7}{100} \times \frac{1000}{18} \times 44 = 448 \text{ mg/kg steam}$$

and in the total discharge (T.D.) of the well

$$C_{CO_2} = 448 \times 0.3 = 135 \text{ mg/kg}$$

$$\text{or} = 3.04 \text{ millimoles/kg}_{T.D.}$$

Table 2.5 -- Gas concentrations in steam separated from geothermal well discharges

	Sample separation pressure b.g.	Enthalpy J/gm	Steam Fraction	total gas in steam millimoles mole^{-1} steam	CO_2 in total discharge mg/kg	CO_2	H_2S	CH_4	H_2	N_2* (+Ar)	NH_3
						millimoles/ mole total gases					
Wairakei (average)	1	≈1135	.3	0.2	134.5	917	44	9	8	15	6
Tauhara 1	8.8	1120	.2	1.2		936	64	na	na	na	na
Broadlands 22	10	1169	.19	10.04		956	18.4	11.8	1.01	8.89	4.65
Ngawha 4	1.87	968	.19	24.8		945	11.7	28.1	3.0	2.1	10.2
Cerro Prieto 19A	6.6	1182	.289	5.88		822	79.1	39.8	28.6	5.1	23.1
Tongonan 103	7.6	1615	.414	2.95		932	55	4.1	3.6	1.2	4.3
Reykjanes 9	19.0	1154	.135	.248		962	29	1	2	6	na
Salton Sea IID1	0	1279	.38	.507		957	43.9	na	na	na	na
Dominant Systems											
Geysers	-	-	1.0	20.0		941	16	12	23	8	na
Larderello	-	-	1.0	5.9		550	48	95	150	125	na

na = not analysed

* In this data set N_2 and Ar are not distinguished although this is quite practicable if a suitable gas chromatograph is available. Many geothermal analyses are recorded in this way.

Problem: Recalculate the data of Table 2.5 to obtain the CO_2 concentration of the discharges (i.e., underground fluid compositions) and list the systems in order of increasing deep system CO_2 concentration.
$

SUMMARY PROBLEMS

Now lets see how these relationships can help in understanding the hydrology of some geothermal systems.

$ 1. Examine the spring and well analyses from the Tauhara field (Tables 2.2 and 2.3) and plot a mixing diagram as follows: (H = enthalpy. Sample locations are shown in Figure 2.8).

   a. Plot H and C_{Cl} for the aquifer fluid represented by the well discharge (see Table 2.3).

   b. Plot C_{Cl} and $H_{spring\ temp.}$ for the Tauhara Springs.

   c. Using your quartz (adiabatic) geothermometer find the composition and temperature of the Fissure Spring water prior to boiling in its feeding vent.

$ 2. How many ways can you find to account for the Fissure Spring composition and temperature?

$ 3. What problems have arisen in using your silica geothermometer and how do you account for them?

$ 4. How do you account for the occurrence of 'steaming-ground' in the center of the cross-section?

$ 5. Draw a schematic hydrological flow and underground temperature diagram on the cross section provided (Fig. 2.9). You can use the boiling depth relation (Fig. 2.2b) to obtain minimum depths for the silica temperatures you have obtained. You may wish to refine your hydrological model when you learn more about geothermometry in Chapter 3. Additional help is given by the resistivity-sounding data shown in Fig. 2.8;

the sharp resistivity gradient at the 'boundary' of the field locates the area underlain by conductive chloride water. Also note how the cross-section is aligned relative to the lake and to high ground.

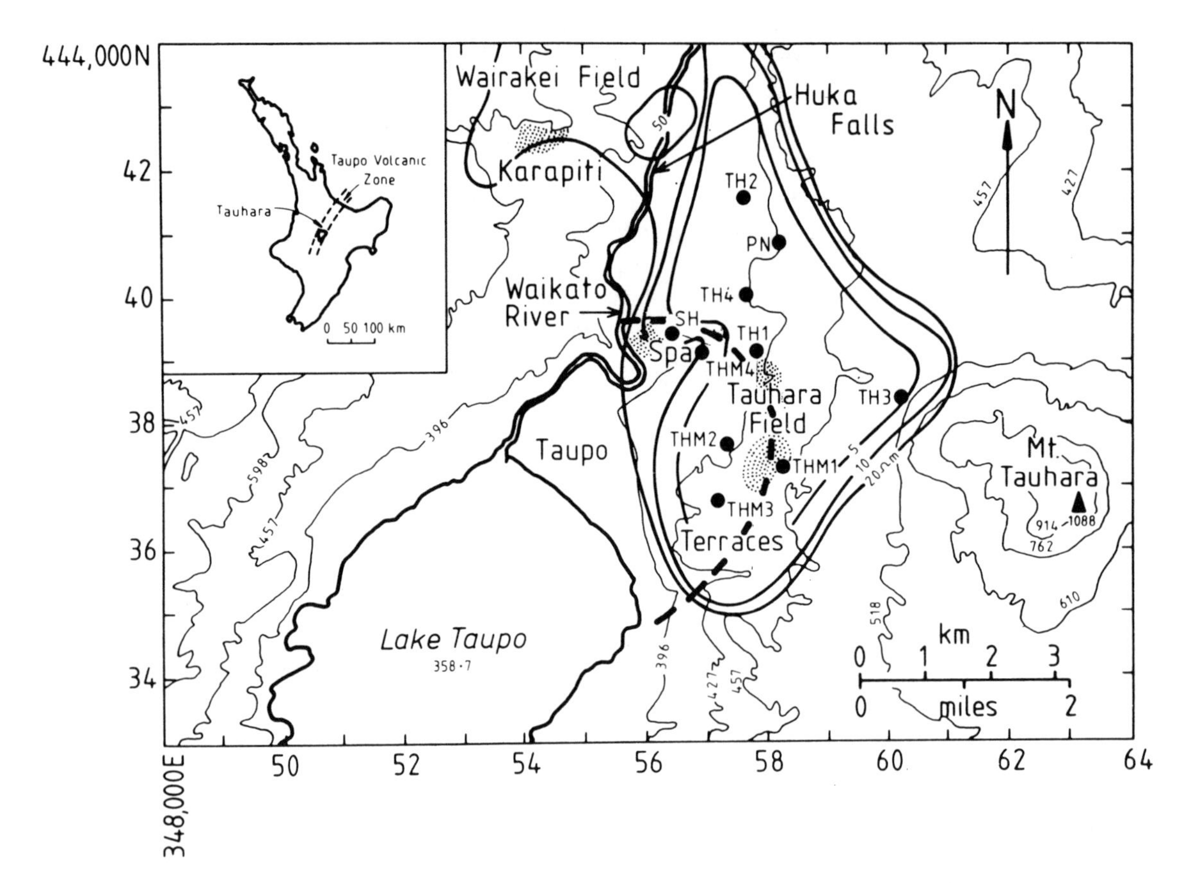

Figure 2.8. Map of the Tauhara geothermal field, showing altitude (meters), resistivity contours, exploration wells, areas of steaming ground (stippled) and other physiographic features. The broken, curved line provides the cross-section line for the exercise. (With permission, from Henley and Stewart, 1983.)

As another example draw a mixing diagram to account for the origin of the Ohaaki Pool (Table 2.2) at Broadlands. Well data are given in Table 2.3 and were used in our first steam fraction calculation. The well is about 500 m from the Ohaaki Pool. This pool is of special interest because of the occurrence of ore grade gold and silver in precipitates on its margin (Chapter 9)

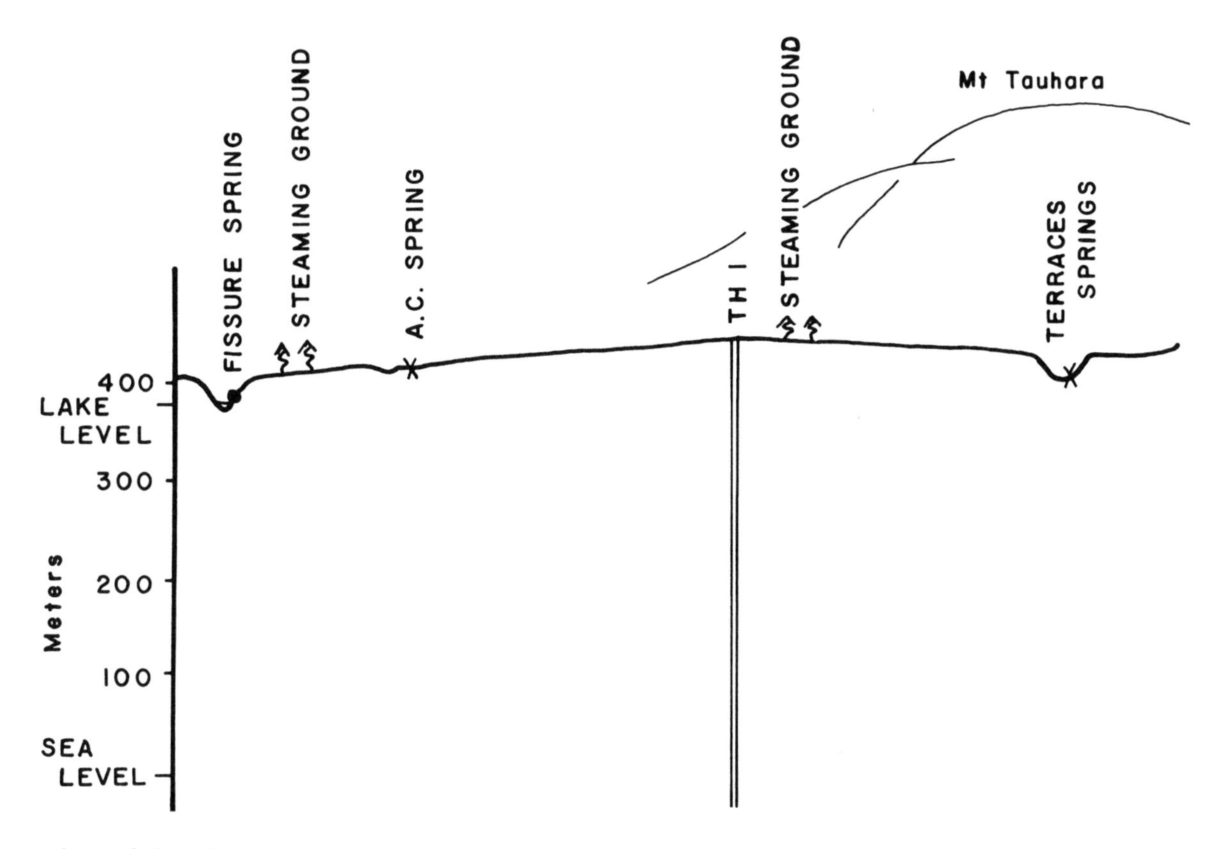

Figure 2.9. Sketch cross-section of the Tauhara field, looking north-east.

REFERENCES

D'Amore, F., and Celati, C., 1983, Methodology for Calculating Steam Quality in Geothermal Reservoirs: Geothermics, v. 12, no 2/3, p. 129-140.

Ellis, A.J., and Mahon, W.A.J., 1977, Chemistry and Geothermal Systems: Academic Press, 392 p.

Elder, J.W., 1981, Geothermal Systems: Academic Press, London, 508 p.

Giggenbach, W.F., 1975, A simple method for the collection and analysis of volcanic samples: Bulletin Volcanologique, v. 39, p. 132-145.

Giggenbach, W. F., 1980, Geothermal gas equilibria: Geochimica et Cosmochimica Acta, v. 44, p. 2021-2032.

Grant, M. A., Donaldson, I. G., and Bixley, P. F., 1982, Geothermal Reservoir Engineering, Academic Press, New York, 369 p.

Haas, J. L., 1971, The effect of salinity on the maximum thermal gradient of a hydrothermal system at hydrostatic pressure: Economic Geology, v. 66, p. 940-946.

Henley, R. W., and Ellis, A. J., 1983, Geothermal Systems Ancient and Modern, a geochemical review: Earth Science Reviews, v. 19, p. 1-50.

Henley, R. W., and Stewart, M. K., 1983, Chemical and isotopic changes in the hydrology of the Tauhara Geothermal Field due to exploitation at Wairakei: Journal Volcanology and Geothermal Research, v. 15, p. 285-314.

James, R., 1962, Steam-water critical flow through pipes: Proceedings of the Institute of Mechanical Engineers, v. 176, p. 741-748.

Klyen, L.E., 1982, Sampling Techniques for Geothermal Fluids: DSIR New Zealand, Chemistry Division Report CD2322, 88 p.

Kieffer, S. W., 1982, Dynamics and Thermodynamics of Volcanic Eruptions; implications for the plumes of Io; in Morrison, D. (ed.) Satellites of Jupiter, Univ. Arizona Press, Tucson, p. 647-723.

Mahon, W.A.J., 1961, Sampling of geothermal drill hole discharges: Proceedings of U.N. Conference New Sources of Energy, Rome, v. 2, p. 269-278.

Mahon, W. A. J., Klyen, L. E., and Rhode, M., 1980, Natural sodium-bicarbonate-sulphate hot waters in geothermal systems: Chinetsu (Journal of the Japanese Geothermal Energy Association) v. 17, p. 11-24.

Rybach, L., and Muffler, L.J.P. 1981, Geothermal Systems - Principles and Case Histories, Wiley, New York, 359 p.

Rybach, L., 1981, Geothermal Systems; Definition and Clasification; in Rybach, L. and Muffler, L.J.P.(eds), Geothermal Systems, Wiley, New York, p. 3-36.

Thompson, J.M., 1975, Selecting and collecting thermal spring waters for chemical analysis a method for field personnel: U.S. Geological Survey Open File Report 75-68.

Truesdell, A. H., and White, D. E., 1973, Production of superheated steam from vapor-dominated geothermal reservoirs: Geothermics, v. 2, no. 3-4, p. 154-173.

Truesdell, A.H., and Nehring, N., 1978, Collection of chemical, isotope and gas samples from geothermal wells: Proceedings 2nd Workshop on Sampling Geothermal Effluents, Environmental Protection Agency Report 600-7-78-121, p. 130-140.

White, D. E., Muffler, L. J. P., and Truesdell, A. H., 1971, Vapor-dominated hydrothermal systems compared with hot-water systems: Economic Geology, v. 66, no. 1, p. 75-97.

White, D. E., Fournier, R. O., Muffler, L. J. P., and Truesdell, A. H., 1975, Physical results of research drilling in thermal areas of Yellowstone Park, Wyoming: U.S. Geological Survey Professional Paper 892, 70 p.

Plate 2. The Pink Terrace, Lake Rotomahana, New Zealand (c. 1800 A.D.); an example of the discharge and precipitation of silica from a geothermal system. This and the White Terrace were destroyed in 1886 by hydrothermal eruptions initiated by the violent eruption of Mt. Tarawera, 2-3 km to the northeast. Photo courtesy of the Rotorua Museum.

Chapter 3
CHEMICAL GEOTHERMOMETERS FOR GEOTHERMAL EXPLORATION

In Chapter 2 chemical data from a set of geothermal wells were used to derive an empirical geothermometer based on silica content. In this section we shall derive this geothermometer by an independent method based on experimental data and examine some other empirical geothermometers which are related to mineral equilibria in the system. During exploration the chemical geothermometers provide rare and valuable windows into the deep system through which we see "as through a glass darkly" (St Paul), the choice and interpretation of geothermometer data being the art of the exploration geochemist.

SILICA GEOTHERMOMETERS

Quartz occurs in cuttings and cores from most geothermal fields. You look up the experimental solubility of quartz and find this data (Table 3.1) from Fournier and Potter (1982):

Table 3.1

t°C	1/T°K	SiO_2 mg/kg	QA	t°C	1/T°K	SiO_2 mg/kg	QA
0		2		175		190	
25		7		200		271	
50		13		225		367	
75		25.5		250		471	
100		46		275		571	
125		79		300		660	
150		126		325		738	

Now complete the following exercises:

$1. Write an equation for the solution of quartz to give SiO_2(aq). Write the equilibrium constant. Using the program LIN (Appendix I) evaluate the constants in log K = a + b (1/T), use mg/kg as activity units for SiO_2(aq) and use only data from 125° to 250°C for the best fit. Rearrange this into a geothermometer equation.*

---

* To avoid cumbersome sub- and superscripts we shall write geothermometer reactions without valence or concentration symbols (e.g. Na/K, SiO_2 not $CSiO_2$).

$2. Calculate the water fraction (x = 1-y) for a fluid that cools by adiabatic single stage steam loss from 250° to 100°C. Use this to calculate the silica content of this water at 100°C if it was saturated with quartz at 250°C (abbreviated QA for quartz saturated, adiabatically cooled). Write this value next to the quartz solubility at 250°C in the table.

For initial temperatures between 125° and 275°C and single stage steam separation, the water fraction at 100°C is related to initial temperatures by the equation (good to 0.6% or better)

$$x = 1.2148 - 0.002044 \ t \ (°C)$$

3. Use this formula and the tabulated quartz solubilities to calculate the rest of the silica contents of water cooled adiabatically to 100°C from temperatures of 125° to 275°C. Add these data to Table 3.1 in the column headed QA.

$ 4. Use your calculators (as above) to obtain the equation of log (SiO_2 QA) vs 1/T°K. Rearrange into a geothermometer equation. Why do your equations differ from those in Table 3.2?

Problems in the use of silica geothermometers: steam separation and dilution affect the concentration of silica in spring samples and sometimes in geothermal well samples. We have seen how maximum steam loss effects may be taken into account, but in many situations the rate of upflow is unknown so that it is impossible to estimate precisely the proportion of steam lost. Thus a range of temperatures obtained from the quartz geothermometers must be used in system modelling. In some cases where sufficient data are available mixing models may be applied.

For most deep wells quartz saturation may be assumed since quartz is abundant in the altered rocks, but in systems where temperatures are less than about 190°C it is sometimes found that silica contents reflect equilibrium with chalcedony rather than with quartz. Consequently spring waters for which independent methods suggest source temperatures less than 190°C should be also considered on the basis of the chalcedony geothermometer (Table 3.2).

In the case of the Tauhara springs discussion in the first section, both the amorphous silica and quartz geothermometers coincidentally give satisfactory temperature estimates for the Terraces dilute chloride spring, while steam-heated waters, which have extensively interacted with the local pumice breccias, give good amorphous silica temperatures when compared to their sampling temperatures. Evidently great care needs to be exercised when using the silica geothermometer in spring surveys where a range of independent methods should also be used. In deep wells, however, the quartz thermometer is found to give uniformly excellent results - some of the problems of "excess enthalpy" wells are discussed later.

Problems

1. Use your quartz solubility equation and the equations for chalcedony and amorphous silica solubility in Table 3.2 to draw a plot of silica (range 0-1500 mg/kg) against temperature (0 to 350°C). Plot also the experimental values for quartz solubility at 275°, 300° and 325°C.

2. If amorphous silica precipitation is essentially instantaneous what is the upper temperature limit of application of the quartz conductive and adiabatic geothermometers to hot springs near sea level and to hot springs at El Tatio, Chile where, due to the elevation, the local boiling temperature is 85°C?

3. Using your diagram as a mixing model, plot paths for dilution by cold water (10°C) and evaporation for a water initially saturated with quartz at 250°C.

Table 3.2 -- Equations expressing the temperature dependence of selected geothermometers. All concentrations are in mg/kg. (adapted from Fournier, 1981)

	Geothermometer	Abbreviation*	Equation	Restrictions**
a.	Quartz-no steam loss	TQC	$t°C = \frac{1309}{5.19 - \log SiO_2} - 273.15$	t = 0-250°C
b.	Quartz-maximum steam loss	TQA	$t°C = \frac{1522}{5.75 - \log SiO_2} - 273.15$	t = 0-250°C
c.	Chalcedony	TCH	$t°C = \frac{1032}{4.69 - \log SiO_2} - 273.15$	t = 0-250°C
d.	α-Cristobalite		$t°C = \frac{1000}{4.78 - \log SiO_2} - 273.15$	t = 0-250°C
e.	β-Cristobalite		$t°C = \frac{781}{4.51 - \log SiO_2} - 273.15$	t = 0-250°C
f.	Amorphous silica	TAM	$t°C = \frac{731}{4.52 - \log C} - 273.15$	t = 0-250°C
g.	Na/K (Fournier)	TNAK-F	$t°C = \frac{1217}{\log (Na/K) + 1.483} - 273.15$	t > 150°C
h.	Na/K (Truesdell)	TNAK-WE	$t°C = \frac{855.6}{\log (Na/K) + 0.8573} - 273.15$	t > 150°C
i.	Na-K-Ca	TNaKCa	$t°C = \frac{1647}{\log (Na/K) + \beta [\log(Ca/Na) + 2.06] + 2.47} - 273.15$	
j.	$\Delta\, ^{18}O(SO_4^{=} - H_2O)$		$1000 \ln \alpha = 2.88(10^6\, T^{-2}) - 4.1$, where $\alpha = \frac{1000 + \delta^{18}O(HSO_4)}{1000 + \delta^{18}O(H_2O)}$ and T in °K	

* These abbreviations are used in the text and in the program BAL (Appendix I).

** These equations do not describe silica solubilities above 250°C.

ALKALI GEOTHERMOMETERS

In a later section you will show that alkali concentrations (Na, K, Ca) and pH are related through a few relatively simple mineral equilibria between feldspars, mica and calcite. Field data (Figure 3.1) show that the atomic ratio (Na/K), above about 180°C, is a very useful geothermometer which is presumably controlled by mineral equilibria related to feldspar alteration. Assigning the empirical data to a specific mineral assemblage is a problem because of uncertainties in thermodynamic data and in the choice of polymorph (e.g., low albite vs high albite).

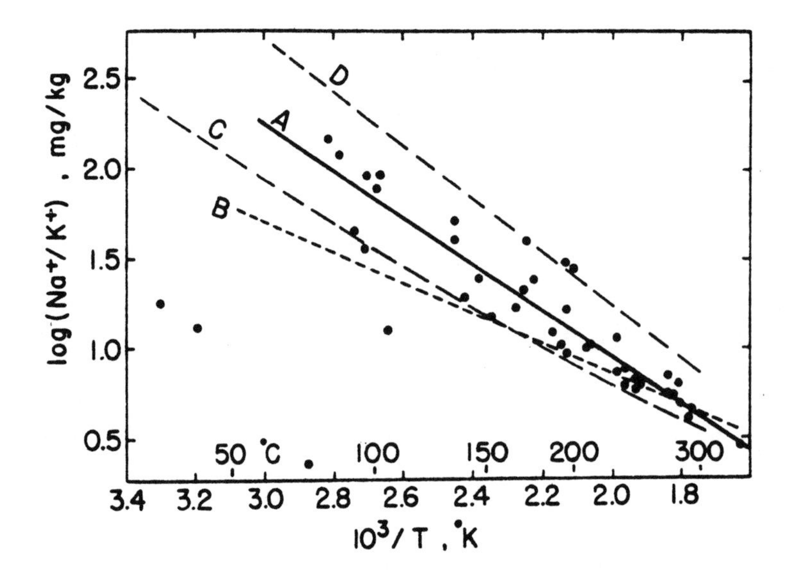

Figure 3.1. Na/K atomic ratios of well discharges plotted at measured downhole temperatures. Curve A is the least squares fit of the data points above 80°C. Curve B is another empirical curve (from Truesdell, 1976). Curves C and D show the approximate locations of the low albite-microcline and high albite-sanidine lines derived from thermodynamic data. (From Fournier, 1981)

In many areas the simple Na/K geothermometer (equation g or h in Table 3.2) gives unreasonably high temperature estimates due to ion exchange reactions with clay minerals particularly where deep temperatures are less than 200°C. Our study of mineral alteration equilibria (Chapter 6) suggests that calcic minerals (plagioclase, epidote, calcite) also play an important role in controlling fluid chemistry. The empirical NaKCa thermometer (equation i in Table 3.2) derived by Fournier and Truesdell (1973) must be based on equilibria involving these mineral phases, and, although some care is required in its use, is calibrated to give more acceptable temperatures over the temperature range 100°-300°C.

The rules for using the NaKCa thermometer are complex; use $\beta = 4/3$, if $t < 100°C$ and $[\log(\sqrt{Ca}/Na) + 2.06] > 0$ but if t with $\beta = 4/3 > 100$ or the above function is negative use $\beta = 1/3$ to obtain the temperature. A hand calculator program is invaluable in saving time. It is also important to notice that because $\sqrt{Ca}/Na$ is involved, the thermometer is not independent of steam loss or dilution - as you can show yourself. (The program listed in Appendix I may be used for your geothermometry problems.)

At low temperatures -- where calcite solubility is relatively high -- the P_{CO_2} of the solution may markedly affect the results. Also, although solubilities of magnesium silicates are very low at high temperatures, high magnesium contents at low temperatures affect the geothermometer and an empirical correction must be applied (see Figure 3.2) -- care must be taken however in case Mg has increased in the near surface due to mixing or rapid reactions. Notice that the geothermometer with its Mg correction is probably based on a succession of different mineral equilibria -- the Mg-Na-K-Ca relations of seawater for example show that low temperature reactions may be important.

Geothermal waters are generally relatively enriched in other alkalis -- Li, Rb, Cs -- and these may have some potential as geothermometers as is evident by inspection of Table 3.3. In the case of these elements however the controlling equilibria are probably cation exchange reactions with clays and zeolites rather than formation of discrete mineral phases. Fouillac and Michard (1981) proposed an empirical Na/Li geothermometer for waters where $Cl^- < 11000$ mg/kg (molal concentrations).

$$\log Na/Li = 1000/T - 0.38$$

and for higher salinity waters,

$$\log Na/Li = 1195/T + 0.13$$

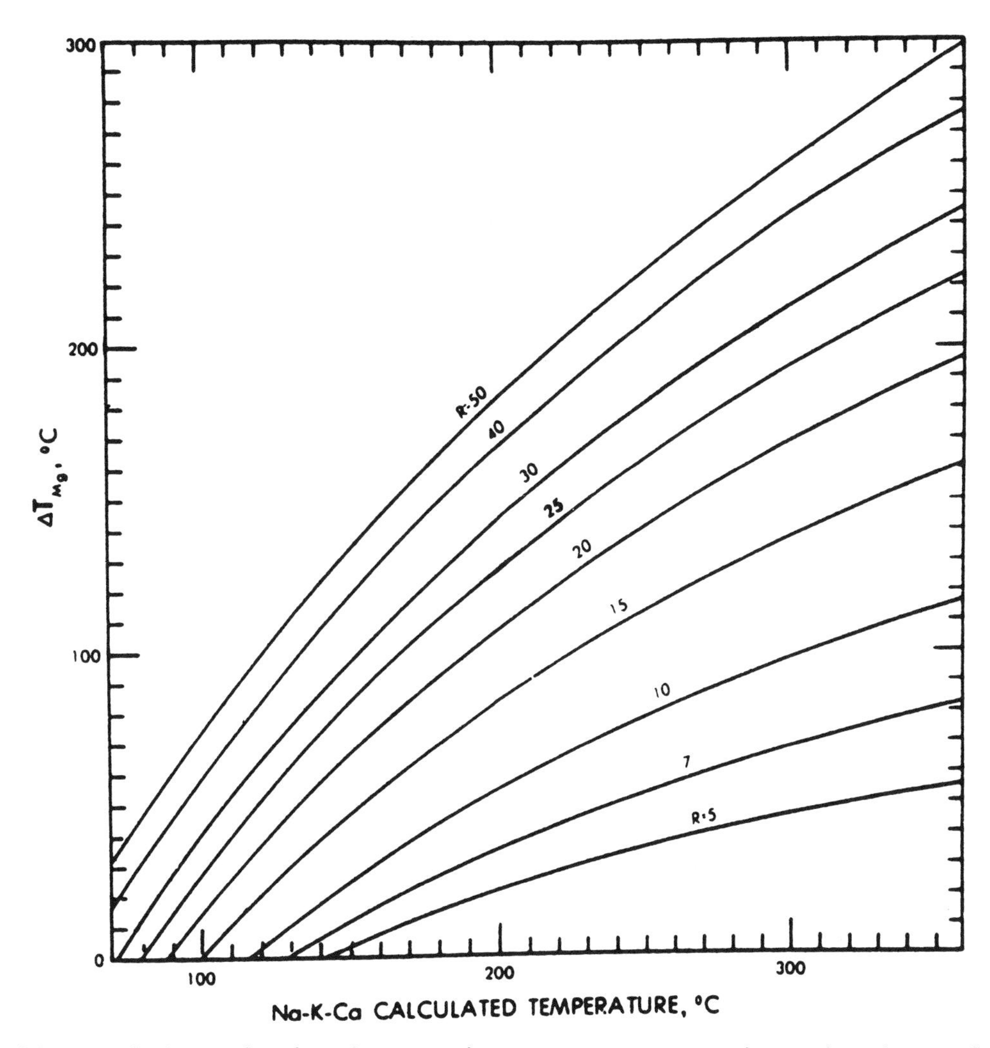

Figure 3.2. Graph for estimating the magnesium temperature correction to be subtracted from Na-K-Ca calculated temperature. R = 100 Mg/(Mg+Ca+K) with concentrations expressed in equivalents, i.e,, molality/ionic charge (From Fournier, 1981)

Table 3.3 -- Relationship between Na/K and Na/Li ratios and maximum downhole temperatures for waters discharged from Broadlands wells.

Well Number	Temp. (°C)	Na/K	Na/Rb	Na/Li
19	267	7.8	1605	25
2	262	8.0	1775	27
17	255	8.2	2000	30.2
18	240	8.5	2240	29.9
16	184	11.4	4155	30.2

Giggenbach et al. (1983) have proposed a K-Mg geothermometer (concentrations in mg/kg) which is useful in situations when Na and Ca do not equilibrate rapidly enough (e.g. seawater in a low temperature aquifer). Try this geothermometer on the following problems.

$$t_{KMg} = 4410/[13.95 - \log (K^2/Mg)] - 273.15$$

Problems

1. Is it logical to apply chemical geothermometry to analyses of steam-heated water?

2. Apply the quartz, Na/K and NaKCa geothermometers to the spring and well data given in Chapter 2 and discuss your results. Why is it that quartz temperatures are often about 30°C less than the alkali thermometer estimates?

3. We have the following (largely fictitious) water analyses from Israel where hot springs occur close to the coast (in mg/kg).

Table 3.4a

	Source	Temp.	Na	K	Ca	Mg	SiO_2
1.	Seawater	10°C	10500	380	400	1350	6
2.	Banias	14	22	0.2	45	10	4
3.	Tiberias	60	6990	350	3460	651	
4.	Golan	70	378	39	29	8	235
5.	Jordan	100	5439	210	200	-	120
6.	Haifa	74	141	13	40	9	81

Routine application of the geothermometry equations gives the data shown in the table below. Using the criteria discussed earlier circle the apparently most reliable results.

From the occurrence and chemistry we suspect that Golan is a high temperature water and that Jordan and Haifa are mixtures with seawater and rainwater (like Banias) respectively. If you calculate the mixing proportions based on Na and the resulting temperature you find that the geothermometers give mixed results. Can you describe the causes of these results? Which geothermometer would you use in a coastal geothermal system? What is the origin of the Tiberias spring? How would the Mg-correction affect the result? Plot the temperatures and silica contents on your silica-temperature diagram. Which analyse(s) could be used in mixing calculations?

Table 3.4b

					TNaKCa		Tsilica		
		Spring Temp. °C	log $\sqrt{Ca}$/Na + 2.06	R_{Mg}	β =4/3	β =1/3	Quartz	Chalcedony	Amorphous Silica
1.	Seawater	10	-0.7	79	270	173	24	-11	-77
2.	Banias	14	1.5	27	-22	55	12	-23	-87
3.	Tiberias	60	0.02	23	166	164	-	-	-
4.	Golan	70	0.2	21	167	194	188	171	67
5.	Jordan	100	-0.5	-	244	171	143*	121	26
6.	Haifa	74	0.7	24	97	167	124	96	7

* adiabatic cooling.

ISOTOPE METHODS

Various isotope geothermometers have been developed for exploration and are discussed in Chapter 10. Of these the exchange of ^{18}O between sulfate and water is currently the most useful (Truesdell and Hulston, 1980). The utility of the exchange reaction is based on its slow rate, so that deep system conditions may be preserved through to the surface outflow. There are however problems associated with data interpretation as described by Fournier (1981, p. 122):

> "If steam separation occurs during cooling, the oxygen isotopic composition of water will change. The fractionation of ^{16}O and ^{18}O between liquid water and

steam is temperature dependent and comes to equilibrium almost immediately at temperatures as low as 100°C. Therefore, if a water cools adiabatically from a high temperature to 100°C, the liquid water remaining at the termination of boiling will have a different isotopic composition depending on whether steam escaped continuously over a range of temperatures or whether all the steam remained in contact with the liquid and separated at the final temperature, 100°C (Truesdell, et al., 1977).

Although boiling makes interpretation more complex, it does not preclude use of the $\delta^{18}O$ ($SO_4^{=}$-H_2O) geothermometer. McKenzie and Truesdell (1977) showed that sulfate/oxygen isotope geothermometer temperatures could be calculated for three end-member models: (1) conductive cooling, (2) one-step steam loss at any specified temperature, and (3) continuous steam loss. Where water is produced from a well and steam is separated at a known temperature (or pressure), there is no ambiguity in regard to which model to use. Likewise, conductive cooling is assumed for springs hat emerge well below boiling.

The validity of temperatures calculated by the $\delta^{18}O$($SO_4^{=}$-H_2O) method is adversely affected by mixing of different waters (generally hot and cold) unless corrections are made for changes in isotopic composition of both the sulfate and water that result from that mixing. Even if the cold component of the mixture contains no sulfate, calculated sulfate/oxygen isotope temperatures will be in error unless the isotopic composition of the water is corrected back to the composition of the water in the hot component prior to mixing. Examples of calculated corrections for boiling and mixing effects applied to waters from Yellowstone National Park and Long Valley, California, were given by McKenzie and Truesdell (1977) and by Fournier et al. (1979).

The formation of sulfate by oxidation of H_2S at low temperatures is a particularly difficult problem to deal with when applying the sulfate/oxygen isotope geothermometer. [If there is little sulfate of deep origin,] a small amount of low-temperature sulfate can cause a large error in the geothermometer result. If analytical data are available for only one or two springs, addition of sulfate by H_2S oxidation may go unnoticed unless pH values are abnormally low. Where analytical data are available for several springs and they all have the same Cl/SO_4 ratio, oxidation of H_2S is probably unimportant. When variations in Cl/SO_4 are found in hot spring waters from a given region, the water with the highest Cl/SO_4 ratio has the best chance of being unaffected by H_2S oxidation."

CHEMISTRY AND GEOTHERMOMETRY OF SHOSHONE HOT SPRINGS

Shoshone Geyser Basin is a small hot spring area in Yellowstone Park. We will use it to illustrate the use of geothermometers and mixing models. Figures 3.3 (from Truesdell and Thompson, 1982) and 3.4 show the distribution of thermal features in the area, and Tables 3.5-3.7 give water and gas analyses and geothermometry data calculated using the program GEOTHERM (Truesdell, 1976b).

$ 1. Using a colored pen or pencil note the chloride contents of each hot spring on map (Fig. 3.4a) and contour the values. Do the same on the other maps for HCO_3/Cl and CO_2/residual gases. What is the connection between these quantities?

$ 2. On graph paper draw axes for Cl (0 to 400 mg/kg scale in the x direction) vs enthalpy (0 to 2800 J/g in the y direction). For the springs a and bb plot surface chloride and enthalpy (from Table 3.7) and draw a line from this point to the composition of steam at 93°C (surface boiling at Shoshone). Use the quartz adiabatic geothermometer temperature to calculate the enthalpy of the water before boiling. Plot this enthalpy along the line. Check the computer calculation of aquifer chloride (CLAQ) and enthalpy (EAQ) for these spring waters.

$ 3. On the same graph plot the aquifer compositions (CLAQ and EAQ) for all the other springs using the computer calculated values. Draw a boiling line from the highest chloride surface spring composition (E and CL) through the aquifer composition (CLAQ and EAQ) to the composition of steam (assume steam has no chloride). Draw the most likely dilution lines and estimate the range of enthalpy (and temperature) of the parent hot water. Cold (10°C) water at Shoshone has 4 mg/kg Cl.

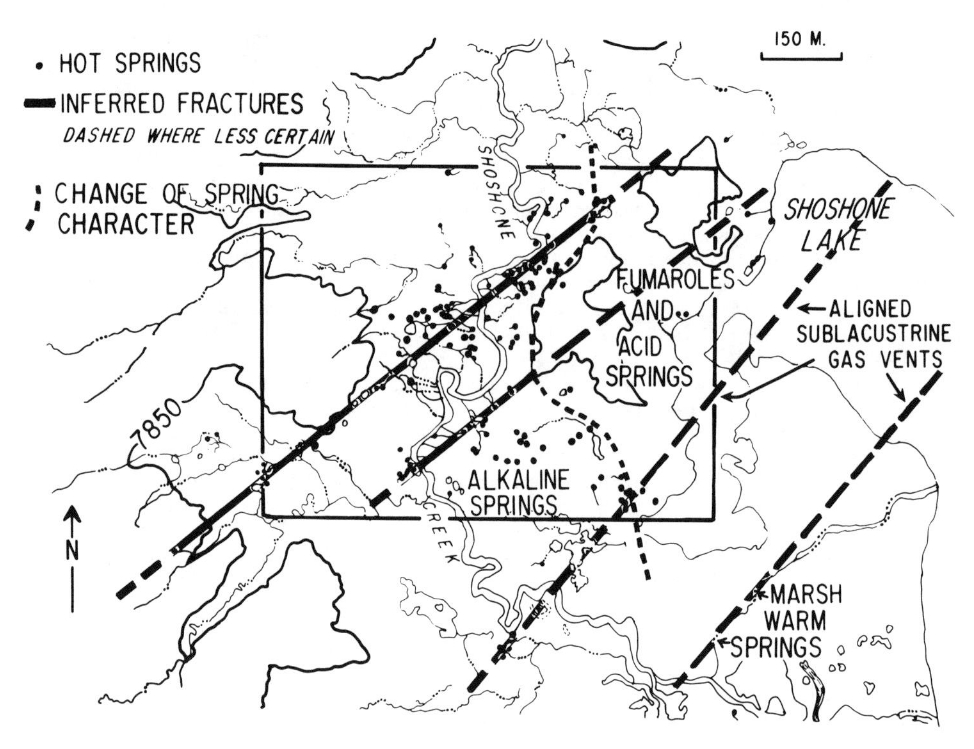

Figure 3.3. Shoshone Geyser Basin, Yellowstone National Park, showing location of hot-springs and inferred fractures. The box outline shows the area enlarged in Figure 3.4.

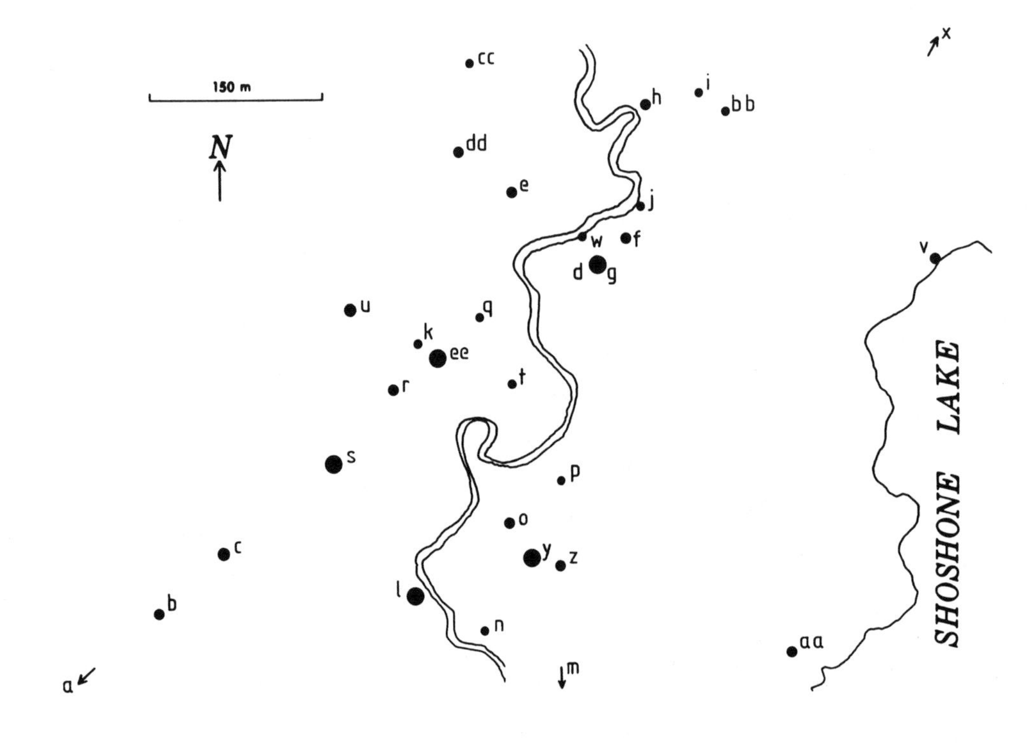

Figure 3.4a–c. Three maps of the distribution of hot springs in Shoshone Geyser Basin for the contouring of Cl, HCO_3/Cl and CO_2/R. Size of dots indicates relative discharge rates.

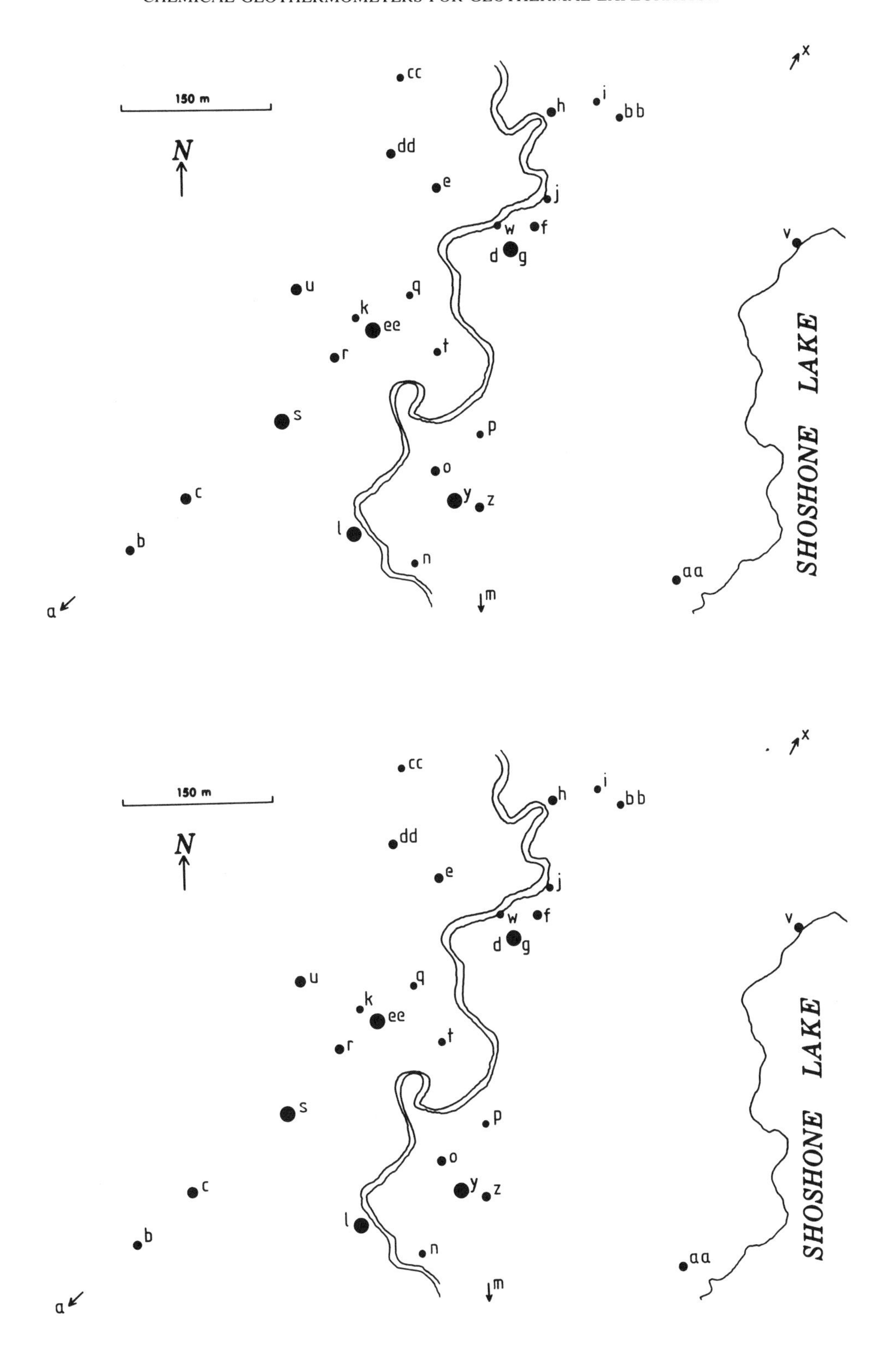
150 m
N
SHOSHONE LAKE
a
b
c
d
e
f
g
h
i
j
k
l
m
n
o
p
q
r
s
t
u
v
w
x
y
z
aa
bb
cc
dd
ee

$ 4. Using what you know, draw a hypothetical cross section from SE to NW of the Shoshone system perpendicular to the spring alignments. Assume fractures carrying only steam are vertical. Use the boiling curve equation (R. James, personal communications, 1976) to calculate the depth, h (in meters), of first boiling.

$$h = (t°C/69.56)^{4.7962}$$

Draw reasonable slopes for fractures carrying hot water and draw a shallow mixing reservoir. If you were drilling a 1 km exploratory hole to attempt to find the parent "reservoir" where would you drill it?

5. List the reactions suggested by the chemical data. Indicate on your cross section the direction of subsurface flow of hot water and cold water.

$ 6. Spring bb had $\delta^{18}O(H_2O) = -17.4$ and $\delta^{18}O(SO_4) = -11.1$.

Using the equation (Table 3.2)

$$\delta^{18}O(SO_4) - \delta^{18}O(H_2O) \simeq 10^3 \ln\alpha = 2.88 \times (10^6/T^2) - 4.1$$

with T in °K, calculate the sulfate isotope temperature for this spring and compare with results from other geothermometers and mixing calculations.

Table 3.5 -- Water analyses of Shoshone Springs (J.M. Thompson, analyst)

Code	Name	t°C	SiO_2	Cl	Na	K	Ca	HCO_3	SO_4	B	Mg
a	Boiling Spring	94	170	135	295	11.3	4.1	456	36	1.7	0.05
b	Boiling Caldron	94.5	260	175	325	10.7	0.5	416	43	2.1	0.05
c	Velvet	92.8	250	183	330	13.7	0.9	438	45	2.5	0.05
d	Gourd	93	256	155	315	12.4	1.5	435	46	2.3	0.05
e	Unnamed	93	260	165	315	13.3	1.1	425	50	2.3	0.05
f	Little Bugler	87	256	157	300	11.7	2.0	425	55	2.0	0.05
g	Shield	92.5	250	156	300	12.4	2.5	437	46	2.0	0.05
h	Little Giant	92.5	292	168	300	18.4	0.9	399	46	2.2	0.05
i	Unnamed	94	308	278	350	15.6	1.1	326	50	3.6	0.05
j	Black Sulfur	93.8	244	133	275	11.6	1.2	345	65	2.0	0.05
k	Bead	90.5	278	174	330	15.5	1.1	437	47	2.3	0.0
l	Unnamed	---	242	179	280	11.7	1.6	322	40	2.2	0.05
m	"	86	344	215	350	12.5	0.9	393	43	2.7	0.05
n	Washtub	81	328	238	365	16.0	0.4	406	48	2.8	0.05
o	Unnamed	94.5	316	238	375	10.3	0.7	419	52	2.9	0.05
p	Taurus	94.5	296	200	340	13.3	0.6	434	53	2.6	0.05
q	Pearl	90	266	178	315	12.2	1.1	427	49	2.2	0.05
r	Unnamed	93	266	175	315	14.8	1.0	432	46	2.2	0.05
s	Coral	80.5	286	193	325	11.3	1.3	437	48	2.0	0.05
t	Bronze	93	246	167	315	12.4	1.1	424	50	2.3	0.05
u	Glen	94.5	268	170	315	15.6	1.1	433	49	2.3	0.05
v	Unnamed	93	404	60	60	32	4.3	0	436	0.9	0.49
w	"	93.5	256	153	295	11.7	1.4	410	57	2.0	0.05
x	"	83	286	92	170	15.5	0.7	213	61	1.1	0.14
y	Union	91.5	352	242	380	11.0	0.3	445	42	2.8	0.04
z	Unnamed	81	349	243	380	10.0	0.3	445	46	2.8	0.01
aa	"	---	262	174	260	19.0	0.6	267	54	1.9	0.03
bb	"	89.5	350	323	350	13.0	0.3	230	70	3.7	0.01
cc	"	83.5	299	158	300	20.0	0.5	435	46	1.4	0.01
dd	"	93	300	123	250	25.0	0.6	428	34	2.0	0.01
ee	"	80	310	183	310	12.0	0.5	438	43	0.7	0.01

Table 3.6 -- Analyses of gases from Shoshone Springs (in mole %; R = residual gases [N2, H2, Ar...])

Spring	CO_2	H_2S	R	CO_2/H_2S	CO_2/R
b	76.0	<.1	24.0	>760	3.2
c	82.3	0.2	17.5	410	4.7
i	80.9	0.6	18.5	135	4.4
j	93.8	0.3	5.9	310	16
l	82.0	0.2	17.8	410	4.6
m	90.5	0.2	9.3	450	9.7
u	79.9	0.1	20.0	800	4.0
p	90.4	0.3	9.3	300	9.7
q	83.0	0.1	16.9	830	4.9
v	94.2	0.7	5.1	135	18
x	88.8	6.4	4.8	14	19
aa	97.3	0.8	1.9	122	51
bb	91.0	4.8	4.2	19	22
dd	93.4	<.1	6.5	>930	14
ee	90.8	0.2	9.1	450	10

Table 3.7 -- Shoshone geyser basin, Yellowstone Park, Wyoming

NONMIXED CL = 323 T = 89.5 COLD CL = 2 T = 4

CODE	SPRING	NAME	T	E	CL	SIO2	TQA	TQC	TCH	T13	T43	XHOT	THOT	CLAQ	EAQ	EHOT	CLHOT	TMIXI	XCLD	CLHCO3	CLSO4	CLB
a	T7201	BOILING	94.0	394	135	170	161	169	146	154	166	0.56	272	118	679	1194	208	238(25)	0.63	0.51	10.16	24.22
b	T7202	BOILING	94.5	396	175	260	185	199	180	166	250	0.68	265	145	787	1157	214	0(75)	0.00	0.72	11.02	25.41
c	T7203	VELVET	92.8	389	183	250	183	196	177	173	240	0.69	256	152	776	1116	219	0(75)	0.00	0.72	11.01	22.32
d	T7204	GOURD	93.0	390	155	256	184	198	179	165	210	0.63	279	128	782	1232	203	0(75)	0.00	0.61	9.12	20.55
e	T7205	UNNAMED	93.0	390	165	260	185	199	180	171	228	0.65	272	136	786	1196	208	0(75)	0.00	0.67	8.94	21.88
f	T7206	LITTLE B	87.0	364	157	256	184	198	179	161	194	0.63	278	129	779	1223	202	0(75)	0.00	0.64	7.73	23.94
g	T7207	SHIELD	92.5	387	156	250	183	196	177	162	189	0.63	277	129	776	1220	204	0(75)	0.00	0.61	9.18	23.79
h	T7208	LITTLE G	92.5	387	168	292	192	207	190	193	259	0.66	277	136	817	1221	204	0(75)	0.00	0.72	9.89	23.29
i	T7209	UNNAMED	94.0	394	278	308	195	211	195	176	241	0.91	214	224	832	917	248	0(75)	0.00	1.47	15.06	23.55
j	T7210	BLACK SU	93.8	393	133	244	181	194	175	168	212	0.58	296	111	770	1319	190	0(75)	0.00	0.66	5.54	20.28
k	T7211	BEAD	90.5	379	174	278	189	204	186	178	239	0.67	269	142	803	1182	209	0(75)	0.00	0.69	10.03	23.07
l	T7212	UNNAMED	0.0	379	179	242	0	0	0	166	202	0.00	0	0	0	0	0	0(0)	0.00	0.96	12.12	24.81
m	T7213	UNNAMED	86.0	360	215	344	201	220	206	166	235	0.77	255	168	858	1110	218	0(75)	0.00	0.94	13.54	24.28
n	T7214	WASHTUB	81.0	339	238	328	198	216	201	185	296	0.81	239	186	842	1030	228	0(75)	0.00	1.01	13.43	25.92
o	T7215	UNNAMED	94.5	396	238	316	197	213	198	156	235	0.82	236	192	839	1020	233	0(75)	0.00	0.98	12.39	25.03
p	T7216	TAURUS	94.5	396	200	296	193	208	192	174	258	0.74	255	163	821	1111	220	0(75)	0.00	0.79	10.22	23.46
q	T7217	PEARL	90.0	377	178	266	186	200	182	167	222	0.68	264	146	791	1153	213	0(75)	0.00	0.72	9.84	24.67
r	T7218	BUNNAMED	93.0	390	175	266	187	200	182	178	239	0.68	266	144	792	1164	212	0(75)	0.00	0.70	10.30	24.26
s	T7219	CORAL	80.5	337	193	286	189	206	189	161	211	0.71	258	154	805	1122	215	0(75)	0.00	0.76	10.89	29.43
t	T7220	BRONZE	93.0	390	167	246	182	195	175	168	223	0.65	267	139	772	1170	211	0(75)	0.00	0.68	9.04	22.14
u	T7221	GLEN	94.5	396	170	268	187	201	183	180	238	0.67	270	140	795	1186	210	0(75)	0.00	0.68	9.40	22.54
v	T7222	UNNAMED	93.0	390	60	404	212	235	222	286	192	0.46	373	46	908	1963	99	0(75)	0.00	0.00	0.37	20.33
w	T7223	UNNAMED	93.5	392	153	256	184	198	179	165	208	0.63	281	127	782	1241	201	0(75)	0.00	0.64	7.27	23.33
x	T7305		83.0	348	92	286	190	206	189	205	242	0.49	343	74	806	1613	147	0(75)	0.00	0.74	4.08	25.51
y	T2313	UNION	91.5	383	242	352	203	222	208	166	282	0.83	241	191	867	1042	229	0(75)	0.00	0.94	15.60	26.36
z	T7314		81.0	339	243	349	201	221	207	161	274	0.83	240	188	859	1036	227	0(75)	0.00	0.94	14.31	26.47
aa	T7316		0.0	339	174	262	0	0	0	204	278	0.00	0	0	0	0	0	0(0)	0.00	1.12	8.73	27.93
bb	T7319		89.5	375	323	350	203	221	208	178	293	1.00	203	254	865	865	254	0(75)	0.00	2.42	12.50	26.62
cc	T7327		83.5	350	158	299	192	209	193	203	297	0.64	286	126	818	1268	195	0(75)	0.00	0.63	9.30	34.42
dd	T7328		93.0	390	123	300	194	209	193	223	300	0.57	316	99	824	1437	173	0(75)	0.00	0.49	9.80	18.76
ee	T7329		80.0	335	183	310	194	212	196	173	257	0.69	270	144	826	1182	207	0(75)	0.00	0.72	11.52	79.73

The table is an example of output from the program, GEOTERM (Truesdell, 1976). Abbreviations: TQA, quartz saturation with adiabatic cooling; TQC, quartz saturation with conductive cooling; TCH, chalcedony saturation with conductive cooling; T13, the Na-K-Ca geothermometer with beta = 1/3; T43, the Na-K-Ca geothermometer with beta = 4/3; TMIX1, the warm springs mixing model; THOT, the boiling springs mixing model. The number in parentheses after the warm-springs mixing temperature is the silica content in mg/kg assumed for the cold component of the mixture and this is followed by the fraction of the cold component, XCLD; From the boiling springs mixing model the hot component temperature, THOT, and fraction, XHOT are calculated, as are the aquifer chloride content and enthalpy after mixing (CLAQ and EAQ) and the chloride and enthalpy of the hot component of the mixture (CLHOT and EHOT) for assistance in plotting the data. The printout also includes the surface temperature, enthalpy, chloride and silica contents of the hot spring waters as well as the atomic ratios of chloride to total bicarbonate- CLHCO3, sulfate- CLSO4, and boron- CLB. All temperatures are in °C, enthalpies in Joules/gm and concentrations in mg/kg.

REFERENCES

Fournier, R. O., 1981, Application of water geochemistry to geothermal exploration and reservoir engineering; Chapt. 4 in Geothermal Systems: Principles and Case Histories, L. Ryback and L.J. P. Muffler eds., Wiley New York, p.109-143.

Fournier, R. O., and Potter, R. W., 1982, A revised and expanded silica (quartz) geothermometer: Geothermal Research Council Bulletin, v. 11, p. 3-9.

Fournier, R. O., and Truesdell, A. H., 1973, An empirical Na-K-Ca geothermometer for natural waters: Geochimica et Cosmochimica Acta, v. 37, p. 1255-1275.

Fournier, R. O., Sorey, M. L., Mariner, R. H., and Truesdell, A. H., 1979, Chemical and isotopic prediction of aquifer temperatures in the geothermal system at Long Valley, California: Journal of Volcanology and Geothermal Research, v. 5, p. 17-34.

Fouillac, C., and Michard, G., 1981, Sodium/lithium ratio in water applied to the geothermometry of geothermal waters: Geothermics, v. 10, p. 55-70.

Giggenbach, W. F., Gonfiantini, R., Jangi, B.L., and Truesdell, A. H., 1983, Isotopic and chemical composition of Parbati Valley geothermal discharges, NW Himalaya, India: Geothermics, v. 12, p. 199-222.

McKenzie, W. F., and Truesdell, A. H., 1977, Geothermal reservoir temperatures estimated from the oxygen isotope compositions of dissolved sulfate and water from hot springs and shallow drillholes: Geothermics, v. 5, p. 51-62.

Truesdell, A. H., 1976a, Summary of section III, Geochemical techniques in exploration: Proc. 2nd U.N. Symp. on the Development and Use of Geothermal Resources, San Fran., 1975, v.1, p. liii-lxxix.

Truesdell, A. H., 1976b, GEOTHERM, a geothermometric computer program for hot spring systems: Proc. 2nd U.N. Symp. on the Development and use of Geothermal Resources, San Francisco, 1975, v. 1, p. 831-836.

Truesdell, A. H., and Hulston, J. R., 1980, Isotopic evidence on environments of geothermal systems; Chap. 5 in Handbook of Environmental Isotope Geochemistry, Vol. 1 The terrestrial environment, P. Fritz and J. Ch. Fontes, eds Elsevier, Amsterdam, p. 179-226.

Truesdell, A. H., and Thompson, J. M., 1982, The geochemistry of Shoshone Geyser Basin, Yellowstone National Park: Guidebook 33rd Ann. Field Conf., Wyoming Geol. Assoc. p. 153-159.

Truesdell, A. H., Nathenson, M., and Rye, R. O., 1977, The effects of subsurface boiling and dilution on the isotope compositions of Yellowstone thermal waters: Journal of Geophysical Research., v. 82, p. 3694-3704.

Plate 3. Fumaroles and intense steaming ground, with advanced argillic alteration, in the Karapiti Thermal Area; part of the Wairakei Geothermal Field, New Zealand. This is an impressive example of steam and gas discharge from a geothermal system.

Exploitation of the Wairakei Field for electric power production led to a major increase in the heat flow of this area as aquifer water flashed to steam. Prior to this, surface activity was confined to minor steaming ground and a single fumarole (off RIGHT). The present intense activity includes several new fumaroles, mud-pots and hydrothermal blowouts due to high temperature steam condensing into shallow ground-water.

Chapter 4
GASEOUS COMPONENTS IN GEOTHERMAL PROCESSES

Gas concentrations in fluids encountered during drilling of geothermal fields range from 0.05 wt% (Wairakei, Ahuachapan) up to about 1 wt% (Ngawha, Broadlands). We discovered in Chapter 2 that carbon dioxide is the dominant gas in geothermal systems and, as we shall see later, plays an important role in controlling the pH of the aquifer fluid. The ratios of the principal gases (e.g., CO_2, H_2, CH_4) are controlled by reactions such as

$$CO_2 + 4H_2 = CH_4 + 2H_2O$$

and may therefore be used as geothermometers in the same way as we have used alkali ion ratios. The development of gas geothermometers is discussed in a later chapter; at this stage we will examine the behaviour of gases when phase separation occurs from an initially single-phase geothermal fluid. This is important when we consider the recalculation of analyses of steam samples separated at the surface to determine aquifer dissolved gas compositions.

Gas pressures are also important in reservoir modelling studies as well as in a number of engineering problems associated with geothermal field development; in studies of fossil hydrothermal systems -- ore deposits -- the constraints imposed by gas contents are just as important and deserve much more attention.

DISTRIBUTION OF GASES BETWEEN LIQUID AND VAPOR

When steam separates from a liquid during boiling, gases like H_2, N_2, and CH_4 move preferentially into the vapor phase. The more soluble gases (CO_2, H_2S, NH_3) are partially retained in the residual liquid. It is possible to calculate gas concentrations in both the vapor and liquid phases resulting from such a process. These concentrations depend on the initial enthalpy of the fluid, the final temperature, the gas distribution coefficient (B) and the composition of the initial liquid phase.

The gas distribution coefficient, B, is defined as the concentration of gas in vapor divided by the concentration of gas in liquid. Because B is a ratio, any consistant molal or related concentration units may be employed. At low gas concentrations mole fractions may be used without introducing any significant error.

From experimental gas solubility data, Giggenbach (1980) derived the following regression equations valid from 100 to 340°C,

Table 4.1

$$\log B_{NH_3} = 1.4113 - .00292t \qquad (\text{ t in °C })$$

$$\log B_{H_2S} = 4.0547 - .00981t$$

$$\log B_{CO_2} = 4.7593 - .01092t$$

$$\log B_{CH_4} = 6.0783 - .01383t$$

$$\log B_{H_2} = 6.2283 - .01403t$$

$$\log B_{N_2} = 6.4426 - .01416t$$

Problem: Calculate B values for CO_2, H_2S and H_2 at 300 and 150°C. Which gas is the most soluble in the liquid phase?

Single Step Steam Separation

Consider a deep fluid with an initial CO_2 content, C_o, and initial temperature of 260°C. If a fraction of steam (y) separates adiabatically and remains in contact with the liquid until it reaches 250°C, what is the concentration of CO_2 in the water phase, C_l, and in the steam phase, C_v, in terms of the original concentration?

To solve this we need an equation, giving C_l in terms of C_o, y and B. The first step is to write the mass balance equation giving C_l and C_v in terms of C_o and y.

$$C_o = C_l(1 - y) + C_v(y) \qquad (1)$$

Dividing through by C_l will now allow us to eliminate C_v by substituting our B expression and the rest is straightforward.

$$C_o/C_l = (1 - y) + B\,y \quad \underline{\text{or}} \quad 1 + y(B - 1) \qquad (2)$$

Now rewrite the equations in terms of C_l/C_o and C_v/C_o. The separated steam fraction, y, depends on the initial fluid enthalpy and the temperature interval over which boiling occurs. (Notice that for most purposes we do not need to consider the heat of solution of gas in water when calculating y).

Consider a fluid containing 500 millimoles CO_2/kg solution which is initially at 260°C and is separated into water and steam fractions at the temperatures shown. Complete the following table.

Table 4.2

t°C	Σy	B_{CO_2}	$\dfrac{1}{1 + y(B_{CO_2} - 1)}$	C_l	C_v	C_l/C_o
260	-	83	1.000	500.0	-	1.000
256	.01	92	.524	262.0	24104.0	.524
253	.02	99	.337	168.9	16721.1	.337
250	.03	107				
240	.057					
220	.103					
200	.146					
150	.238					

$ What range of steam sampling pressures do you recommend for the determination of the CO_2 content of a geothermal discharge? (i.e., under what conditions can most of the originally dissolved gas be collected in a single steam sample?)

The process you have considered is termed SINGLE STEP STEAM SEPARATION. In some cases you will find it referred to as 'closed-system boiling'. The process is broadly analogous to the formation of fully equilibrated phenocrysts from a magma.

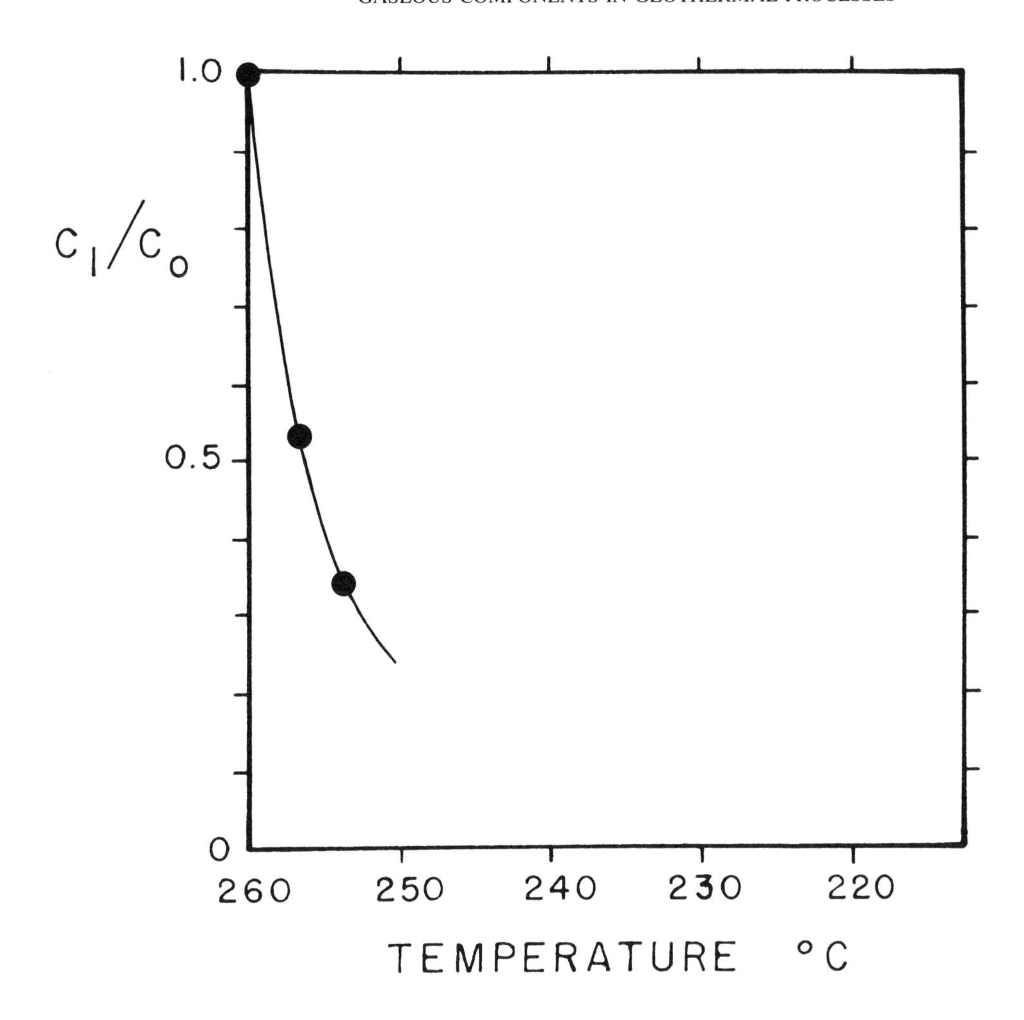

Figure 4.1. Decrease of the CO_2 content of residual water relative to the initial CO_2 content following single-step steam loss in the temperature range 260 to 220°C.

Figure 4.1 shows the CO_2 data from Table 4.2; notice that C_1/C_0 is independent of C_0. Sketch in equivalent data for H_2S and H_2 using appropriate values of B calculated from Table 4.1.

In rigorous calculation of total discharge compositions, the gas contents of both the separated vapor and liquid must be combined. Write an expression which will allow you to correct a steam analysis for H_2S to allow for H_2S partitioned into the liquid phase. Often this calculation is required because liquid phase samples are not available or suitable for independent CO_2 and H_2S analysis.

Multistep Steam Separation

Suppose that now you remove each successive steam fraction as it forms. In Table 4.2 we carefully chose temperatures so that between 250 and 260° steam fractions were constant over each interval; now complete Table 4.3 for this new process of MULTISTEP STEAM SEPARATION. (First inspect the table to find the fraction of steam removed at each step.)

Table 4.3

t,°C	Σy*	B_{CO_2}	$\frac{1}{1 + y(B-1)}$	mmoles CO_2/kg $CO_{2(l)}$	$CO_{2(v)}$	$CO_{2(l)}/CO_{2(o)}$
260	.00	83	1.000	500.0	-	1.000
256	.01	92	.524	261	24104	.523
253	.02	99	.505	132	13100	.264
250	.03	107				

* Σy is the cumulative steam fraction removed during the process considered and not the steam fraction removed at each successsive step.

Notice that for a given temperature interval much more gas is lost from solution by this multistep process than for the single step method. Over a small temperature interval we may insert an average distribution coefficient, B', so that for n steps, each removing a fraction of steam, y,

$$C_l/C_o = [1/\{1 + y(B'-1)\}]^n \qquad (3)$$

Now add curves for multistep CO_2, H_2S and H_2 loss to Figure 4.1.

Continuous Steam Separation

As the steam fraction becomes smaller, equation (4) approaches a limiting condition for the multistep process, termed CONTINUOUS STEAM SEPARATION. In the present case this process may be expressed by a Rayleigh-type equation in which B' is the average distribution coefficient for the small temperature interval considered and y is the cumulative steam fraction removed.

$$C_l/C_o = e^{-B'y} \qquad (4)$$

You can justify this relationship yourself using a table like Table 4.3 and operating with progressively smaller y values until a limiting condition is found. Add curves of C_l/C_o vs t for CO_2, H_2S and H_2 to Figure 4.1.

Both multistep and continuous steam separation are 'open-system processes'; a petrologic analogy is the formation of zoned phenocrysts.

> In many hydrothermal problems, stable isotopes of hydrogen and oxygen may be used to trace or identify these steam separation processes in underground systems. See Chapter 10.

A Chemical Method for Measuring Discharge Enthalpy

Where physical measurements for a discharging well have not been completed, or an independent estimate is required, the enthalpy of the discharge may be determined from gas analyses. Two steam samples are required at different separation pressures; this is achieved by collecting samples upstream and downstream from a valve or orifice plate and using a mini-separator for steam separation.

If the ratio of the gas concentrations in the two steam samples is r, and sample conditions are such that equation (3) of Chapter 2 holds, derive an expression for the discharge enthalpy of the well (Mahon, 1966).

GEOTHERMAL GASES - CONCENTRATION UNITS

Units such as millimoles/100 moles, gas fraction, etc., will be unfamiliar to most readers and so we will digress for a moment to consider some of the ways in which gas analyses are presented in the literature.

Gas concentrations in water or steam may be expressed in the same units as any other dissolved constituents e.g., mg/kg, wt%, etc., but because they are measured most frequently on separated steam samples and are often used directly in calculations involving the Gas Laws, molal units are the most convenient, e.g. millimoles gas/100 moles steam (n.b. this unit is approximately 10^5x mole fraction).

For the BR22 analysis, data may be reported as follows:

		Enthalpy J/gm	Sampling Pressure b.g.	Gas concentration in steam millimoles gas/100 moles steam					
				CO_2	H_2S	CH_4	H_2	$(N_2 + Ar)$	NH_3
BR22	1/6/80	1160	10	959.4	18.5	11.82	1.02	8.93	4.67

(Note that the total of the non-condensible gases may be obtained from their gas pressure, without the necessity for individual component analyses)

To obtain the CO_2 concentration in the total discharge (TD) (neglecting CO_2 dissolved in the separated water) multiply the steam fraction by the gas concentration in the steam,

$$yC_{sample} = C_{TD}$$

$y = .190$ $CO_2 = 101.2$ millimoles/kg and $H_2S = 1.96$ millimoles/kg

or $CO_2 = 4456$ mg/kg and $H_2S =$ ______ mg/kg

Assuming for the purpose of illustration, that the other gas concentrations are negligible, then in the steam sample, converting millimoles/100 moles to gm/kg,

$$\text{Total gas} = (.9549 \times 44 + 0.0185 \times 34)/1.8 \text{ gm gas/kg steam}$$

$$= 2.380 \text{ wt\% gas in steam}$$

and this is equivalent to 0.45 wt% gas in the total discharge.

An alternative method of presentation is often used which is derived as follows:

(a) of the total gas present (959.4 + 18.5 + 26.4) millimoles/100 moles steam, 959.4 millimoles are CO_2

... on a dry gas basis

$$CO_2 = \frac{959.4 \text{ millimoles}}{1004.3 \text{ millimoles total gas}} = 955 \text{ millimoles/mole total gas}$$

$$H_2S = \qquad = 18.4 \text{ millimoles/mole total gas}$$

Just to confuse matters, note that CO_2 could also be given as 95.5 mole % or, on an ideal gas basis, 95.5 volume %.

(b) The gas "fraction", x_g, is given in units of millimoles gas per mole of steam. x_g is sometimes referred to as the gas to steam molar ratio.

$\therefore$ Total gas = $CO_2 + H_2S + CH_4$ etc millimoles gas/100 moles steam

= 1004.3 millimoles gas/100 mole steam

= 10.04 millimoles/mole steam

(c) Finally the data is reported as follows -- and note the inclusion of y for convenience

	Enthalpy (j/gm)	Sampling Pressure b.g.	Steam Fraction (y)	m.moles gas / mole steam	millimoles gas/mole total gas: CO_2	H_2S	Other gases
BR22 1/6/80	1160	10	.19	10.04	956	18.4	25.6

From data presented in this way, the concentration of say CO_2 in the total discharge is obtained from the product of x_g, y, and the CO_2 concentration in the gas.

NOTE THAT IN TABULATING GAS DATA IT IS ESSENTIAL TO CAREFULLY SPECIFY EXACTLY WHAT MEASUREMENT UNITS ARE EMPLOYED.

Examples:

1. Data from some Cerro Prieto wells are reported by Nehring and D'Amore (1984) as follows:

	Steam Fraction (y)	Molar Gas/steam ratio (x_g)	Non condensible gases in moles: CO_2	H_2S	H_2	CH_4	Ar	N_2	NH_3
CPM5 4/77	.287	4.46×10^{-3}	78.86	7.63	4.66	5.18	0.014	.60	2.28
CPM8 4/77	.284	9.24×10^{-3}	92.02	2.59	2.33	1.51	0.0037	.15	1.47

The mole fraction (X_i) of a dissolved gas can be calculated from the steam fraction (y), the gas/steam ratio (x_g) and the % of a gas in the total gas. Using the expression (derive this yourself if you wish),

$$X_i = \frac{(\%/100)x_g}{1/y + x_g}$$

calculate the mole fraction of dissolved gases (other than NH_3) in the total discharge from Cerro Prieto well M-5. (Assume that all gas was dissolved in aquifer liquid).

2. In the 10 b.g. separated water at BR22, NH_3 = 6.1 mg/kg, calculate the NH_3 concentration of the total discharge in units of millimoles NH_3 kg^{-1} total discharge. (Don't forget to add in the NH_3 in the separated steam.)

Notice that, in many cases, analytical data may not be present for the NH_3 content of both phases. In these instances the total discharge concentration can still be estimated by assuming an equilibrium distribution of the gas between the two phases at the sampling pressure and allowing for ionized NH_4^+ (pH dependent) in the liquid. The calculations involved are also discussed in Chapters 5 and 7. In most cases where sampling pressures are low you can assume that most of the gases like CO_2 and H_2S (but not NH_3) partition strongly to the steam phase but in rigorous work or when dealing with reservoir boiling, corrections should be made to allow for the gases remaining in solution in the residual liquid as, for example, HCO_3^- and HS^- (Chapter 13). A similar situation exists for boron which partitions strongly to the liquid but has significant volalility.

GAS PRESSURES

The partial pressure (P_{gas}) of a gas dissolved in a liquid is given by Henry's Law

$$P = K_H X \qquad (5)$$

where K_H is the Henry's Law coefficient for the gas and X is the mole fraction of the gas in solution. Strictly, fugacities should be used but in most geothermal studies, fugacity coefficients are >0.98 so that pressures may be used directly in gas calculations. This equation applies whether or not there is a vapor phase. (In general, the sum of the partial pressures of all components, including H_2O, must equal the total pressure, for a vapor phase to exist.) Table 4.4 gives values for K_H for carbon dioxide in water and these are also shown in Figure 4.2. Figure 4.2 also includes data for the solubility of CO_2 in salty solutions. Is CO_2 more or less soluble in seawater (m_{Cl} = 0.05) than in pure water?

Table 4.4 -- (Data recalculated from Ellis and Golding, 1963; and Glover, 1982)

t°C	K_H (bars mole fraction^{-1})		
	CO_2	H_2S	NH_3
25	1660	551	1.0
50	2880	899	2.7
100	5245	1555	14.5
125	6150	1703	28.1
150	6670	1829	49.1
175	6860	1959	78.8
200	6620	2042	119.0
250	5340	1934	233.0
300	3980	1645	413.0

Notice that the Henry's Law coefficient for CO_2 passes through a maximum at about 175°C. In this chapter we will focus on CO_2 but we shall use K_H values again in later chapters dealing with gas geothermometry, redox reactions and the chemistry of steam condensates.

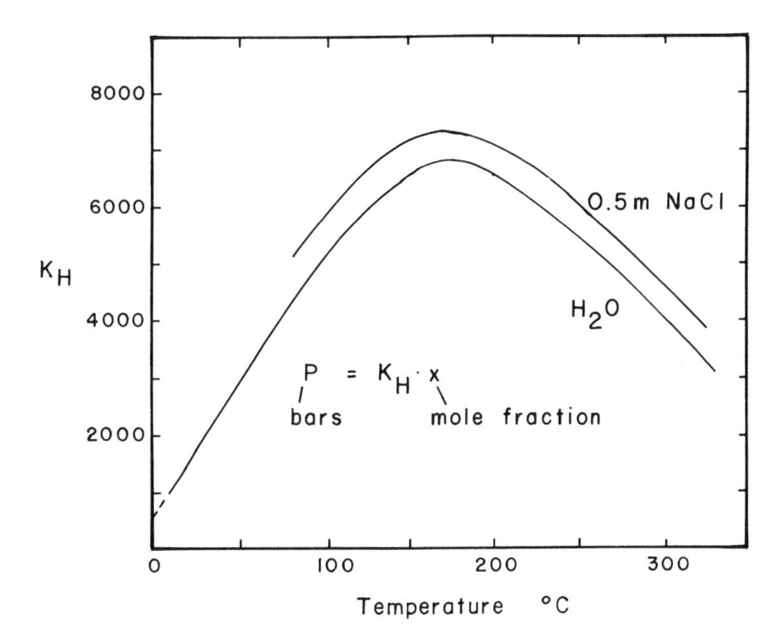

Figure 4.2. Values of the Henry's Law Constant for CO_2 in water and in a 0.5m NaCl solution. (Redrawn from Ellis and Golding, 1963)

Consider water containing 1.5 wt% CO_2 - what would its CO_2 gas pressure and total pressure be at 300°C and 200°C respectively ?

Gas pressures can be dangerous. In a number of geothermal wells gas pressures have become sufficiently high that explosions have occurred. At Dieng, Indonesia, one particularly violent explosion occured with the loss of three lives just as an overpressured valve was being cracked open. The effect is particularly common where fluid flow can occur between a well and the reservoir when the well-head valve is closed. Boiling and cooling through the ambient temperature gradient leads to the formation of a high pressure cold gas cap within the well so that deep fluid pressures are transmitted to the well-head. Natural hydrothermal eruptions such as those at Kawerau (Nairn and Wiradiradja, 1980) may have occurred in the same way when silica deposits locally sealed off spring vents (Fig. 4.3). Breakage of the seal by some mechanism such as seismic shock triggers the hydrothermal eruption which in some cases may involve rock transport from depths of up to 200 meters. The associated eruption breccias mantle a wide area and similar deposits are now well known in a number of hot spring-type epithermal deposits (e.g., Round Mountain, Nevada, McLaughlin, California; Berger and Eimon, 1982; Henley and Ellis, 1983) as well as in some polymetallic massive sulfide deposits (e.g. Buchans, Newfoundland and Kosaka, Japan; Henley and Thornley, 1978).

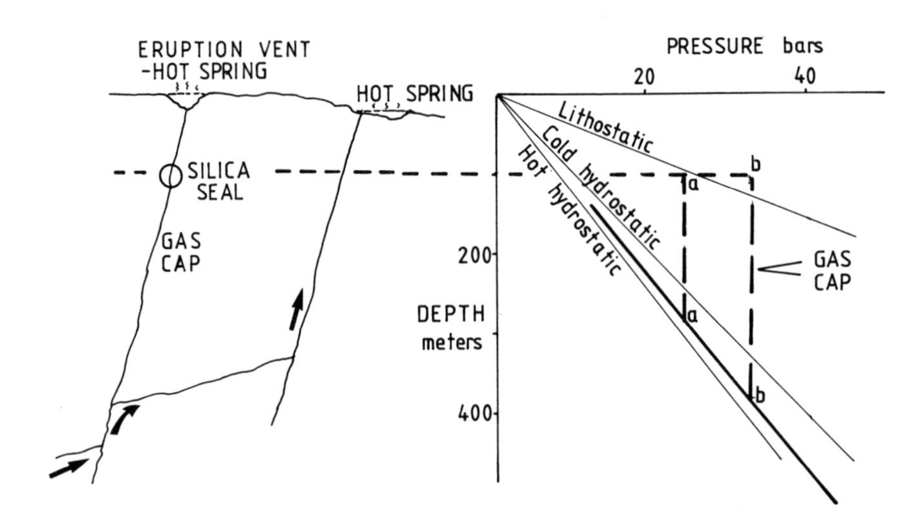

Figure 4.3. Pressure vs Depth relation for the evolution of hydrothermal eruptions in hot-spring systems and geothermal wells. The formation of a local seal in the system (e.g., through silica deposition, closure of well head valve) allows gas, exsolved during flow through the temperature gradient of the underlying fissure system, to accumulate as a "gas-cap". Prior to sealing, pressures in the flowing system are just above hot hydrostatic but as the gas accumulates, progressively greater aquifer pressures are transmitted to the seal by the gas-cap. Eventually the transmitted pressure exceeds lithostatic and an eruption may occur when triggered by seismic shock or opening of the well-head valve. Much of the succeeding eruption and crater formation results from erosion of the fissure walls by the expanding gas + steam + water mixture. The destruction of the Pink Terrace of Lake Rotomahana (Plate 2) may have resulted from hydrothermal eruptions triggered in this way.

$ What procedure might be followed to minimize the threat of explosion at a geothermal well-head?

In some systems hydrothermal eruptions may be caused by removal of overburden (e.g. drainage of glacial lakes, Muffler et al.,1971) or local changes in near surface heat flow (Elder, 1981), both mechanisms leading to relatively shallow eruptions.

Gas Solubility Expressions

The solubility of gases may be expressed in several ways. In this chapter we have considered it as a distribution factor (B) without reference to pressure, and as a pressure variable (K_H) for a given gas component. These quantities are related through the following expression:

$$K_H = B(P_v/Z_v) \qquad (6)$$

where P_v is the water vapor pressure, in bars, and Z_v the compressibility factor for pure saturated steam (Z allows for deviations of the PVT relations of steam from those of an ideal gas)*.

---

* Drummond (1981) provides the more rigorous equation, $B = K_H\underline{V}_v/\gamma_{CO2}\ Z_{CO2}\ RT$ where γ and Z are the fugacity coefficient and compressibility of CO_2 gas. For most geothermal applications both factors are close to unity.

$$Z_v = P_v\underline{V}_v/RTn \qquad (7)$$

where n is the number of moles, $\underline{V}_v$ is the molar volume of steam in m^3, T is in °K and the gas constant, R, is in units of 8.309×10^{-5} m^3bar/deg mole.

For the gas concentrations encountered in most geothermal problems the equivalent compressibility factor for CO_2 is not required. The expression may also be written

$$K_H = nBRT/\underline{V}_v \qquad (8)$$

For example: For CO_2 at 150°C we have $K_H = 6670$

$$B = \frac{K_H\underline{V}_v}{nRT} = \frac{6670 \times 392.8 \times 18.02}{1 \times 83.09 \times 423.15}$$

$B_{CO_2} = 1343$ R is given in cc bars/deg mole

Note that the equation given in Table 4.1 gives B = 1322, a value reflecting the smoothing involved in the regression. The difference is negligible.

Problem. Using equation (8) and data from Steam Tables (Appendix III) explain why K_H should pass through a maximum value whereas B is a simple exponential function of t°C. Do the solubility minima (i.e. K_H maxima) of other gases occur at the same or at different temperatures?

Aquifer Gas Pressure

Gas dissolved in an aquifer fluid contributes to the total fluid pressure within the system.

For fields like Wairakei where gas contents are low, temperatures and pressures are constrained by a simple boiling point - depth relationship close to that for pure water. In other systems, like Broadlands, gas contents significantly affect pressure - temperature patterns in the field. Let's draw the boiling point - depth relation for a low salinity water with initial temperature 300°C and gas content as CO_2 = 4.4 wt% ($X_{CO2} = 1.8 \times 10^{-2}$).

The boiling point - depth curve represents the maximum temperature which a liquid (with or without vapor) may have at a given depth. For pure water, the total pressure, P_{TOT}, at depth is given by the integral $\int\rho gh$ (Chapter 2) where ρ is the density of water at the boiling point (steam saturated) at all depths from the water table to the depth h and the corresponding temperature is obtainable from steam tables for saturated (2 phase) conditions.

In a system with significant gas concentrations, the total pressure is given by

$$P_{TOT} = \int\rho gh = P_w + P_{gas} \qquad (9)$$

where P_w is the pressure of steam-saturated liquid water and P_{gas} is the contributed gas pressure calculated through Henry's Law -- equation (5).*

---

* In this equation, the density is no longer that of pure, steam-saturated water, but must include the effect of dissolved gases and lower temperature. To a first approximation, these effects are negligible.

Consider an isolated box of fluid (X_{CO2}= 1.8×10^{-2}) rising adiabatically through a geothermal system. As it rises the pressure of the overlying fluid decreases and eventually a point is reached where boiling commences, with the formation of a gas-rich vapor phase. Specifying a closed system, we can calculate the mass fraction of steam formed and the proportion of CO_2 remaining in the liquid at selected temperatures. Once this is known we can obtain P_{CO2} and, through equation (9), the minimum pressure required to prevent such boiling. The total pressure is then converted to an equivalent depth of hot water using the data of Haas(1971) reproduced in Table 4.6.

To simplify the calculation we may, in the temperature range considered, use the regression equation for $K_{H,CO2}$ given in Table 5.3., and use the equation to convert pressures to depths for temperatures from 220° to 300°C.

$$\text{depth(m)} = 7.9697 \text{ bars}^{1.1031} \qquad (10)$$

Now complete Table 4.5 to obtain the boiling point-depth relation for fluid of the given CO_2 content. Plot the data onto Figure 2.2b to compare the pure water boiling point - depth curve with that of the gaseous system.

Table 4.5

t°C	K°_{H}	B	y	X_{CO2} mole fraction in liquid phase	P_{CO2} bars	P_{H2O} bars	P_{tot} bars	Depth to Boiling Point m
300	3745	31	0	1.8×10^{-2}	71.6	86	157	2088
290	3968	40	.037	7.4×10^{-3}	29.2	74	103	1332
280	4205	51	.070	4.0×10^{-3}	16.8	64		
270	4455	66						
260								
250								
240								
230								
220								

Giggenbach (1981) shows some boiling point depth curves constructed in this way, while Sutton and McNabb (1977) provided a more rigorous analysis of a flowing system based on pressure-temperature data actually measured in the Broadlands field. The example you have just completed is based on their analysis. Notice how, as temperature decreases the gas curve approaches that of pure water and that "boiling" occurs to much greater depths than you would anticipate on the basis of water alone. At Wairakei ($X_{CO2,250}$ = 2×10^{-4}) boiling occurs to a depth of only about 500 m but at Broadlands drilling to over 2500 m has not penetrated beneath the boiling zone.

Steam (and gas) may be lost discontinuously from a rising fluid depending upon the permeability structure of the system, while mixing with cooler gas-free waters may also occur. At Broadlands, where X_{CO2} is up to 1.8×10^{-2} in the deep fluid, at 300°C pressure-temperature data diverge from the b.p.d. curve at about 230°C suggesting that these other effects become important at that level.

Table 4.5 was constructed assuming closed system adiabatic boiling during fluid ascent. What would be the effect of continuous steam loss on the boiling-point depth relation of the same starting fluid that we considered above?

Table 4.6 — Thermal profile for H_2O liquid with corresponding vapor and density. (from Haas, 1971)

Temp. (°C)	Depth (metres)	Pressure (bars)	Density (g/cm^3)
100	0.0	1.0	0.958
110	4.5	1.4	0.951
120	10.4	2.0	0.943
130	18.2	2.7	0.935
140	28.2	3.6	0.926
150	40.9	4.8	0.917
160	56.8	6.2	0.907
170	76.4	7.9	0.897
180	100.5	10.0	0.887
190	129.7	12.6	0.876
200	164.7	15.6	0.865
210	206.8	19.1	0.853
220	256.4	23.2	0.840
230	314.9	28.0	0.827
240	383.3	33.5	0.814
250	462.9	39.8	0.799
260	555.2	46.9	0.784
270	661.8	55.1	0.768
280	784.6	64.2	0.751
290	925.6	74.4	0.732
300	1088.	85.9	0.712
310	1273.	98.7	0.691
320	1487.	122.9	0.667
330	1732.	128.6	0.640
340	2017.	146.1	0.609
350	2350.	165.4	0.573
360	2746.	186.7	0.525
370	3243.	210.5	0.446

REFERENCES

Berger, B.R., and Eimon, P. I., 1983, Conceptual models of epithermal precious metal deposits; in Shanks, W.C. (ed.), Cameron Volume on Unconventional Mineral Deposits: AIME, p. 191-206.

Drummond, E.J.S., 1981, Boiling and mixing of hydrothermal fluids: chemical effects on mineral precipitation, PhD Thesis, Pennsylvania State University, 380 p., University Microfilms International, Ann Arbor, Michigan.

Elder, J.W., 1981, Geothermal Systems: Academic Press, London, 508 p.

Ellis, A.J., and Golding, R.M., 1963, The solubility of carbon dioxide above 100°C in water and in sodium chloride solutions: American Journal of Science, v. 261, p. 47-60.

Giggenbach, W.F., 1980, Geothermal gas equilibria: Geochimica et Cosmochimica Acta, v. 44, p. 2021-2032.

Giggenbach, W.F., 1981, Geothermal mineral equilibria: Geochimica et Cosmochimica Acta, v. 45, p. 393-410.

Glover, R.B., 1982, Calculation of the chemistry of some geothermal environments: DSIR New Zealand, Chemistry Division Report CD2323.

Haas, J.L., 1971, Effect of salinity on the maximum thermal gradient of a hydrothermal system at hydrostatic pressure: Economic Geology, v. 66, p. 940-946.

Henley, R.W., and Ellis, A.J., 1983, Geothermal Systems Ancient and Modern: a geochemical review: Earth Science Reviews, v. 19, p. 1-50.

Henley, R.W., and Thornley, P., 1979, Some geothermal aspects of polymetallic massive sulfide formation: Economic Geology, v. 74, p. 1600-1612.

Mahon, W. A. J., 1966, A method for determining the enthalpy of a steam/water mixture discharged from a geothermal drill hole: New Zealand Journal of Science, v. 9, p. 791-800.

Muffler, L.J.P., White, D.E., and Truesdell, A.F., 1971, Hydrothermal explosion craters in Yellowstone National Park: Geological Society of America Bulletin, v. 82, p. 723-740.

Nairn, I. A., and Wiradiradja, S., 1980, Late Quaternary Hydrothermal Explosion Breccias at Kawerau Geothermal Field, New Zealand: Bulletin Volcanologique, v. 43-1, p. 1-13.

Nehring, N. L., and D'Amore, Franco, 1984, Gas chemistry and thermometry of the Cerro Prieto geothermal field: Geothermics, v. 13, p. 75-90.

Sutton, F.M., and McNabb, A., 1977, Boiling curves at Broadlands, New Zealand: New Zealand Journal of Science, v. 20, p. 333-337.

Chapter 5
MORE MILEAGE FROM YOUR GAS ANALYSES: THE GAS GEOTHERMOMETERS

In this chapter we will consider equilibria among gases in hydrothermal fluids from hot water geothermal reservoirs which produce from a liquid phase at depth. We leave until Chapter 11 consideration of gas equilibria in vapor-dominated reservoirs and hot-water reservoirs with aquifer steam, because discharges from these reservoirs contain gases of mixed origin from vapor and liquid and require more complicated computational methods.

The calculations in this chapter will introduce us to the use of thermochemical (free energy) data to derive equilibrium constants and to the conversion of measured gas concentrations in the liquid phase into reservoir gas pressures.

SIMPLE GAS GEOTHERMOMETERS

Two gas reactions exist for which we have all the reactants and products in our gas samples:

Methane breakdown geothermometer

The first is the methane breakdown reaction,

$$CH_4 + 2H_2O = CO_2 + 4H_2. \qquad (1)$$

(All equations are written with the species favored by high temperature as products.)

We are given the following thermodynamic data (from Robie et al., 1979):

Table 5.1

	$\Delta G°_f$ (J/mole)		
Temp °K	400	500	600
CH_4	-41984	-32667	-22799
H_2O	-223882	-219035	-213987
CO_2	-394660	-394920	-395129
H_2	0	0	0

To derive a geothermometer equation based on this reaction, first calculate the free energy change of the reaction through the equation,

$$G^\circ_R = \Sigma\Delta G°_{f,products} - \Sigma\Delta G°_{f,reactants} \qquad (2)$$

$$\Delta G°_R = \Delta G°_{CO_2} + 4\ \Delta G°_{H_2} - \Delta G°_{CH_4} - 2\ \Delta G°_{H_2O} \qquad (3)$$

Complete column (1) of the following table.

Table 5.2

T°K	$\Delta G°_R$	log K	1/T
400			
500			
600			

The mass action constant K expressed in gas partial pressures, is written as follows,

$$K = (P_{CO_2} P^4_{H_2})/(P_{CH_4} P^2_{H_2O}) \qquad (4)$$

$ Calculate log K at the 3 temperatures using $\Delta G°_R = -RT \ln K$ (R = 8.3144 joules/deg mole; 2.303 R = 19.148 joules/deg mole). Add these values to Table 5.2.

These data are formed into a gas geothermometer as follows:

$ 1. Use your calculator to find the coefficients of log K = a + b (1/T). The result should be near

$$\log K = + 10.278 - 9082/T \qquad (5)$$

The mass action constant contains P_{H2O}. We don't usually determine P_{H2O} directly for the reservoir but we have it as a function of temperature in steam tables. Giggenbach (1980) gives the useful equation

$$\log P_{H_2O} = 5.51 - 2048/T \qquad (6)$$

$ 2. Combine this equation with your log K equation to eliminate P_{H2O} (so you can concentrate on CH_4, CO_2 and H_2) and now you have a gas geothermometer.

This gas geothermometer is for equilibration in a pure vapor. Some vapor dominated reservoirs may locally contain only vapor but most steam from these fields is derived by evaporation of reservoir liquid. Giggenbach (1980) and D'Amore and Celati (1983) have suggested equations to calculate the temperature and fraction of vaporized liquid in steam of mixed vapor + liquid origin. These equations are discussed in Chapter 11.

Fortunately, however, most hot water systems have only liquid in the reservoir and your methane gas geothermometer can be used without corrections for mixed steam origin but the pressures must be converted to concentrations of gas in solution.

The following equations (most accurate from 200 to 350°C) for Henry's Law constants $K_H(=P/X)$ can be derived from expressions for B given by Giggenbach (1980) and the equation $K_H/B = RT/V_v$ (Chapter 4). Lower temperature K_H data for CO_2, H_2S and NH_3 are listed in Chapter 4.

Table 5.3

$\log K_H(CO_2)$	=	5.0149 - 0.002515 T	(T in °K, 473 - 623°K)
$\log K_H(H_2)$	=	7.3334 - 0.005625 T	
$\log K_H(CH_4)$	=	7.1287 - 0.005425 T	
$\log K_H(N_2)$	=	7.5832 - 0.005755 T	
$\log K_H(NH_3)$	=	-0.5183 + 0.005485 T	
$\log K_H(H_2S)$	=	4.0071 - 0.001405 T	

$ 3. Use these equations to rewrite your methane gas geothermometer in mole fractions of gas dissolved in water so it can be applied to hot water systems. Recast the equation into logarithms of the mole fractions of CO_2, H_2 and CH_4.

Your equation for the methane geothermometer in terms of log mole fractions (log X) CO_2, H_2, and CH_4 will be near

$$\log X(CO_2) + 4 \log X(H_2) - \log X(CH_4) = - 5.922 - 13178/T + 0.01959 T \qquad (7)$$

Mole fractions in the total fluid can be calculated from (Chapter 4)

$$X_i = (\%i/100)x_g/(x_g+1/y) \qquad (8)$$

where x_g is the gas/steam ratio and y is the steam fraction.

Some gas analyses of steam from wells at Cerro Prieto (Mexico) from Nehring and D'Amore (1984) are given in Chapter 4. In an earlier exercise you recalculated these data to give mole fractions of each gas in the total discharge. For well M-5 tabulate the logs of these mole fractions and for future use calculate log X_{N2}.

Table 5.4a

		CO_2	H_2	CH_4	N_2
CPM 5	X_i	1.01×10^{-3}	5.96×10^{-5}	6.62×10^{-5}	
$	$\log X_i$				

$ Calculate the methane temperature of the fluid feeding M-5.

$ Compare this with the silica, NaKCa and other cation geothermometer temperatures using the following analyses of the water flashed to one atmosphere (in mg/kg)

Table 5.4b -- Composition of brine flashed to 1 atm from well M-5 discharge (Truesdell et al., 1981)

pH	Li	Na	K	Rb	Cs	Mg	Ca	Mn	Fe
6.5	19.5	7745	1860	8.1	1.4	0.5	440	1.0	0.3

F	Cl	Br	HCO_3	SO_4	B	SiO_2	H_2S
2.1	14380	48	48	5.6	16	880	5.6

Why do these temperatures differ? The measured reservoir temperature is about 300°C; The flowing bottomhole temperature is lower due to near-well boiling. Which geothermometer equilibrates most rapidly? Which least rapidly? What special information do we obtain from having different geothermometer temperatures?

Ammonia geothermometer

The second gas geothermometer for which we have all the reactants and products in our gas analysis is ammonia breakdown,

$$2NH_3 = N_2 + 3H_2 \quad (9)$$

The thermodynamic data are (in Joules/mole)

Table 5.6

T	400	500	600
$\Delta G°_f\ (NH_3)$	-5984	+4760	+15841
$\Delta G°_f\ (N_2)$	0	0	0
$\Delta G°_f\ (H_2)$	0	0	0
ΔG_R			
log K			

You can use the procedure we followed for the methane geothermometer to derive the ammonia geothermometer yourself. The steps are as follows:

(a.) Fill in Table 5.6. Why are $\Delta G^\circ_f (N_2)$ and $\Delta G^\circ_f (H_2)$ equal to zero? Calculate the dependence of log K on 1/T.

Your equation should approximate

$$\log K = 11.375 - 5179/T \quad (10)$$

(b.) Use the log KH expressions given earlier to derive a geothermometer equation useful for hot water systems. Your equation should be near

$$\log K = -19.245 - 5179/T + 0.0336\,T \quad (11)$$

Ammonia is a very water soluble gas and we cannot assume (as we did for CO_2, H_2 and CH_4) that all NH_3 is in the separated steam. We must use a mass balance equation and the gas distribution constants (B) introduced in Chapter 4.

$$(n_{NH_3}/n_{H_2O})_{total} = y(n_{NH_3}/n_{H_2O})_v + (1-y)(n_{NH_3}/n_{H_2O})_l \quad (12)$$

$$B,NH_3 = (n_{NH_3}/n_{H_2O})_v/(n_{NH_3}/n_{H_2O})_l \quad (13)$$

combining and rearranging

$$(n_{NH_3}/n_{H_2O})_{total} = (\%NH_3/100)(x_g)(y + (1-y)/B,NH_3) \quad (14)$$

For the low gas contents of most geothermal fluids n_{NH3}/n_{H2O} is essentially identical to X_{NH3} and can be used directly in the geothermometer equations.

$ Use the analyses of Cerro Prieto M-5 steam to calculate the temperature of ammonia equilibrium in the reservoir.

$ Compare this temperature with the methane, NaKCa and silica temperatures you calculated earlier. What other geothermometer has about the same rate of equilibration as ammonia?

Hydrogen-Carbon Dioxide Geothermometer

Other gas geothermometers depend on the assumed presence of certain minerals in the aquifer.* In many cases the aquifer mineral assemblage is known from cores or cuttings.

---

* Most water geothermometers also depend on the presence of certain aquifer minerals e.g. quartz for silica geothermometer and feldspars for the Na/K geothermometer.

At Cerro Prieto the sedimentary aquifer rocks contain elemental carbon (graphite) as coal. This allows us to use CO_2 and H_2 as a geothermometer.

We can write $C + O_2 = CO_2$ and $2H_2O = 2H_2 + O_2$. (15a,b)

Combining these equations we obtain,

$$2H_2O + C = 2H_2 + CO_2 \quad (16)$$

Again you can derive the geothermometer yourself:

(a.) The necessary thermodynamic data were given earlier ($\Delta G°_f$ of carbon is zero at all temperatures). Construct a table of $\Delta G°_f$ at 400, 500 and 600°K, and calculate $\Delta G°_R$ and log K at each temperature

(b.) Now calculate the dependence of log K on 1/T. You should obtain

$$\log K = 5.278 - 4886/T \quad (17)$$

(c.) Again add the expression for 2 x log $P(H_2O)$ to obtain a water-free geothermometer as,

$$\log P_{CO_2} + 2 \log P_H = 16.298 - 8982/T \quad (18)$$

(d.) Now since our geothermometer is to be used at Cerro Prieto we have to add the proper combination of log K_H (Henry's Law) expressions (given earlier) to obtain a geothermometer equation for dissolved gases.

You should obtain something like,

$$\log X(CO_2) + 2 \log X(H_2) = -3.384 - 8982/T + 0.01377\ T. \quad (19)$$

$ Calculate the hydrogen-carbon dioxide temperature for M5 at Cerro Prieto.

Tabulate the geothermometer temperatures you have for this well discharge in 1977. The measured downhole temperature in 1976 was 299°C. The measured flows from the separator (at 6.72 bars absolute) were 50 tonnes/hr steam and 124 tonnes/hr water. Can you calculate the enthalpy of discharge and the temperature of aquifer water with this data? Is it useful to have so many ways of measuring the aquifer temperature?

AN EMPIRICAL GAS GEOTHERMOMETER

Although earlier we have used only experimentally based data to obtain gas geothermometers, it is also useful to consider the empirical gas geothermometer proposed by D'Amore and Panichi (1980). Empirical geothermometers (such as NaKCa) are based on observed fluid compositions and observed subsurface temperatures. They cannot be entirely empirical - the form of the equation must be suggested by theory.

D'Amore and Panichi made 4 assumptions:

1) free carbon, carbon dioxide and hydrogen react to form methane

$$C + CO_2 + 6H_2 = 2CH_4 + 2H_2O \quad (20a)$$

$$\log K = 6.82 + 11801/T - 7.11 \log T \quad (20b)$$

2) anhydrite and pyrite react to form hydrogen sulfide

$$CaSO_4 + FeS_2 + 3H_2O + CO_2 = CaCO_3 + 1/3\ Fe_3O_4 + 3H_2S + 7/3\ O_2 \quad (21a)$$

$$\log K = 23.68 - 62220/T \quad (21b)$$

3) the oxygen partial pressure is empirically related to temperature by

$$\log P(O_2) = 8.20 - 23643/T \quad (22)$$

4) the CO_2 pressure is related to the relative amount of CO_2 in the gas

$$P_{CO_2} = 0.1 \text{ if } \%CO_2 < 75$$

$$P_{CO_2} = 1.0 \text{ if } \%CO_2 > 75$$

$$P_{CO_2} = 10 \text{ if } \%CO_2 > 75 \text{ and}$$

$$CH_4 > 2H_2 \text{ and}$$

$$H_2S > 2H_2$$

These assumptions are rather arbitrary and cannot be fully accepted for all fields, but through their use D'Amore and Panichi have produced a gas geothermometer that can be used on fumarole and hot spring gases. The geothermometer equation is,

$$t = 24775/(2 \log (CH_4/CO_2) - 6 \log (H_2/CO_2) - 3 \log (H_2S/CO_2) + 7 \log P(CO_2) + 36.05) - 273 \qquad (23)$$

$ Apply this equation to the Cerro Prieto M-5 gas analyses. Use the assumption 4) above to determine $P(CO_2)$. Is the temperature reasonable?

COMPARISON OF PUBLISHED GEOTHERMOMETRY EQUATIONS

As a postscript to this exercise in the derivation and use of gas geothermometers it should be noted that:

1) In general, gas geothermometers depend on a knowledge of the gas/steam and (for hot water systems) the steam/water ratios. These ratios usually cannot be determined for fumarole and hot spring samples because gas and water usually do not reach the surface together. The D'Amore - Panichi empirical geothermometer is an exception which may be used on natural discharges.

2) The thermodynamic data used here are from one source. Other data sources may differ. For example, the methane breakdown equation in pressure units is given by other authors as,

$\log K = -2145.73 - 196187/T + 5969000/T^2 + 24428/\sqrt{T} + 228.01 \ln T$ — Arnorsson et al (1982)

$\log K = +10.76 - 9323/T$ — Giggenbach (1980)

$\log K = +21.78 - 13419/T$ — D'Amore and Nehring (1984)

compared with

$\log K = +10.266 - 9075/T$ — Our equation

Let's calculate some values

T	423	473	523	573
t	150	200	250	300
log K				
Giggenbach	-11.27	-8.94	-7.06	-5.51
D'Amore	-9.93	-6.58	-3.87	-1.63

Arnorsson	-9.62	-6.34	-3.67	-1.43
Us	-11.18	-8.91	-7.08	-5.57

Evidently we agree with Giggenbach but the others disagree -- or do they? Let's try subtracting 2 log (H_2O) (from the equation or the steam tables) from the values of D'Amore and Arnorsson,

T	423	473	523	573
- 2 log P(H_2O)	-1.34	-2.36	-3.19	-3.87
log K (D'Amore) - 2 log P(H_2O)	-11.27	-8.94	-7.06	-5.50
log K (Arnorsson) - 2 log P(H_2O)	-10.96	-8.70	-6.86	-5.30

Thus we find that D'Amore and Arnorsson had already added 2 log P(H_2O) before they presented their equations. Now three of the data sets agree closely (ours, Giggenbach's and D'Amore's) and the other is only slightly different. To resolve this difference we would have to examine the primary thermodynamic data and weigh all the possible errors. That would be too much for now .

REFERENCES

Arnorsson, S., Sigurdsson, S., and Svavarsson, H., 1982, The chemistry of geothermal waters in Iceland. I. Calculation of aqueous speciation from 0° to 370°C: Geochimica et Cosmochimica Acta, v. 46, p. 1513 to 1532.

D'Amore, F., and Panichi, C., 1980, Evaluation of deep temperatures of hydrothermal systems by a new gas geothermometer: Geochimica et Cosmochimica Acta, v. 44, p. 549-556.

D'Amore, F., and Celati, R., 1983, Methodology for calculating steam quality in geothermal reservoirs: Geothermics, v. 12, p. 129-140.

D'Amore, F., and Nehring, N.L., 1984, Gas chemistry and thermometry of the Cerro Prieto geothermal field: Geothermics, v. 13, p. 75-90.

Giggenbach, W.F., 1980, Geothermal gas equilibria: Geochimica et Cosmochimica Acta, v. 44, p. 2021-2032.

Robie, R.A., Hemingway, B.S., and Fisher, J.R., 1979, Thermodynamic Properties of Minerals and Related Substances at 298.15K and 1 bar pressure and at higher Temperatures: U.S. Geological Survey Bulletin 1492, 456 p.

Truesdell, A. H., Thompson, J. M., Coplen, T. B., Nehring, N. L., and Janik, C. J., 1981, The Origin of the Cerro Prieto geothermal brine: Geothermics, v. 10, p. 225-238.

Chapter 6
HYDROLYSIS REACTIONS IN HYDROTHERMAL FLUIDS

In previous sections you have considered the components of a geothermal fluid analysis in terms of:

(a.) comparison of conservative element concentrations (like Cl) between wells and effects of boiling and dilution (mixing diagrams).

(b.) relations between component ratios, concentrations and deep temperatures (Na/K, NaKCa, gas and silica geothermometers).

The next step in fully utilizing the chemistry of geothermal discharges is to examine carefully the relations between observed fluid chemistry and alteration minerals occurring in the drillcore. These relations form the basis for chemical geothermometry as well as highlighting some of the important interwoven relationships between the fluid components. To formulate these relations we rely heavily on thermodynamic data which, because of experimental difficulties at high temperatures, may sometimes be suspect -- we can often recognize these cases by using natural fluid-mineral equilibria as a guide.

In studies of mineral deposits, alteration assemblages are frequently used in conjunction with salinity estimates from fluid inclusion data to indicate the pH of ore-forming fluids. The compatibility of fluid chemistry and mineralogy established through geothermal studies (Browne and Ellis, 1971; Arnorsson et al., 1978; Truesdell and Henley, 1982) provides the confidence to apply this approach in ore forming systems.

Since many mineral equilibria may be written involving hydrogen ions

$$\text{e.g. } 3\ KAlSi_3O_8 + 2\ H^+ = KAl_3Si_3O_{10}(OH)_2 + 6\ SiO_2 + 2\ K^+$$

we shall first, using water and steam samples from geothermal wells, apply some approximations to estimate the pH of the geothermal fluid which feeds from the deep geothermal aquifer into an exploration well. Using this pH estimate we may then examine mineral alteration reactions in the system itself. At that stage we shall need to refine our methods for calculating the pH of the reservoir fluid using some complex chemical routines. These methods are readily programable for routine calculations and open the door to tackling a whole range of high temperature chemical calculations related both to alteration reactions deep in the system and to scaling and corrosion problems important to geothermal development.

Figure 6.1 shows the range of measured temperatures over which the most common alteration minerals have been observed in active geothermal systems. Their occurrences are described by a number of workers (Steiner, 1977, Browne, 1978, Hoagland and Elders, 1978; Kristmannsdottir, 1976). The same mineral assemblages occur in fossil geothermal systems i.e. hydrothermal ore deposits where they are often classified as <u>argillic</u> (if kaolinite is present) or <u>propylitic</u> assemblages (Rose and Burt, 1979). Advanced argillic assemblages, characterized by kaolinite, alunite, etc., occur in acid water environments such as are found in the upper levels of geothermal and epithermal ore-depositing systems or in deep magmatic environments such as at Butte, Montana.

Many of the mineral phases listed show extensive compositional variation due to solid solution (e.g. epidote, chlorite). For these, thermodynamic analysis is complex, while the extensive interlayering shown by the clay and clay-chlorite mixed-layer minerals poses additional problems. The emphasis in the calculations discussed below is therefore on reference mineral equilibria involving simple cation exchange.

SIMPLE METHODS FOR THE ESTIMATION OF RESERVOIR FLUID pH

The pH of reservoir fluids cannot be measured directly, so in order to examine fluid-mineral equilibria in active geothermal systems we will first use two simple methods for the estimation of pH from water and steam samples taken from discharging wells. In the next chapter we shall learn a more rigorous method for pH calculation.

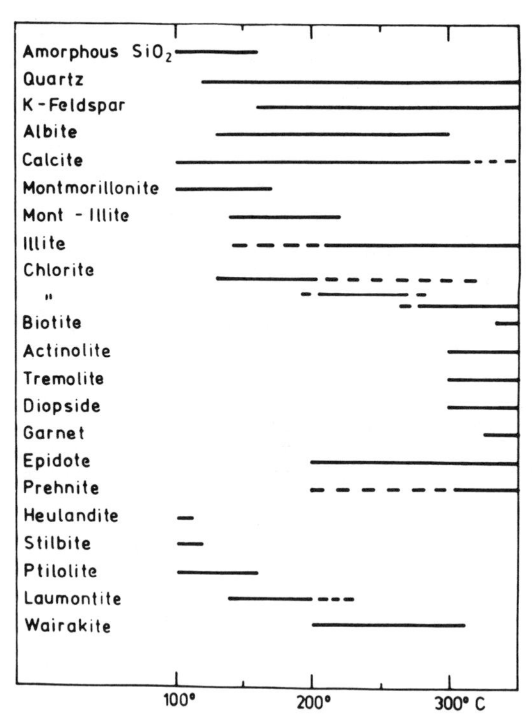

Figure 6.1. Generalized summary of the temperatures over which alumino-silicate alteration minerals have been observed in active geothermal systems. The three chlorite stability ranges indicate the transition from swelling through mixed layer to non-swelling chlorite with increasing temperature. Phases which occur in the surficial sulfate zone are not included; e.g., alunite, kaolinite, cristobalite....(Reproduced with permission from Henley and Ellis, 1983)

The first method is based on the dissociation of dissolved CO_2 to give bicarbonate and hydrogen ions.

$$H_2CO_3 = HCO_3^- + H^+$$

$$K_{H_2CO_3} = \frac{a_{HCO3^-} \, a_{H^+}}{a_{H_2CO_3}} \qquad (1)$$

$$\log K_{H_2CO_3} = \log \gamma_{HCO_3} / \gamma_{H_2CO_3} + \log m_{HCO_3} / m_{H_2CO_3} - pH \qquad (2)$$

γ_i is the individual ion activity coefficient of the dissolved species, i, which we may calculate through the Debye-Hückel equation.

We can illustrate the method using an analysis of fluid from well WR27 at Wairakei, given below as steam and water sample analyses.

Date	Enthalpy J/gm	Separation Pressure b.g	$pH_{20°}$	Na	K	Ca	Mg	Cl	SO_4	B	SiO_2	NH_3	HCO_3	CO_2
				mg/kg										
(Water sample)														
8/60	1085	0	8.10*	1320	220	16.8	.05	2198	31	28	690	.2	33	0.4
			millimoles/kg											
(Steam sample)			CO_2	H_2S										
8/60	1085	14.8	67.2	2.3										

* NOTE: As in previous tables this is the pH of the surface sample and not that of the reservoir fluid.

Using the heat balance equation discussed in Chapter 2, recalculate these data to pre-flash conditions to obtain the composition of the fluid feeding the well.

To justify this recalculation compare the enthalpy of water at the silica temperature calculated from the total discharge data and the discharge enthalpy measured for the well (1085 J/gm) - comment on the assumptions you make. Complete the tables below.

Total discharge composition. (mg/kg)

Na	K	Ca	Mg	Cl	SO_4	B	SiO_2	NH_3	HCO_3	CO_2	H_2S
		11.8		1549					23.2	348.4	

To use these data in thermochemical calculations we need to recalculate them to molal units, as follows.

Total discharge composition (10^3 x moles/kg = millimoles/kg)

Na	K	Ca	Mg	Cl	SO_4	B	SiO_2	NH_3	HCO_3	CO_2	H_2S
		.30							0.38	7.92	

In order to use these data in equation (2) to obtain pHT, we need γ and K values. For the WR27 fluid we may use $\gamma_{HCO_3^-} = 0.7$ and $\gamma_{H2CO3} = 1.0$. Values of log K_1 are provided in thermodynamic tables - which we shall discuss later. At 250°C log $K_1 = -7.75$ for this reaction.

If for our introductory calculations we assume that $m_{HCO_3^-}$ of the total discharge is essentially that of the deep fluid, we may then substitute values into equation (2) as follows as a means of estimating deep system pH:

$$-7.75 = -0.155 + (-1.32) - pH_{250°}$$

$$\text{and } pH_{250°} = \underline{6.27}$$

Notice that this is almost 2 pH units lower than the pH of the weirbox water sample.

The dissociation constant for water at 250°C is $10^{-11.2}$, compared with 10^{-14} at 25°C, so that the neutral pH of high temperature water is 5.6. The Wairakei fluid is therefore slightly alkaline at depth.

During flow up the well and flashing to lower temperatures some HCO_3^- converts to CO_2 through reactions with silica, boron and ammonia species such as

$$H_2O + CO_2 + H_3SiO_4^- = HCO_3^- + H_4SiO_4$$

In the next chapter we shall learn to refine our calculation procedure to deal with this.

Show that without these reactions the proportion of HCO_3^- that would be transformed to CO_2 during flash is negligible.

Another first order calculation procedure is open to us if the drillcore contains calcite, as the fluid would be expected to be close to equilibrium with this mineral. Although calcite is not abundant at Wairakei, its solubility may be used as a constraint on the pH of the deep fluid.

$$CaCO_3 + 2H^+ = Ca^{++} + H_2O + CO_2 \qquad (3)$$

Equilibrium constants for this reaction are given in Table 7.3.

At 250°C $\log K_C$ = 6.46 and we may write

$$\log K_C = 6.46 = \log \gamma_{Ca^{++}} + \log (m_{Ca} m_{CO_2}) + 2pH \qquad (4)$$

so that with $\gamma_{Ca^{++}}$ = 0.27

we find that pH_t = 6.33

Individual Ion Activity Coefficients

For calculations on the relatively dilute solutions described in this and the next chapter the following activity coefficients may be used. In more saline fluids, such as the Salton Sea, activity coefficient calculation is more complex - for a review of such procedures see Pitzer (1981).

	25°C		260°C	300°C
	Wairakei	Cerro Prieto	Wairakei	Cerro Prieto
I (moles x 10^3)	65.8	403	44.6	209
γ_{H^+}	.85	.80	.76	.56
$\gamma_{HCO_3^-}$	.81	.7	.70	.42
$\gamma_{H_4SiO^-}$	.81	.69	.70	.40
γ_{HS^-}	.81	.69	.70	.40
$\gamma_{BO_2^-}$	.81	.69	.70	.40
γ_{Na^+}	.81	.69	.70	.40
γ_{K^+}	.79	.65	.68	.36
$\gamma_{Ca^{++}}$	.46	.28	.27	.05

Problem: Using the first order methods described above, estimate the pH of the deep fluid at Hatchobaru, Japan from the data below.

(Water sample)

Enthalpy J/gm	S.P. b.g.	$pH_{20°}$	Li	Na	K	Ca	Mg	Cl	SO_4	B	SiO_2	NH_3	HCO_3	CO_2
1270	0	8.15	1.11	1396	289	9.9	.16	2327	98	32	1076	0.1	58	0.6

(Steam Sample) CO_2 in total discharge = 184.2 millimoles/kg

The silica temperature is 295°C and $\log K_{1,H_2CO_3} = -8.22$, $\log K_C = 5.91$,

$$\gamma_{Ca^{++}} = 0.21, \quad \gamma_{HCO_3^-} = 0.66$$

$ $CO_2 - HCO_3$ method pH =

$ Calcite method pH =

Of course the pH values we obtain are strongly dependent on the accuracy of the thermodynamic data, the assumptions we make and the analysis particularly of $m_{Ca^{++}}$, but the agreement between these two simple methods is satisfying.

This may not always be the case. If we apply the same methods to well M19A at Cerro Prieto we obtain pH = 5.7 (bicarbonate) but pH = 4.9 (calcite). The problem may be related to processes occurring in the feed zones of the well. In many high temperature wells (> 275°C) or where extensive production has occurred, excess steam may enter the well giving both an artifically high enthalpy and gas content to the discharge. In other cases mixing of the fluids from different temperature zones may occur so that the "mixture" sampled is not truly representative of the fluid actually present in the system. In Chapter 11 we shall examine some methods to handle these complex situations.

ACTIVITY DIAGRAMS

Silicate - Water Reactions

Most of the alteration reactions whose products we observe in drillcore involve aluminosilicate minerals (e.g. replacement of feldspar by clay or mica). The calculation of aluminosilicate solubilities however presents problems. If microcline, for example, dissolves in a geothermal fluid and we assume we know the solution species

$$KAlSi_3O_8 + 8H_2O = K^+ + Al(OH)_4^- + 3H_4SiO_4$$

we may then compare our geothermal fluid composition with the mineral solubility, calculated from available thermodynamic tables, just as we did for calcite. We are able then to discover whether our geothermal fluid is under, over, or just saturated with microcline. The difficulties come in analyzing for aluminium which is usually a trace constituent (< 1 mg/kg) and in calculating the distribution of aluminium solution species at high temperatures for which little reliable data are available.

Usually in drillcore we see feldspars being replaced by mica or by clay minerals and quartz. We may write replacement reactions such as this so that the problem of dissolved aluminium is avoided.

$$3KAlSi_3O_8 + 2H^+ = KAl_3Si_3O_{10}(OH)_2 + 6SiO_2 + 2K^+$$

This is an example of an incongruent mineral solubility or hydrolysis reaction.

What is the justification for writing silica as quartz rather than as a dissolved species?

Written in this form we see that the coexistence of K-feldspar and Kmica implies, at quartz saturation, a relationship between m_{K^+} and pH. Similar reaction equations may be given for other feldspar end-members. Provided thermodynamic data are available, alteration assemblages may be related to the fluid discharged from geothermal wells. This is usually tackled by constructing 'activity' diagrams.

We shall now construct a simple activity diagram using primary experimental data and from there proceed to the use of tabulated thermodynamic data.

The Na_2O - K_2O - Al_2O_3 - SiO_2 - H_2O system

Let's start by listing some mineral formulae: (i = Na or K)

		i/Al
K-feldspar (Kspar)	$KAlSi_3O_8$	1.0
Kmica	$KAl_3Si_3O_{10}(OH)_2$	0.3
Kaolinite	$Al_2Si_2O_5(OH)_4$	0.0
Na-montmorillonite	$Na0_{.33}Al_{2.33}Si_{3.66}O_{10}(OH)_2$	0.14
Na-feldspar	$NaAlSi_3O_8$	1.0

It is apparent from these ratios that at quartz saturation these minerals will be ordered similarly with respect to solution composition; K-feldspar will be in the most K^+ rich solutions and kaolinite in the least. The same will be true of sodic minerals.

If we write the reaction between K-feldspar and Kmica and assume that aluminium is conserved* in the solid phases,

$$\underset{\text{Kspar}}{3KAlSi_3O_8} + 2H^+ = \underset{\text{Kmica}}{KAl_3Si_3O_{10}(OH)_2} + \underset{\text{Quartz}}{6SiO_2} + 2K^+$$

we find that the only solution species are K^+ and H^+ and we can write the equilibrium constant as aK^+/aH^+. For sodium minerals the constants will be aNa^+/aH^+ and for reactions of potassium and sodium minerals the constants will be combinations of both ratios. These ratios represent phase boundaries under the specified conditions of temperature, pressure and, as discussed later, silica activity. Usually the silica activity is determined by the solubility of quartz in high temperature fluids.

---

* By this we do not mean that aluminum does not go into solution at all. Actually there is a constant small amount of aluminum in solution so that any incremental conversion of Kspar to Kmica involves no net increase of dissolved aluminum.

Phase boundaries such as these need to be experimentally determined. These experiments are not simple; they require special high temperature techniques and a constant awareness of the possibilities of non-attainment of equilibrium. At least three studies have been published on the Kspar-Kmica equilibrium and these are relatively consistent. The diagram below (Fig. 6.2) shows experimental data from Hemley and coworkers, who measured the pH and $m\Sigma K$ of alkali chloride solutions quenched from the experimental temperature after prolonged reaction with finely ground Kmica-Kspar-quartz mixtures, or an equivalent mixture containing kaolinite or montmorillonite (Hemley, 1959). These experiments do not give equilibrium constants directly but by making allowance for the increased association of Na^+, K^+, H^+ and Cl^- as NaCl, KCl and HCl at high temperatures, equilibrium constants may be calculated. For now lets ignore these reactions and use the the experimental measurements of total potassium and of total available hydrogen ion directly as a measure of aK^+/aH^+. You can, or will be able to, evaluate the reliability of this yourself. This opportunity is provided as a problem at the end of the next chapter.

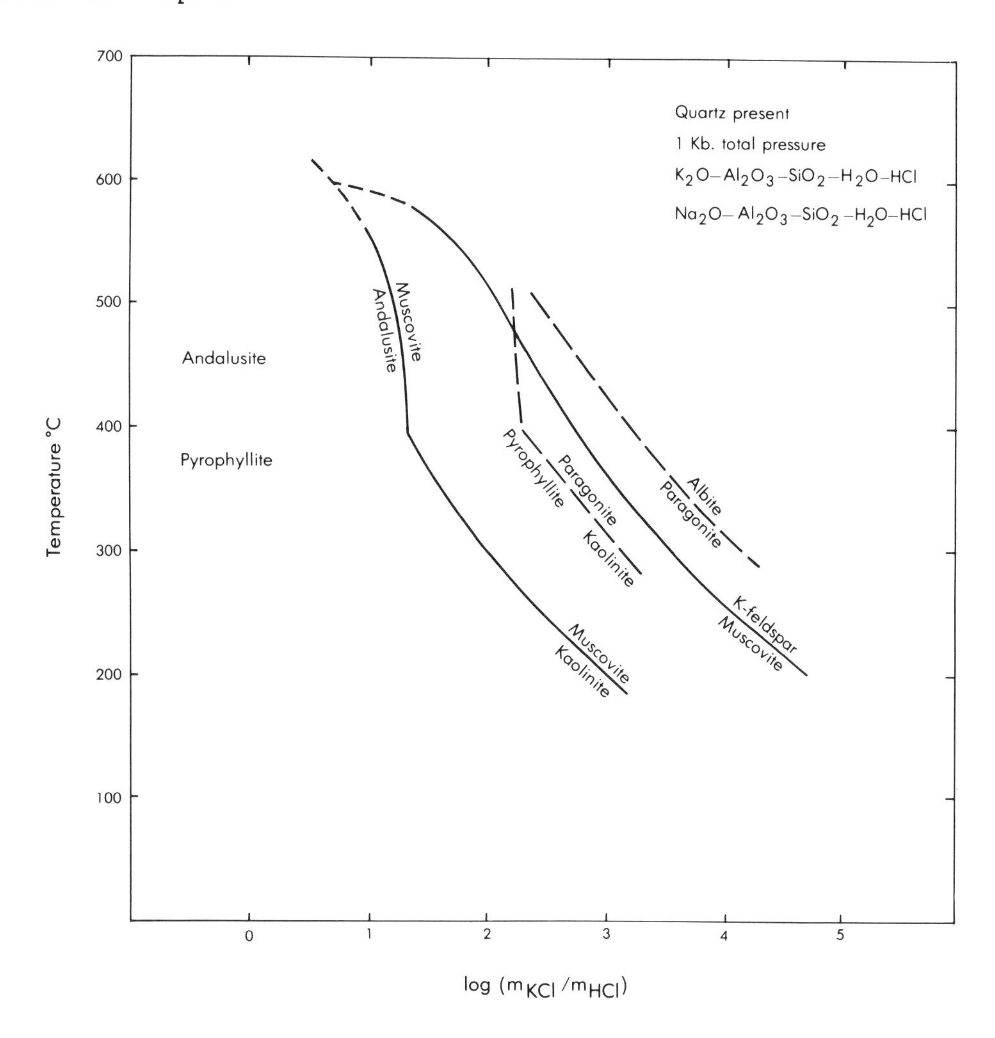

Figure 6.2. Concentration and calculated activity ratio curves for the systems $K_2O-Al_2O_3-SiO_2-H_2O$ and $Na_2O-Al_2O_3-SiO_2-H_2O$. (Redrawn from Montoya and Hemley, 1975)

From the data of Figure 6.2, read values of log $m\Sigma K/(mH^+ + mHCl)$ and log $m\Sigma Na/(mH^+ + mHCl)$ for the following mineral pairs at 250°C:

	log $m\Sigma K/(mH^+ + mHCl)$ or log $m\Sigma Na/(mH^+ + mHCl)$
Kaolinite - Kmica	
Kmica - K-feldspar	

Kaolinite - Na-montmorillonite

Na-montmorillonite - albite

____

Now you can draw an activity diagram with, on the x-axis, log a_{K^+}/a_{H^+} from 0 to +5 and, on the y-axis, log a_{Na^+}/a_{H^+} from 0 to +7.

In addition to these boundaries we have others involving both Na and K such as that between montmorillonite and Kmica.

$$2.3\ \text{Kmica} + Na^+ + 1.3\ H^+ + 4SiO_2 = 3\ \text{Na-mont.} + 2.3\ K^+$$

$$K = a_{K^+}^{2.3}/(a_{Na^+}\ a_{H^+}^{1.3})$$

$$\log K = 2.3 \log (a_{K^+}/a_{H^+}) - \log(a_{Na^+}/a_{H^+}) \qquad (5)$$

We do not need to calculate the log K value of this reaction because we know that the boundary extends from the intersection of the kaolinite-Kmica and kaolinite-Na-montmorillonite boundaries and that it has a slope of 2.3. Similarly we can write a reaction for Kmica-albite and deduce the boundary slope for that reaction.

What is the slope of the K-feldspar-albite reaction boundary?

The most convenient form for storage of information about mineral-mineral or mineral-fluid reactions is as tabulated thermodynamic data but you should always remember that these numbers are (or should be) derived from experimental data and are not independent. Many of these data have not yet been obtained. For this reason many of the phase boundaries of interest to us are not sufficiently well known to be plotted with great confidence on activity diagrams. In these cases it is often possible, using equilibria indicated by petrographic data, to estimate phase boundaries - and therefore thermodynamic data - from the activity ratios calculated in the fluid produced from a geothermal well. The phase boundaries for wairakite or epidote solid solution are examples to which we will return later.

One of the most useful achievements of recent years has been the development by Helgeson and his coworkers of the computer program SUPCRT (Helgeson et al., 1978). This contains an enormous amount of self-consistent data for minerals and aqueous species over a wide temperature range. Most stability data for solution species are obtained by extrapolation from low temperature and therefore need experimental verification. We have extracted (in Table 6.1) a sample of thermodynamic data for a number of the minerals and aqueous species of interest to us. A useful compilation of thermodynamic data, not including aqueous species, is provided by Robie et al. (1978), and much of the data in this source may easily be made consistent with those in the Helgeson compilation (see box). Another useful data source is Haas and Fisher (1976).

Because absolute values of the free energies of substances cannot be measured it is necessary to adopt an arbitrary zero. The free energies of elements are defined as zero in their standard state (generally 298.15°K and 1 bar pressure). In the compilation of Robie et al. (1978), the standard state adopted for elements is 1 bar absolute at any temperature and therefore the free energies of elements are zero at all temperatures. A different convention is adopted by Helgeson et al. (1978), in the SUPCRT file; the free energies of elements are defined as zero only at 1 bar absolute and 298.15°K but are adjusted for temperature through the relation, $\Delta G = \int c_p dT$ where c_p is the specific heat of the element. To combine Robie et al. (1978) and Helgeson et al. (1978) data the easiest path is to use their tabulated values of log K (of formation from the elements) rather than $\Delta G°$ values. This avoids the difference in standard state conventions. Because of these differing conventions one must be very cautious in mixing thermodynamic data from different sources.

For ionic species a further convention is necessary because the free energies of individual ions are unobtainable. The convention adopted is that $\Delta G°_{H^+}$ is zero at all temperatures (Helgeson, 1969).

Table 6.1 -- Free Energies of formation (ΔG_f^o) of Minerals and Aqueous Species (from Helgeson et al., 1978) in kJ/mole. at the saturated vapor pressure of pure water.

	Temperature (°C)					
	100	150	200	250	300	350
Quartz	-859.8	-862.7	-865.7	-869.0	-872.4	-876.5
Albite	-3725.9	-3739.3	-3754.3	-3770.6	-3787.8	-3805.8
K-feldspar	-3763.9	-3777.7	-3793.2	-3810.0	-3827.5	-3845.9
Muscovite	-5615.8	-5639.2	-5656.8	-5680.6	-5706.1	-5732.9
Kaolinite	-3806.6	-3820.4	-3836.3	-3853.5	-3872.3	-3892.4
Calcite	-1137.6	-1143.9	-1150.2	-1157.3	-1164.4	-1172.4
Wairakite	-6215.8	-6235.0	-6274.7	-6307.4	-6342.5	-6378.9
Zoisite	-6515.3	-6534.6	-6556.3	-6580.2	-6606.1	-6633.7
H^+	0	0	0	0	0	0
Na^+	-266.9	-270.7	-274.9	-279.1	-283.7	-286.2
K^+	-290.4	-295.8	-301.7	-307.5	-312.5	-317.6
Ca^{++}	-548.9	-546.0	-543.5	-540.6	-535.6	-523.8
H_2O liquid	-243.1	-247.7	-252.7	-258.2	-264.0	-269.9
H_2O vapor	-243.1	-253.1	-263.2	-273.2	-284.1	-294.6
CO_2, gas	-410.9	-422.2	-433.5	-445.2	-457.3	-469.4

Drawing an Activity Diagram from Thermodynamic Data

To examine fluid-mineral equilibria at Wairakei we wish to calculate phase boundaries at the aquifer temperature (250°C) so that first we obtain thermodynamic data from Table 6.1 at this temperature. (For other temperatures, say 260°C, it might be necessary to use a linear regression or simple graph to obtain the required data).

For the reaction

$$3\text{Kspar} + 2H^+ = \text{Kmica} + 6\text{quartz} + 2K^+$$

we obtain the standard free energy change ($\Delta G°_R$) by summing the standard free energies of the reactants and products

$$\Delta G°_R = (\Delta G°\text{Kmica} + 6\,\Delta G°\text{quartz} + 2\,\Delta G°K^+) - (3\,\Delta G°\text{Kspar} + 2\,\Delta G°H^+)$$

$$= [(-5680.6) + (-5214.1) + (-615)] - [(11429.9) - (0)]$$

$$= -79.8 \text{ kJ}$$

Then $\Delta G°_R = -RT \ln K = -RT\ 2.303 \log K$

With R = 8.3143 J/deg. mol, $\log K = 2 \log (a_{K^+}/a_{H^+}) = 7.98$

so that for our 250°C activity diagram we have $\log a_{K^+}/a_{H^+} = 3.99$

How does this log activity ratio compare with that which you read directly from Hemley's experimental data (Fig. 6.2)?

Similarly equilibrium constants may be obtained for the following reactions

$$2KAl_3Si_3O_{10}(OH)_2 + 3H_2O + 2H^+ = 3Al_2Si_2O_5(OH)_4 + 2K^+$$

Kmica Kaolinite

$$2NaAlSi_3O_8 + H_2O + 2H^+ = Al_2Si_2O_5(OH)_4 + 4SiO_2 + 2Na^+$$

Albite Kaolinite quartz

but the absence of reliable data for the clay minerals does not allow us to derive activity ratios for these important minerals.

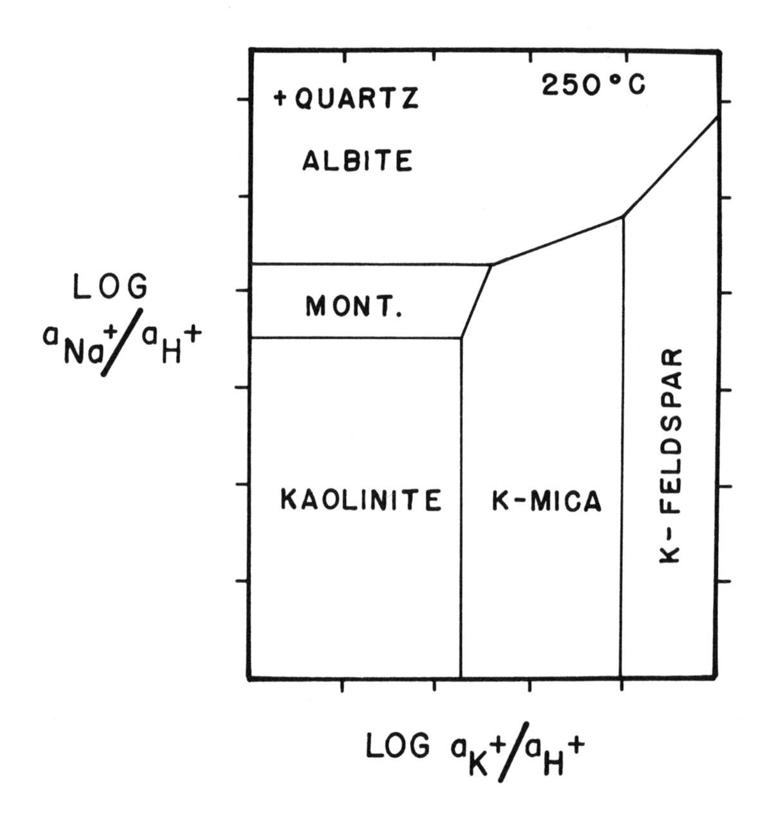

Figure 6.3. Activity Diagram for the principal phases in the system Na_2O-K_2O-Al_2O_3-H_2O at 250°C. Insertion of appropriate axial scales forms an exercise in the text.

Figure 6.3 is an activity diagram for the system K_2O-Na_2O-Al_2O_3-SiO_2-H_2O at 250°C. We have omitted actual numbers from the axes so that you can complete the diagram using thermodynamic data as described above.

		Enthalpy J/gm	Na	K	Ca	Mg	Cl	SO_4	B	SiO_2	ΣHCO_3	CO_2	ΣH_2S
TH1	10/64	1098	891	156	10	.10	1552	21	27	507	16	476	26 mg/kg

BR22	6/80	1160	705	152	1.0	.08	1160	1.6	34	571	27	4437	66

1. The Tauhara Geothermal Field is roughly 8 km south east of Wairakei (Fig. 2.8). The chemistry of the fluids from the two fields are similar but temperatures in some Tauhara wells are a little higher than those in the production wells at Wairakei. The total discharge composition of well TH1 is shown in the table above.

Remembering to convert to appropriate units, plot the composition of this fluid on your activity diagram, the deep pH (calculated as discussed in the next chapter) is 6.0. Incidentally check that t_{SiO2} and t_{NaKCa} for this fluid allow you to use the 250°C diagram (Fig. 6.3).

What alteration minerals would you expect to find in core from this well?

2. The total discharge composition of well BR22 at Broadlands is shown above. What is the principal difference between the fluid compositions of this well and those from Wairakei and Tauhara?

Using the two methods discussed at the beginning of this chapter calculate pH_t and plot activity ratios for this discharge on your activity diagram.

Adding Other Components to the Diagram - The Use of Field Data

We can draw diagrams for any pair of major components we like provided that good thermodynamic or experimental data are available to calculate phase boundary positions. As stressed by Helgeson et al. (1978), data for the aluminosilicates like kaolinite and for calcic phases like wairakite are conflicting, and are often subjectively assessed. We also have problems of solid solution and, in the case of the zeolite group, significant variation in physical HI properties with respect to temperature and pressure. These problems apply particularly to aluminosilicate minerals containing calcium and magnesium, the most important of which in terms of geothermal alteration are epidote (generally represented by clinozoisite, $Ca_2Al_3Si_3O_{12}(OH)$), chlorite, wairakite ($CaAl_2Si_4O_{12},2H_2O$) and the clay minerals.

In such cases chemical data from geothermal systems may be the most reliable guide to phase equilibria. Browne and Ellis (1970) used this approach to map out $a_{Ca^{++}}/a^2_{H^+}$ and $a_{Mg^{++}}/a^2_{H^+}$ activity diagrams to discuss differences in the alteration mineralogy at Broadlands and Wairakei. At Wairakei plagioclase commonly alters to epidote and wairakite at temperatures around 250°C, while at Broadlands chlorite and calcite are the dominant secondary phases with less common wairakite and rare epidote at 260°C. The principal difference in the fluids of the two systems is the concentration of CO_2: about 0.01 m on average at Wairakei, but about 0.15 m at Broadlands.

Ca/Al ratios for the secondary calcium aluminosilicates are as follows:

Clinozoisite	0.66
Wairakite	0.5
Kaolinite	0

so that an appropriate activity diagram may be sketched (Fig. 6.4);

Since reliable thermodynamic data are not available we cannot quantify the $a_{Ca^{++}}/a^2_{H^+}$ axis on the diagram but we observe from field data that calcite displaces epidote as the stable calcic phase at Broadlands. In low grade metamorphic rocks a similar trade-off between calcite and calcic zeolites is also seen and is generally discussed in terms of the activity of CO_2 in the accompanying "metamorphic" fluid - water:rock ratios being much lower than in geothermal systems.

Why is the activity ratio written using $a^2_{H^+}$ for these minerals? Try writing a clinozoisite-wairakite reaction with conservation of aluminium and find $a_{Ca^{++}}/a^2_{H^+}$ for this reaction using the thermodynamic data from Table 6.1.

Calcite

The solubility of calcite was used earlier as a basis for the estimation of pH_t from the chemistry of well discharges. In the examples discussed we found that an assumption of calcite saturation was reasonable for most systems - to within say ± 0.2 pH units.

Writing calcite dissolution as follows:

$$CaCO_3 + 2H^+ = Ca^{++} + CO_{2,aq} + H_2O$$

so that

$$\log K = \log(a_{Ca^{++}}/a^2_{H^+}) + \log m_{CO2,aq}$$

The equilibrium constant that we need was given earlier for 250°C (log K = 6.46). We find for the Wairakei fluid, where $m_{CO2} = 0.01$

$$\log(a_{Ca^{++}}/a^2_{H^+}) = 6.46 + 2 = 8.46$$

and for Broadlands (m_{CO2} = 0.15) $\log(a_{Ca^{++}}/a^2_{H^+}) = 7.28$

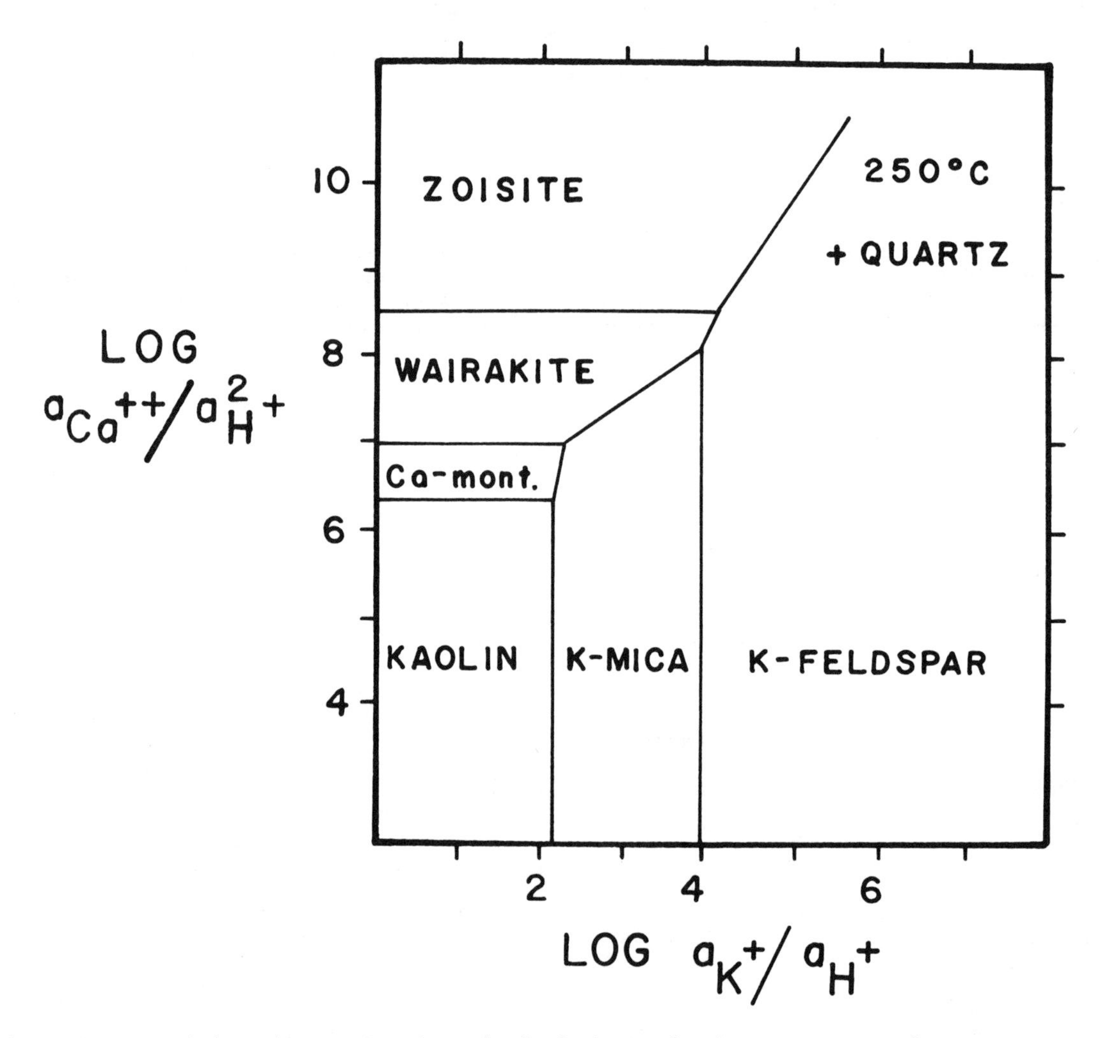

Figure 6.4. Activity Diagram for the principal phases in the system $CaO-Al_2O_3-K_2O-H_2O$ at 250°C.

Draw these calcite solubility boundaries on Fig. 6.4. In order to account for the different alteration assemblages seen at Broadlands and Wairakei the phase boundaries for wairakite-zoisite must be in the approximate positions shown relative to the calcite solubility boundary. Here we assume that the alteration reactions progress close to an equilibrium value, and that other components (such as Fe^{3+}) do not significantly change the activity of zoisite end-member in the epidote solid solution series. Ellis and Browne (1970) provided diagrams at 260°C but these do not differ significantly from that calculated here at 250°C. The method we used above to estimate an equilibrium constant for the wairakite-zoisite reaction could be used to refine our free energy values for one of the two minerals concerned provided that the values for all other phases are sufficiently well known.

For example, we estimated at 250°C $\log (a_{Ca^{++}}/a^2_{H^+}) \simeq 8.5$, so that

$$\Delta G^\circ_R = 2\Delta G^\circ_{zoisite} + 6\,\Delta G^\circ_{quartz} + 4\,\Delta G^\circ_{H2O} - 3\,\Delta G^\circ_{wairakite}$$

$$+ 2\,\Delta G^\circ_{H^+} - \Delta G^\circ_{Ca^{++}} = 85 \text{ kJ}$$

Data from Table 6.1 give ΔG°_R = 55.6 KJ and the difference in these two estimated ΔG° values is equivalent to an experimental error of only 0.5% in the determination of ΔG° for wairakite. In using activity diagrams we are putting some severe constraints on the thermodynamic data. The approach to the problem using field data is limited since extrapolation to other temperatures is precluded without a reliable determination of the specific heat of the reaction and therefore the specific heats of the individual phases. The relationship between these various quantities is discussed by Helgeson (1969).

Bird and Helgeson (1980) have shown how activities of mineral phases may be calculated and applied to alteration equilibria. Using their methods we find that a decrease of a_{mica} from 1 (assumed in all our calculations above) to 0.5 moves the Kmica-Kspar (a_{K^+}/a_{H^+}) - phase boundary by about +0.15 units.

The solubilities of magnesium bearing minerals are extremely small at high temperatures, hence the low magnesium concentrations in dilute geothermal waters. The principal magnesium-bearing phase is chlorite which, with epidote, may be important in limiting redox conditions in hydrothermal systems due to the presence of Fe^{2+} in solid solution.

<u>Effect of silica activity</u>

In all the calculations above quartz was assumed to be the stable silica phase. In general this assumption is valid since the quartz geothermometer is demonstrably successful in geothermal fields at all temperatures above 200°C. It is instructive to consider deviation from this limiting condition as follows:

Reaction	$\log K_{250^\circ}$
3 Kspar $+2H^+$ = Kmica + 6 quartz $+2K^+$	= 7.98
6 quartz + $12H_2O$ = 6 H_4SiO_4	5.32
6 H_4SiO_4 = $12H_2O$ + 6 cristobalite	-6.54
3Kspar $+2H^+$ = Kmica + 6 cristobalite + $2K^+$	6.76

(log K for the quartz and cristobalite reactions are easily obtained from Table 3.2)

Therefore $\log a_{K^+}/a_{H^+}$ at cristobalite saturation = 3.38 compared to about 4 at quartz saturation

By now you will have realized that the calculations involved here are simple but tedious. Computer programs like SUPCRT routinely calculate these equilibrium constants. Even so it is still somewhat daunting to be obliged to specify reactions and proceed through these routines every time you need an activity diagram. The set of equilibria with which we are concerned in geothermal systems is relatively small, so that to save time we could calculate their equilibrium constants and then find regression equations to describe their variation with respect to

1/T. Arnorsson et al. (1982) have done this (see Table 6.2) and presented a set of useful regression equations for a large number of reactions. Their table is largely based on Helgeson et al. (1978), but some data are modified on the basis of field observations. An example of the use of these equations follows - but remember that the regression equations are merely time saving aids and may produce numbers slightly inconsistent with other thermodynamic data.

Table 6.2 -- Equilibrium constants for some selected geothermal fluid-mineral reactions. (Reproduced from Arnorsson et al., 1982).

Equilibrium constants for the disproportionation of selected geothermal minerals

MINERAL	REACTION	TEMPERATURE FUNCTION (°K)
401 ADULARIA	$KAlSi_3O_8 + 8H_2O = K^+ + Al(OH)_4^- + 3H_4SiO_4^\circ$	$+38.85 -0.0458T -17260/T +1012722/T^2$
402 LOW-ALBITE	$NaAlSi_3O_8 + 8H_2O = Na^+ + Al(OH)_4^- + 3H_4SiO_4^\circ$	$+36.83 -0.0439T -16474/T +1004631/T^2$
403 ANALCIME	$NaAlSi_2O_6 \cdot H_2O + 5H_2O = Na^+ + Al(OH)_4^- + 2H_4SiO_4^\circ$	$+34.08 -0.0407T -14577/T +970981/T^2$
404 ANHYDRITE	$CaSO_4 = Ca^{+2} + SO_4^{-2}$	$+6.20 -0.0229T -1217/T$
405 CALCITE	$CaCO_3 = Ca^{+2} + CO_3^{-2}$	$+10.22 -0.0349T -2476/T$
406 CHALCEDONY	$SiO_2 + 2H_2O = H_4SiO_4^\circ$	$+0.11 -1101/T$
407 Mg-CHLORITE	$Mg_5Al_2Si_3O_{10}(OH)_8 + 10H_2O = 5Mg^{+2} + Al(OH)_4^- + 3H_4SiO_4^\circ + 8OH^-$	$-1022.12 -0.3861T +9363/T +412.46\log T$
408 FLUORITE	$CaF_2 = Ca^{+2} + 2F^-$	$+66.54 -4318/T -25.47\log T$
409 GOETHITE	$FeOOH + H_2O + OH^- = Fe(OH)_4^-$	$-80.34 +0.099T +20290/T -2179296/T^2$
410 LAUMONTITE	$CaAl_2Si_4O_{12} \cdot 4H_2O + 8H_2O = Ca^{+2} + 2Al(OH)_4^- + 4H_4SiO_4^\circ$	$+65.95 -0.0828T -28358/T +1916098/T^2$
411 MICROCLINE	$KAlSi_3O_8 + 8H_2O = K^+ + Al(OH)_4^- + 3H_4SiO_4^\circ$	$+44.55 -0.0498T -19883/T +1214019/T^2$
412 MAGNETITE	$Fe_3O_4 + 4H_2O = 2Fe(OH)_4^- + Fe^{+2}$	$-155.58 +0.1658T +35298/T -4258774/T^2$
413 Ca-MONTMOR.	$6Ca_{0.167}Al_{2.33}Si_{3.67}O_{10}(OH)_2 + 60H_2O + 12OH^- = Ca^{+2} + 14Al(OH)_4^- + 22H_4SiO_4^\circ$	$+30499.49 +3.5109T -1954295/T +125536640/T^2 -10715.66\log T$
414 K-MONTMOR.	$3K_{0.33}Al_{2.33}Si_{3.67}O_{10}(OH)_2 + 30H_2O + 6OH^- = K^+ + 7Al(OH)_4^- + 11H_4SiO_4^\circ$	$+15075.11 +1.7346T -967127/T +61985927/T^2 -5294.72\log T$
415 Mg-MONTMOR.	$6Mg_{0.167}Al_{2.33}Si_{3.67}O_{10}(OH)_2 + 60H_2O + 12OH^- = Mg^{+2} + 14Al(OH)_4^- + 22H_4SiO_4^\circ$	$+30514.87 +3.5188T -1953843/T +125538830/T^2 -10723.71\log T$
416 Na-MONTMOR.	$3Na_{0.33}Al_{2.33}Si_{3.67}O_{10}(OH)_2 + 30H_2O + 6OH^- = Na^+ + 7Al(OH)_4^- + 11H_4SiO_4^\circ$	$+15273.90 +1.7623T -978782/T +62805036/T^2 -5366.18\log T$
417 MUSCOVITE	$KAl_3Si_3O_{10}(OH)_2 + 10H_2O + 2OH^- = K^+ + 3Al(OH)_4^- + 3H_4SiO_4^\circ$	$+6113.68 +0.6914T -394755/T +25226323/T^2 -2144.77\log T$
418 PREHNITE	$Ca_2Al_2Si_3O_{10}(OH)_2 + 10H_2O = 2Ca^{+2} + 2Al(OH)_4^- + 2OH^- + 3H_4SiO_4^\circ$	$+90.53 -0.1298T -36162/T +2511432/T^2$
419 PYRRHOTITE	$8FeS + SO_4^{-2} + 22H_2O + 6OH^- = 8Fe(OH)_4^- + 9H_2S$	$+3014.68 +1.2522T -103450/T -1284.86\log T$
420 PYRITE	$8FeS_2 + 26H_2O + 10OH^- = 8Fe(OH)_4^- + SO_4^{-2} + 15H_2S$	$+4523.89 +1.6002T -180405/T -1860.33\log T$
421 QUARTZ	$SiO_2 + 2H_2O = H_4SiO_4^\circ$	$+0.41 -1309/T$ (0-250°C); $+0.12 -1164/T$ (180-300°C)
422 WAIRAKITE	$CaAl_2Si_4O_{12} \cdot 2H_2O + 10H_2O = Ca^{+2} + 2Al(OH)_4^- + 4H_4SiO_4^\circ$	$+61.00 -0.0847T -25018/T +1801911/T^2$
423 WOLLASTONITE	$CaSiO_3 + 2H^+ + H_2O = Ca^{+2} + H_4SiO_4^\circ$	$-222.85 -0.0337T +16258/T -671106/T^2 +80.68\log T$
424 ZOISITE	$Ca_2Al_3Si_3O_{12}(OH) + 12H_2O = 2Ca^{+2} + 3Al(OH)_4^- + 3H_4SiO_4^\circ + OH^-$	$+106.61 -0.1497T -40448/T +3028977/T^2$
425 EPIDOTE	$Ca_2FeAl_2Si_3O_{12}(OH) + 12H_2O = 2Ca^{+2} + Fe(OH)_4^- + 2Al(OH)_4^- + 3H_4SiO_4^\circ + OH^-$	$-27399.84 -3.8749T +1542767/T -92778364/T^2 +9850.38\log T$
426 MARCASITE	$8FeS_2 + 26H_2O + 10OH^- = 8Fe(OH)_4^- + SO_4^{-2} + 15H_2S$	$+4467.61 +1.5879T -169944/T -1838.45\log T$

We can add these reactions to obtain the reaction we want and add log K values from Table 6.2 to obtain its equilibrium constant. From the tables we can calculate the following log K values at 250°C; the log K value for water is tabulated in the next chapter.

Equation		log K (250°C)
	Water	-11.13
402	Albite	-13.96
411	K-spar	-15.07
417	Kmica	-17.84
421	Quartz	-2.11

Using equations 411, 417 and 421 we can write,

eq#	Reaction	Multiplying Factor
411	$3\ \text{K-spar} + 24H_2O = 3K^+ + 3Al(OH)_4^- + 9H_4SiO_4$	3
417	$3H_4SiO_4 + 3Al(OH)_4^- + K^+ = \text{Kmica} + 10H_2O + 2OH^-$	-1
421	$6H_4SiO_4 = 6\text{Qtz} + 12H_2O$	-6
	$2H^+ + 2OH^- = 2H_2O$	-2

$$3\ \text{K-spar} + 2H^+ = \text{Kmica} + 6\text{Qtz} + 2K^+$$

$$\log K = 3 \log K(411) - \log K(417) - 6 \log K(421) - 2 \log K(101)$$

$$= 3(-15.06) - (-17.84) - 6(-2.11) - 2(-11.13) \quad 250°$$

$$= 7.58$$

$$\log\ a_{K^+}/a_{H^+} = 3.79$$

Notice that this result is not the same as the value calculated from thermochemical data. This is quite typical of the problems of high temperature solution chemistry. Different data sources give different results and may not agree with field observations. Each investigator must choose stability data carefully and reconcile it with field data. A similar and less confusing method may be to use log K of formation values as suggested earlier in this chapter.

Use Table 6.2 to obtain an estimate of the activity ratio for the wairakite-zoisite phase boundary at 250°C and compare the value you obtain with those you calculated earlier using (a) field observations and (b) available thermodynamic data.

Giggenbach (1981) used a similar approach as the basis of a study of fluid-mineral equilibria in geothermal systems. Arnorsson et al. (1978, 1982, 1983) have examined mineral-fluid equilibria in a set of Icelandic systems using this technique. These papers are recommended for further study particularly as they emphasize the chemical uniformity exhibited by a large number of geothermal systems covering a range of temperatures. It is such uniformity which is the basis for the alkali geothermometers and our overall understanding of these complex systems.

Using Activity Diagrams

Activity diagrams are just one way to depict relationships between minerals and fluids in geothermal systems and to compare the alteration assemblages observed in geothermal systems. They are also useful in discussing the mineralogical effects of cooling or boiling of hydrothermal solutions.

Use the data shown in Fig. 6.2 (or in the thermodynamic table, Table 6.1) to show which way the Kmica-K-feldspar boundary moves on cooling with respect to its location at 250°C. If a solution, initially in equilibrium with this mineral pair at 250°C, cools conductively, which mineral should precipitate?

Boiling -- Boiling removes gases (CO_2, H_2S, etc.) from solution and it is evident from our equilibrium relation

$$HCO_3^- + H^+ = CO_{2,g} + H_2O$$

that this loss of gas is accompanied by a decrease in a_{H^+}. If boiling occurs from 250 to 245°C over 50% of the dissolved CO_2 may be lost to the steam phase with a removal of about 1% steam. Assuming that log K and m_{HCO_3} change insignificantly over this temperature drop (see Chapter 7)

$$pH_1 - pH_2 = \log (CO_{2,250°}/CO_{2,245°}) = \log 2.0 = +0.3$$

i.e. 50% loss of CO_2 has increased pH by about 0.3 units

What would be the pH change due to 90% CO_2 loss over the same temperature interval?

What minerals would you expect to find in vugs or veins if local boiling occurs at Broadlands? Initially the solution which contains CO_2 is close to equilibrium at ≃ 260°C with quartz, Kmica, Kspar, calcite, wairakite and chlorite. The pH increase due to gas loss moves the fluid composition into the K-feldspar field so that adularia deposits, along with calcite and quartz. Some wairakite may also form and with very extensive gas loss due to boiling, epidote may also deposit. Loss of H_2S may also result in the deposition of pyrite and in some systems, base metal sulfides (see Chapter 8). Mineral assemblages are therefore sensitive indicators of local boiling - probably more sensitive than fluid inclusions because trapping of two fluid phases during mineral growth may not always occur.

Fluid-rock ratios -- One factor which we haven't discussed as a control on mineral alteration assemblages and fluid compositions is the water-rock ratio of the hydrothermal system. Clearly the greater the mass ratio of water:rock, the more likely it is that fluid composition controls the mineral assemblage - and vice versa. Isotope studies in a number of geothermal systems have shown that the ratio usually lies in the range 1 to 4. Notice however that the ratio represents the total mass of fluid passing through a given mass of rock over an extended period of time; it is not an instantaneous value, which would be much lower. With such low values it is not too surprising that mineral alteration reactions tend to strongly influence fluid chemistry.

The alteration pattern depends as well on the concentration of reacting solutes. In near surface zones where H^+ from oxidizing H_2S is abundant, argillization and dissolution dominate, although water/rock ratios may differ little from deeper zones.

Activity diagrams also help identify the suite of reactions which form the basis of some of the geothermometry techniques. In the Na-K-Al-Si-O-H system, geothermal fluid compositions lie close to the albite-Kspar-Kmica triple point but reflect the alteration of primary plagioclase. The sodic component alters to Kmica (or clay) and Na+ according to reactions like,

$$\text{Na-plagioclase} + K^+ + H^+ = \text{Kmica} + 2Na^+$$

while the calcic component alters to calcite, wairakite or epidote depending on the carbon dioxide content of the fluid. These complementary reactions are the basis of the NaK and NaKCa geothermometers.

Figure 6.5 shows stability relationships for alunite ($KAl_3(SO_4)_2(OH)_6$) and other aluminosilicates in the presence of quartz at 200° and 300°C. Write chemical reaction-equations for alunite-Kmica and alunite-kaolinite mineral pairs. At a sulfate + bisulfate activity of 0.005 molal, determine, at both temperatures, the pH range over which these mineral pairs may occur as a function of potassium concentration - in mg/kg.

In what parts of a geothermal system would you expect to find alunite? What would be the effect of an increase in silica concentration above quartz saturation through, for example, the dissolution of primary volcanic glass?

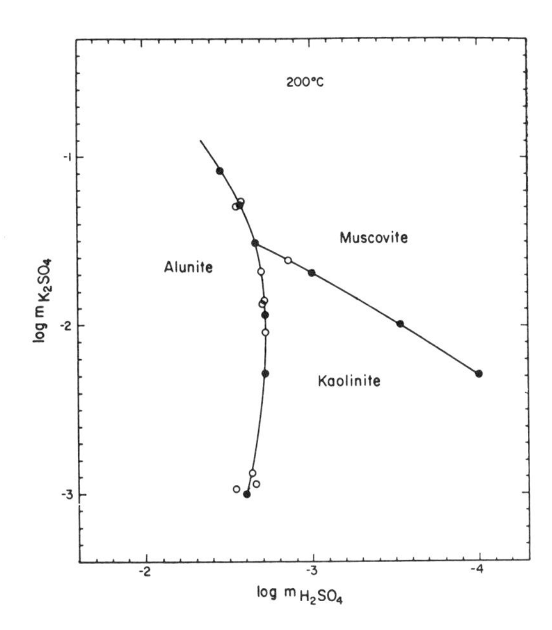

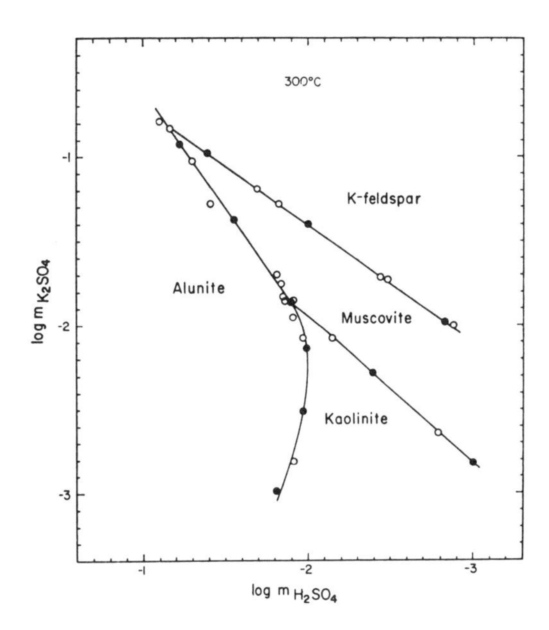

Figure 6.5. Experimentally determined phase relations in the system K_2O-Al_2O_3-SiO_2-H_2O-SO_3. Quartz is present and the total pressure is 15000 psi. The symbols indicate the results of individual experimental determinations. (Reproduced from Hemley et al., 1969)

The kaolinite + Kmica reaction (in marked contrast to that for K-feldspar + Kmica + quartz) has a wide variety of values reported in the literature for the same equilibrium constant (log K = 2 log a_{K^+}/a_{H^+}). At 200°C the values range as follows; 4.48(Helgeson et al., 1978), 5.36(Hemley et al., 1969), 6.20(Montoya and Hemley, 1975) and 7.04(Helgeson, 1969). Can this wide range be the result of disequilibrium or perhaps a failure to characterize the composition or structure of the kaolinite properly? Definitely not! All the values are derived from the same two sets of experiments (Hemley, 1959, for a simple chloride system, and Hemley et al., 1969, for a sulfate-rich system) and so the diverse values come from the diverse ways used to obtain the high temperature pH of the experimental solutions from measurements of pH on samples at laboratory temperature as well as from different sets of constants used in correcting for ion pairing (e.g. $K^+ + Cl^- = KCl$) for the experimental conditions. These corrections still haunt us and a satisfactory reconciliation of all of the data is yet to be found. We shall explore methods for calculating high temperature pH and solute speciation in the next chapter.

REFERENCES

Arnorsson, S., Gronvold, K., and Sigurdsson S. 1979, Aquifer chemistry of four high temperature geothermal systems in Iceland: Geochimica et Cosmochimica Acta, v. 42, p. 523-536.

Arnorsson, S., Sigurdsson, S., and Svarvarsson, H., 1982, The chemistry of geothermal waters in Iceland. I. Calculation of aqueous speciation from 0° to 370°C: Geochimica et Cosmochimica Acta, v. 46, p. 1513-1532.

Arnorsson, S., Gunnlaugsson, E., and Svavarsson, H., 1983, The chemistry of geothermal waters in Iceland. II. Mineral equilibria and independent variables controlling water compositions: Geochimica et Cosmochimica Acta, v. 47, p. 547-566.

Bird, D.K., and Helgeson, H.C., 1980, Chemical interaction of aqueous solutions with epidote-feldspar mineral assemblages in geologic systems 1: American Journal of Science, v. 280, p. 907-941.

Browne, P.R.L., 1978, Hydrothermal Alteration in Active Geothermal Fields: Annual Review Earth and Planetary Sciences, v. 6, p. 229-250.

Browne, P.R.L., and Ellis, A.J., 1970, The Ohaki-Broadlands Hydrothermal Area, New Zealand: mineralogy and related geochemistry: American Journal of Science, v. 269, p. 97-131.

Giggenbach, W.F., 1981, Geothermal mineral equilibria: Geochimica et Cosmochimica Acta, v. 45, p. 393-410.

Haas, J.L., and Fisher, J., 1976, Simultaneous evaluation and correlation of thermodynamic data: American Journal of Science, v. 276, p. 525-545.

Helgeson, H.C., 1969, Thermodynamics of hydrothermal systems at elevated temperatures and pressures: American Journal of Science, v. 267, p. 729-804.

Helgeson, H.C., Delany, J.M., Nesbitt, H.W., and Bird, D.K., 1978, Summary and critique of the thermodynamic properties of rock forming minerals: American Journal of Science, v. 278-A, p. 1-229.

Hemley, J.J., 1959: Some mineralogical equilibria in the System K_2O-Al_2O_3-SiO_2-H_2O: American Journal of Science, v. 257, p. 241-270.

Hemley, J.J., Hostetler, P.B., Gude, A.J., and Mountjoy, W.T., 1969, Some stability relations of alunite: Economic Geology, v. 64, p. 599-612.

Hemley, J.J., Meyer, C., and Richter, D.H., 1961, Some alteration reactions in the system Na_2O-Al_2O_3-SiO_2-H_2O: U.S. Geological Survey Professional Paper 424-D, p. 338-340.

Henley, R.W., and Ellis, A.J., 1983, Geothermal Systems, Ancient and Modern: a geochemical review: Earth Science Reviews, v. 19, p. 1-50.

Hoagland, J.R., and Elders, W.A., 1978, Hydrothermal mineralogy and isotopic geochemistry of Cerro Prieto geothermal field, Mexico, I. Hydrothermal mineral zonation: Geothermal Resources Council Transactions, v. 2, p. 283-286.

Kristmannsdottir, H., 1976, Types of clay minerals in hydrothermally altered basaltic rocks, Reykjanes, Iceland: Jokull, 26, 30-38.

Montoya, J.W., and Hemley, J.J., 1975, Activity relations and stabilities in alkali feldspar and mica alteration reactions: Economic Geology, v. 70, p. 577-582.

Pitzer, K.S., 1981, Characteristics of very concentrated aqueous solutions: in Rickard, D.T. and Wickman, F.E., Chemistry and Geochemistry of Solutions at High Temperatures and Pressures, Pergammon Press, Oxford, p. 249-264.

Robie, R.A., Hemingway, B.S., and Fisher, J., 1978, Thermodynamic properties of minerals and related substances at 298.95°K ... and at higher temperatures: U.S. Geological Survey Bulletin 1452, 456 p.

Rose, A.W., and Burt, D.M., 1979, Hydrothermal Alteration: in Barnes, H.L., Geochemistry of Hydrothermal Ore Deposits: 2nd Edition, Wiley-Interscience, New York, p. 173-235.

Steiner, A., 1977, The Wairakei Geothermal Area, North Island, New Zealand: NZ DSIR Geological Survey Bulletin 90, 135 p.

Truesdell, A.H., and Henley, R.W., 1982, Chemical Equilibria in the Cerro Prieto Hydrothermal Fluid; Proc. 4th Symposium on the Cerro Prieto Geothermal Field. CFE Guadalajara, Mexico, Report (in press).

Chapter 7
pH CALCULATIONS FOR HYDROTHERMAL FLUIDS

In the last chapter the relationship between alteration mineralogy and fluid chemistry was discussed using a greatly oversimplified scheme for estimating the pH_t of deep system fluids. We use this as a starting point to consider some further implications of the fluid-mineral equilibria we have considered and then develop some calculation procedures to improve our calculation of pH_t for high temperature fluids from analytical data.

SALINITY - pH RELATIONSHIPS

Drill core from hydrothermal systems contains a number of secondary minerals such as Kmica, K-feldspar, and epidote that result from the alteration of primary minerals. In many hydrothermal studies it is assumed that fluid composition and pH were controlled--or buffered--by simple alteration phase equilibria. We can examine this proposition as follows. If the pH controlling equilibria are

$$1.5\ KAlSi_3O_8 + H^+ = 0.5\ KAl_3Si_3O_{10}(OH)_2 + 3SiO_2 + K^+ \quad (1)$$

$$3\ NaAlSi_3O_8 + K^+ + 2H^+ = KAl_3Si_3O_{10}(OH)_2 + 6SiO_2 + 3Na^+ \quad (2)$$

We may, from equilibrium constant data for these reactions, obtain a theoretical estimate of pH_t as a function of salinity (approximated as $m_{Na^+} + m_{K^+}$).

'Salinities' of 1.0, 0.1 and 0.01 $m_{(Na + K)}$ cover the range of most geothermal fluids. We first determine m_{K^+} at the temperature of interest by using our empirical NaK geothermometer (m_{Na^+}/m_{K^+}) in reverse (Table 3.2)

Table 7.1 -- Log m_{K^+} as a function of (m_{Na^+} + m_{K^+}) and temperature.

		$m_{Na^+} + m_{K^+}$		
	mNa+/mK+	1.0	0.1	0.01
200°C	15	-1.2	-2.2	-3.2
250°C	10.2	-1.05	-2.05	-3.05
300°C	7.6	-0.93	-1.93	-2.93

Using equilibrium constant values calculated from thermodynamic data we may now calculate pH_t directly from the Kmica-Kspar equilibrium constant.

Table 7.2 -- pH as a function of (m_{Na^+} + m_{K^+}) and temperature.

		$m_{Na^+} + m_{K^+}$			
	log K	1.0	0.1	0.01	$pH_{neutral}$
200°C	4.1	5.44	6.44	7.44	5.65
250°C	4.0	5.18	6.18	7.18	5.6
300°C	3.8	4.86	5.86	6.86	5.65

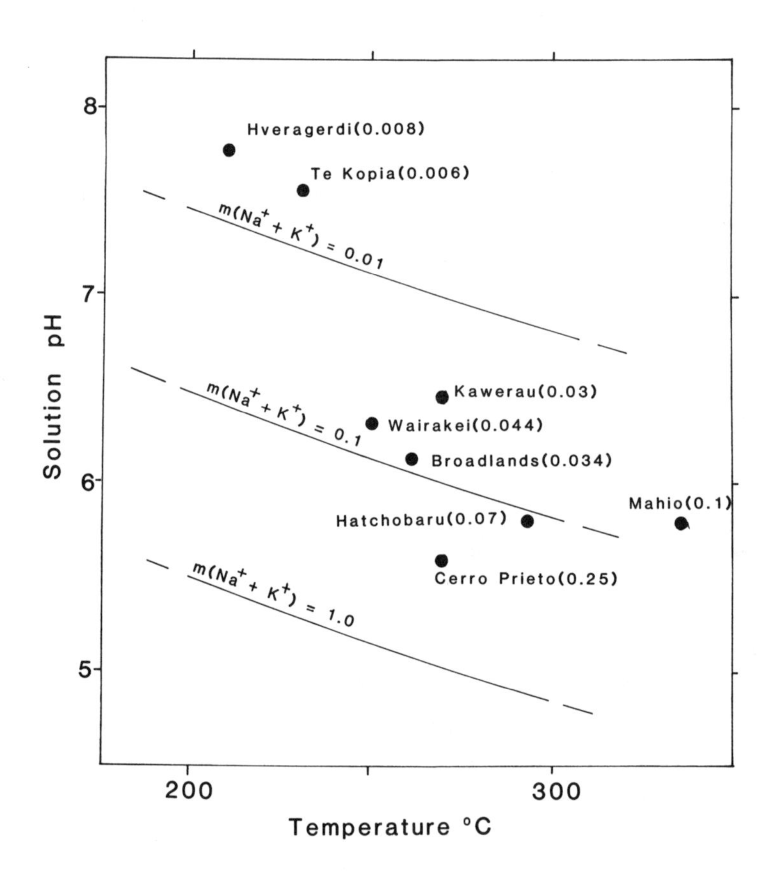

Figure 7.1. Comparison of pH-temperature relations constrained by the Kspar-Kmica-quartz equilibrium at 3 salinities with observed pH in fluids discharged from a number of geothermal fields. Some of the data points are from Ellis (1971). pH-temperature relations are derived from thermodynamic data in combination with empirical Na/K vs temperature relations (see text).

Fig. 7.1 is plotted from the data of Table 7.2 and shows that the pH_t's of dilute geothermal fluids (like Wairakei) are far higher than those of more saline fluids such as in the Tongonan and Cerro Prieto systems, or the Salton Sea system. There are, from this observation, some important implications in assessing potential development problems at an early stage in exploration. The more saline and therefore more acid fluids may involve greater scaling problems, even though, as is the case in the Salton Sea, the scale may have some economic value due to its lead, zinc and other metal contents (see Chapter 8). For metals transported as chloride complexes, as we shall see in Chapter 8, the total chloride and pH are important parameters controlling metal sulfide solubilities. At Creede, Colorado, for example Barton et al (1977) have estimated from fluid inclusion data that the ore-forming fluid contained about 1 mole Cl/kg at 250°C. What is the expected maximum pH_t of this fluid and how would you rate its metal transport capability with respect to Broadlands or Hatchobaru?

Notice that (apart from the high salinity-high temperature solutions) all the values of pH shown in Table 7.2 represent slightly alkaline solutions. For individual fields the pH_t calculated from acid-base equilibria (as discussed below) in the fluid may be compared with values calculated using the assumption of mineral-fluid equilibrium as shown in Figure 7.1. In the fields shown the pH of the reservoir fluid appears to be close to that expected from the silicate equilibrium assumption and our simple calculation.

Our concept of pH buffering by a single aluminosilicate reaction is rather too simple for active geothermal systems. For one thing we have assumed for our own convenience that aluminium is immobile. This is clearly not true and we need to recognize that the alteration parageneses observed represent a stack of incongruent mineral solubility reactions.* Other components must also be brought into the set of pH-controlling reactions which we consider. One of the most important is CO_2 which, as we already have seen, significantly affects temperature-depth relations in geothermal systems and which also participates in pH-dependent mineral-fluid reactions. Remember that we used one as the basis of a simple method for estimating pH_t at the beginning of the last chapter. We also need to recognize that in hydrothermal systems a continuous flux of dissolved components (and heat) occurs so that our field observations really relate to a steady or evolving steady state rather than a simple closed system or local equilibrium.

---

*It would be desirable to routinely obtain analyses of dissolved aluminium in geothermal fluids but analytical problems at low concentrations and the practical problem of sample contamination by particulate alumino-silicates in suspension generally prevent this. Some data for Wairakei fluids are given by Goguel (1977).

Giggenbach (1981) recognized that the mineral assemblages and fluid compositions in a geothermal system resulted from the gradual conversion of mineral phases thermodynamically unstable under geothermal conditions to a stable secondary assemblage. The fluid compositions observed therefore represent a chemical steady state reflecting the attainment to varying degrees of partial or local equilibrium with respect to a small set of controlling reactions. Figures 7.2a and b show data for a number of New Zealand geothermal wells plotted in P_{CO2}-t space. These data show that the pistacite/clinozoisite-epidote solid solution series is the most important secondary calcium alumino-silicate in geothermal systems but that CaAl-silicate-calcite reactions are not controlling P_{CO2}. The distribution of data points does however suggest that a reaction, involving one of the primary minerals may exert such a control, e.g.

$$\text{plagioclase} + CO_2 = \text{clay} + \text{calcite} \qquad (5)$$

[Notice that because of the lack of thermodynamic data for smectites and illite, clay is represented by kaolinite or Kmica.] Both the reactions suggested are consistent with assemblages observed in active and fossil hydrothermal systems.

These data also indicate that at relatively low temperatures (< 250°C) feldspar-mica reactions exerted some control. Giggenbach (1981, p. 399) suggested that the plagioclase/"clay"/calcite reaction predominates in hydrothermal systems but that, depending on temperature a fluid "system may then lock-on to a secondary buffer system involving, at lower temperatures, the potassium phases microcline/muscovite and at higher temperatures low albite and its alteration products or possibly biotite/chlorite."

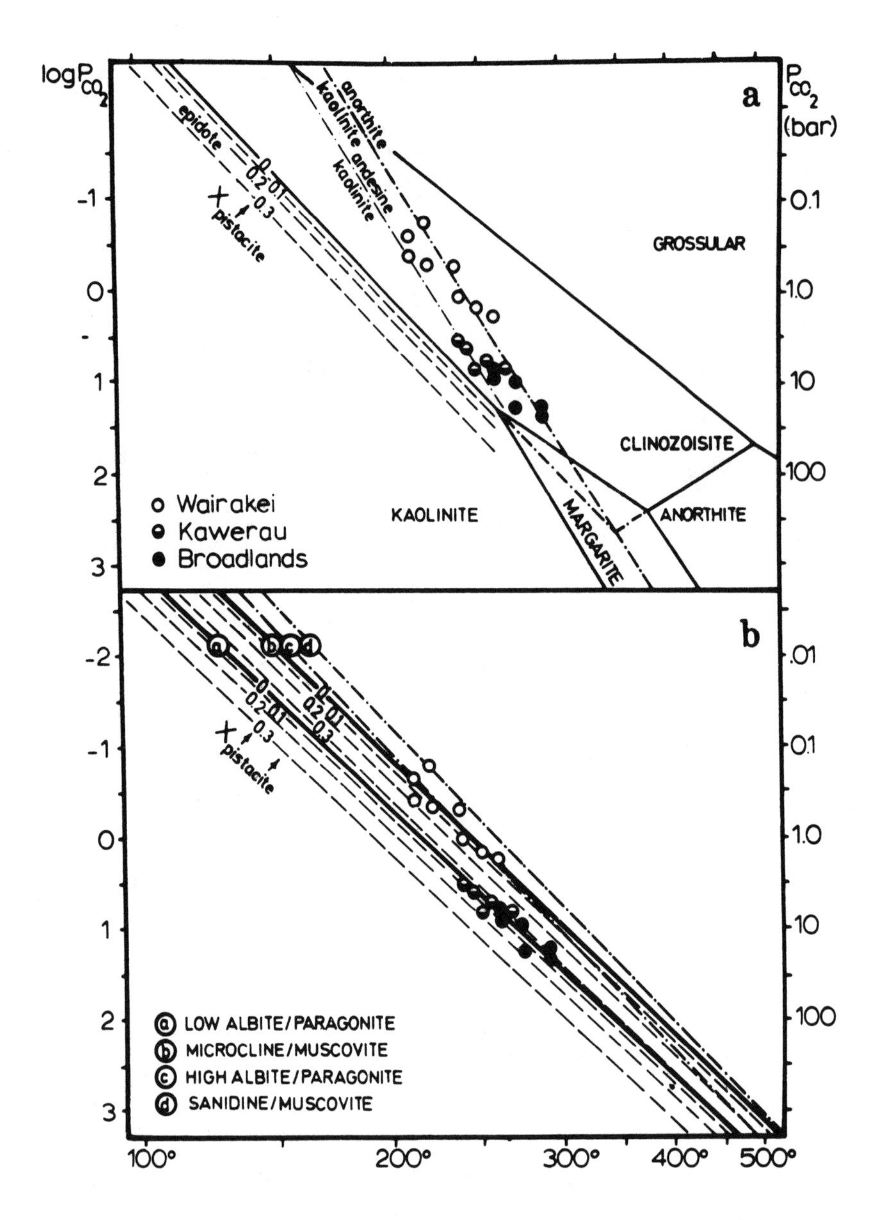

Figure 7.2a and b. Stability diagrams for calcium-aluminium-silicate (a) and feldspar-mica (b) reactions in terms of P_{CO2} and temperature together with data points for well discharges at Wairakei, Kawerau and Broadlands. $X_{pistacite}$ refers to the mole fraction of pistacite $Ca_2Fe_3Si_3O_{12}(OH)$ in epidote. In Fig. 7.2b microcline/muscovite and low albite/paragonite coexistence boundaries in the presence of epidote with $X_{pistacite}$ of 0, 0.1, 0.2 and 0.3 are shown.

If P_{CO2} is controlled through mineral-fluid reactions we may expect to see some relationship between mineral assemblages and our CO_2-H_2O boiling point-depth curves (Chapter 4). Figures 7.2c and d suggest that this generalization may hold true although in any given system a more complex steady state is attained through the kinetics of primary mineral dissolution and secondary mineral deposition, fluid mixing and boiling and, through the permeability and porosity of the system, the degree to which the rising fluid contacts unaltered or partially altered rock.

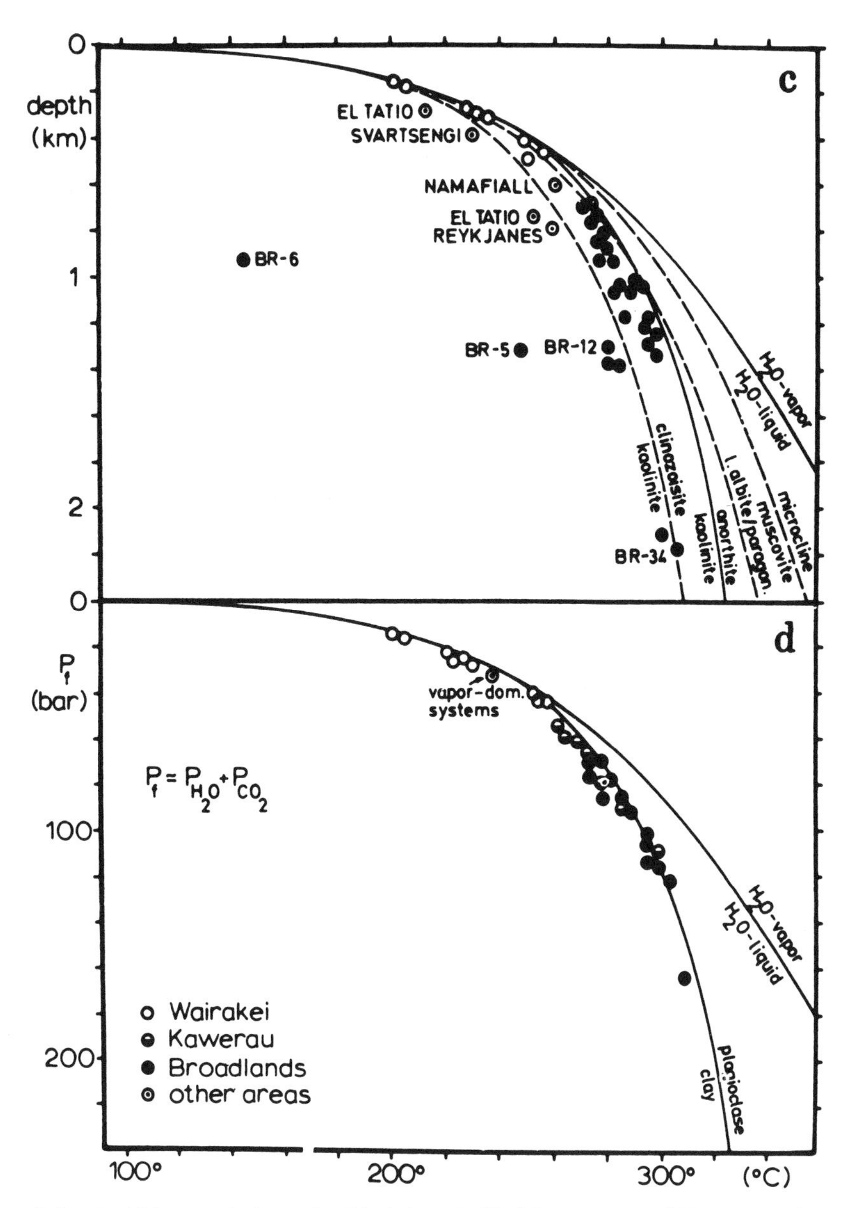

Figure 7.2c and d. Boiling point vs depth (c) and fluid pressure (d) curves for a series of mineral assemblages. The data points for Wairakei, Kawerau, Broadlands, El Tatio and Iceland wells correspond to the point at which measured temperature/depth and temperature/pressure curves change from close to constant temperature with depth behaviour to that indicating boiling point vs depth conditions. (Figures a-d reproduced with permission from Giggenbach, 1981)

In summary, fluid compositions in geothermal systems appear to be constrained by "sliding-scale buffers" with mineral solubilities related to CO_2 partial pressures. The flux of dissolved CO_2 through the system supplies the H^+ responsible for the alteration reactions seen within hydrothermal systems through the production of bicarbonate ion.

Write a reaction equation describing the conversion of plagioclase (represented by anorthite) to Kmica and calcite in terms of bicarbonate ion and with conservation of Al in solid phases. Relate this equation to the activity diagram (Fig. 6.4 discussed in the last chapter.

ION CONCENTRATIONS AND pH AT HIGH TEMPERATURES

In order to examine fluid compositions actually observed in geothermal systems in relation to the equilibrium or steady state models discussed above we need to be able to calculate more rigorously and routinely the concentrations of dissolved species (complexes) and pH in the high temperature fluid. These same high temperature data are also required when we consider the transport of metals in hydrothermal systems and in applied problems such as the control of silica and calcite scaling in geothermal plants.

In order to obtain high temperature pH and dissolved species data, we need some kind of solution model. This involves the same procedure used by geochemists to model the chemistry of seawater at low temperature. However, geothermal fluids are not so user-friendly, and in developing a solution model we must make a number of assumptions (in the derivation of activity coefficients, for example) and a number of approximations.

In any solution the molal sum of all positively charged ions must equal that of the negative ions. This is the 'charge balance' equation. It must be true at any temperature.

$$m_{Na^+} + m_{K^+} + 2m_{Ca^{++}} + 2m_{Mg^{++}} + m_{NH_4^+} + \ldots + m_{H^+}$$

$$= m_{Cl^-} + m_{HCO_3^-} + 2m_{CO_3^=} + 2m_{SO_4^=} + m_{HS^-} + m_{H_2BO_3^-} + m_{H_3SiO_4^-} + m_{NaSO_4^-} + \text{etc.} \qquad (6)$$

The equation has a multitude of terms and should also include all complex ion species e.g. $CaHCO_3^+$. Provided that we have analyzed for every constituent and that we have reliable thermodynamic data it would be possible to calculate every term in the equation. Fortunately we can simplify matters by taking a careful look at the individual terms and realizing that the concentrations of some of the pH-dependent charged species are negligible in certain pH ranges, as well as small relative to analytical errors.

Uncharged solution species are also important particularly at high temperatures. The formation of NaCl ion pairs, for example, reduces the concentration of free Na^+ thereby effecting the ionic strength of the solution and estimates of activity coefficients. So too does the formation of species like $NaSO_4^-$. Notice however that in forming ion pairs like NaCl, concentrations of Na^+ and Cl^- subtract equally from each side of equation (6) so that the charge balance relation and its derivatives are not invalidated by ignoring such neutral species.

Note that analytical concentrations represent the molal sum of all species containing a particular component

e.g. $$m_{\Sigma SiO_2} = m_{H_4SiO_4} + m_{H_3SiO_4^-} + m_{H_2SiO_4^=} \; \cdots$$

$$m_{\Sigma C} = m_{H_2CO_3} + m_{HCO_3^-} + m_{CO_3^=} + m_{CaHCO_3^+}$$

The concentration of total silica in solution is measured using an atomic adsorption or a spectrophotometric technique and is usually recorded as $mgSiO_2/kg$.

The concentration of carbon containing species is determined by pH-titration and is sometimes expressed in terms of individual ion concentrations (CO_2, HCO_3^-, $CO_3^=$) at the analysis temperature. It is preferable however not to break down the titration in this way and to express <u>all</u> carbon containing species as bicarbonate ion. If pH is also recorded for the solution, individual ion concentrations may be determined from the equations developed below.

Neglecting minor species, we can regroup the terms in equation (6) as follows

$$m_{Na^+} + m_{K^+} + 2m_{Ca^{++}} + \ \ldots \ - m_{Cl^-} \simeq m_{HCO_3^-} + 2m_{CO_3^=} + m_{H_3SiO_4^-} - m_{NH_4^+} - m_{H^+}. \quad (7)$$

As we shall find later it is convenient to define a quantity Δ equal to the charge sum on the right hand side of equation (7).

If Na and Cl in dilute solutions remain essentially unassociated the left hand side of the equation is (within the limits of the assumption) constant at all temperatures and is balanced by the pH sensitive ions (HCO_3^- through to NH_4^+). The concentrations of these species and their conjugate acids are given by simple pH dependent equilibrium constants--often called dissociation constants.

For example, for the silicate ion,

$$H_4SiO_4 \ = \ H_3SiO_4^- \ + \ H^+$$

$$K_{H_4SiO_4} \ = \ a_{H^+} \, a_{H_3SiO_4^-} / a_{H_4SiO_4} \quad (8)$$

Now, like all equilibrium constants, K_{H4SiO4}, is temperature dependent so the concentration of silicate ion at the temperature of analysis must be different from that in the same solution at another temperature. Clearly, for a given solution, a change in temperature involves a change in the individual terms on the right side of equation (6), BUT their total charge, Δ, must remain constant. Since each term is dependent on the concentration of the particular component (e.g. total carbonate carbon) which is fixed for a given solution, the relative changes in the individual terms are reflected in a change in pH (i.e. log a_{H^+}) even though the actual concentration of hydrogen ion may contribute little to the charge balance.

This is the basis of one of the procedures (discussed later) for the recalculation of analytical data to high temperature pH and species concentrations. The calculations involved are relatively simple, but because of the iterations required and the number of individual species, they are tedious to do unless you program a hand calculator or larger computer to do them for you. We shall examine the steps required before proceeding to discuss some of the programs currently available for this purpose.

Calculation of Ion Concentrations

An analysis is provided below for a water sample taken at the separator pressure from Broadlands well BR22, the same well we considered in the previous chapter.

Date	Separation Pressure	$pH_{20°}$	Na	K	Ca	Mg	Cl	SO_4	B	SiO_2	NH_3	HCO_3
6/80	10 b.g	7.39	870	188	1.2	0.1	1432	2	42	705	6.1	216

* note that the analytical temperature is specified with the pH.

To use the ion balance equation (7) we need first to calculate the sum of the charged pH sensitive species like $H_3SiO_4^-$ - the silicate ion** - in this solution.

** Notice that we presume that there are no other significant silicate species at the pH and temperature we are considering -- but this assumption should always be checked against available thermodynamic data. pK_2 for silicic acid, for example, is 11.78 and 11.68 at 25° and 260°C, respectively.

Data for K for a range of important constituents are shown in Table 7.3 for the temperature range 0 to 300°C and these are expressed as pK values where pK = -log K.

Table 7.3 -- Values of Weak Acid Dissociation Constants for Various Temperatures at Saturated Vapour Pressures*. (adapted from Glover, 1982, Chem. Div. D.S.I.R. N.Z. Report CD 2323)

	Temperature (°C)													
	0	10	25	50	75	100	125	150	175	200	225	250	275	300
H_2O	14.94	14.54	13.99	13.27	12.71	12.26	11.91	11.64	11.44	11.30	11.22	11.20	11.22	11.30
H_2CO_3	6.57	6.47	6.36	6.29	6.32	6.42	6.57	6.77	6.99	7.23	7.49	7.75	8.02	8.29
H_2S	7.45	7.23	6.98	6.72	6.61	6.61	6.68	6.81	6.98	7.17	7.38	7.60	7.82	8.05
NH_4^+	10.08	9.74	9.24	8.54	7.94	7.41	6.94	6.51	6.13	5.78	5.45	5.15	4.87	4.61
H_4SiO_4	10.28	10.08	9.82	9.50	9.27	9.10	8.97	8.87	8.85	8.85	8.89	8.96	9.07	9.22
H_3BO_3	9.50	9.39	9.23	9.08	9.00	8.95	8.93	8.94	8.98	9.03	9.11	9.22	9.35	9.51
HF	2.96	3.05	3.18	3.40	3.64	3.85	4.09	4.34	4.59	4.89	5.30	5.72	6.20	6.80
HSO_4^-	1.70	1.81	1.99	2.30	2.64	2.99	3.35	3.73	4.11	4.51	4.90	5.31	5.72	6.13
HCl	-0.26	-0.24	-0.20	-0.14	-0.06	0.03	0.14	0.25	0.37	0.50	0.66	0.84	1.06	1.37
HCO_3^-	10.63	10.49	10.33	10.17	10.13	10.16	10.25	10.39	10.57	10.78	11.02	11.29	11.58	11.89

* Expressed as $-\log K_a = pK_a$

Table 7.3 References

H_2O Sweeton, F.H., Mesmer, R.E., and Baes, C.F., 1974, Acidity measurements at elevated temperatures, VII, Dissociation of water: Journal of Solution Chemistry, v. 3, p. 191-214.

H_2CO_3 Read, A.J., 1975, The first ionization constant of carbonic acid from 25 to 250°C and to 2000 bars: Journal of Solution Chemistry, v. 4, p. 53-70.

H_2S Ellis, A.J., and Giggenbach, W.F., 1971, Hydrogen sulfide ionization and sulfur hydrolysis in high temperature solutions: Geochimica et Cosmochimica Acta, v. 35, p. 247-260.

NH_4^+ Hitch, B.F., and Mesmer, R.E., 1976, The ionization of aqueous ammonia to 300°C in KCl media: Journal of Solution Chemistry, v. 5, p. 667-680.

H_4SiO_4 Seward, T.M., 1974, Determination of the first ionization constant of silicic acid from quartz solubility in borate buffer solutions; Geochimica et Cosmochimica Acta, v. 38, p.1651-1664.

Busey, R.H., and Mesmer, R.E., 1977, Ionization equilibria of silicic acid and polysilicate formation in aqueous sodium chloride solutions to 300°C: Inorganic Chemistry, v. 16, p. 2444- 2450.

H_3BO_3 Mesmer, R.E., Baes, C.F., and Sweeton, F.H., 1971, Acidity measurements at elevated temperatures, V, Boric acid equilibria: Inorganic Chemistry, v.11, p.537-543.

HF Truesdell, A. H., and Singers, W., 1971, N.Z.D.S.I.R., Report CD 2136.

HSO_4^- Lietzke, M.H., Stoughton, R.W., and Young, T.F., 1961, The bisulfate acid constant from 25 - 225°C as computed from solubility data: Journal of Physical Chemistry, v. 65, p. 2247-2249.

HCl Johnson, K.S., and Pytkowicz, R.M., 1978, Ion association of Cl^- with H^+, Na^+, K^+, Ca^{2+} and Mg^{2+} in aqueous solutions at 25°C: American Journal of Science, v. 278, p. 1428-47.

Wright, J. M., Lindsay, W. T., and Druga, T. R., 1961, The behavior of electrolytic solutions at elevated temperatures as derived from conductance measurements, U.S.A.E.C. Report WAPD-TM-204.

Pearson, D., Copeland, C.S., and Benson, S.W., 1963, The electrical conductance of aqueous hydrochloric acid in the range 300 to 380°C: Journal of American Chemistry Society, v. 85, p. 1047-1049.

HCO_3 Plummer, L.N., and Busenberg, E., 1982, The solubilities of calcite, aragonite and vaterite between 0 and 90°C, and an evaluation of the aqueous model for the system $CaCO_3$-CO_2-H_2O: Geochimica et Cosmochimica Acta, v. 46, p. 1011-1040.

Activities and concentrations are related via individual ion activity coefficients ($a_i = \gamma_i m_i$) and, as discussed in Chapter 1, these may be calculated for dilute solutions using the Debye-Hückel expression. Substituting in equation (8) we obtain

$$pH = pK + \log (m_{H_3SiO_4^-}/m_{H_4SiO_4}) + \log(\gamma_{H_3SiO_4^-}/\gamma_{H_4SiO_4}) \qquad (9)$$

Insertion of appropriate values of γ gives $\log(m_{H_3SiO_4^-}/m_{H_4SiO_4})$ at any given pH.

$$m_{SiO_2,total} = m_{H_4SiO_4} + m_{H_3SiO_4^-} + \ldots . \qquad (10)$$

Neglecting minor species,

$$m_{SiO_2,total}/m_{H_3SiO_4^-} = \{ m_{H_4SiO_4}/m_{H_3SiO_4^-} + 1 \} \qquad (11)$$

$$\text{and } m_{H_3SiO_4^-} = m_{SiO_2,total}/ \{(m_{H_4SiO_4}/m_{H_3SiO_4^-}) + 1\} \qquad (12)$$

Setting $\gamma_{H4SiO4} = 1$ for dilute solutions, equations (8) and (12) can be combined to give

$$m_{H_3SiO_4^-} = m_{SiO_2,total}/ \{(a_{H^+} \gamma_{H_3SiO_4^-}/K_{H_4SiO_4}) + 1\} \qquad (13)$$

With insertion of appropriate values of γ, K_{H4SiO4} and m_{SiO2}, we can solve equation (13) to obtain the concentration of silicate ion at any pH we choose.

Figure 7.3 is a log m vs pH diagram constructed in this way for the Broadlands sample. (Notice that diagrams like this may be rapidly constructed even without rigorous calculation; $m_{silicate} = m_{silicic\ acid}$ where pH $\simeq$ pK and where pK + 1.5 < pH < pK - 1.5 the concentration slope for silicate ion and silicic acid on the log-log plot = 1. Such geometrical procedures are discussed by Butler, 1964.)

Starting with a calculation flow chart write a program for the calculation of $m_{silicate}$ at any specified pH and temperature. (Assume that you calculate γ_i independently).

The program you have written may be used for any of the monoprotic acids (H_4SiO_4, H_3BO_3, HF, NH_4^+). Diprotic acids, like carbonic acid, which forms both bicarbonate and carbonate ions, are also important and require a more complex calculation.

$$H_2CO_3 = H^+ + HCO_3^- \qquad K_{1,H_2CO_3} \qquad (14)$$

$$HCO_3^- = H^+ + CO_3^= \qquad K_{2,H_2CO_3} \qquad (15)$$

A useful exercise is to write an expression for m_{HCO3^-} as a function of total carbonate concentration and to write a program for its solution which may then be generally applicable to other mono- and diprotic acids (see Appendix I).

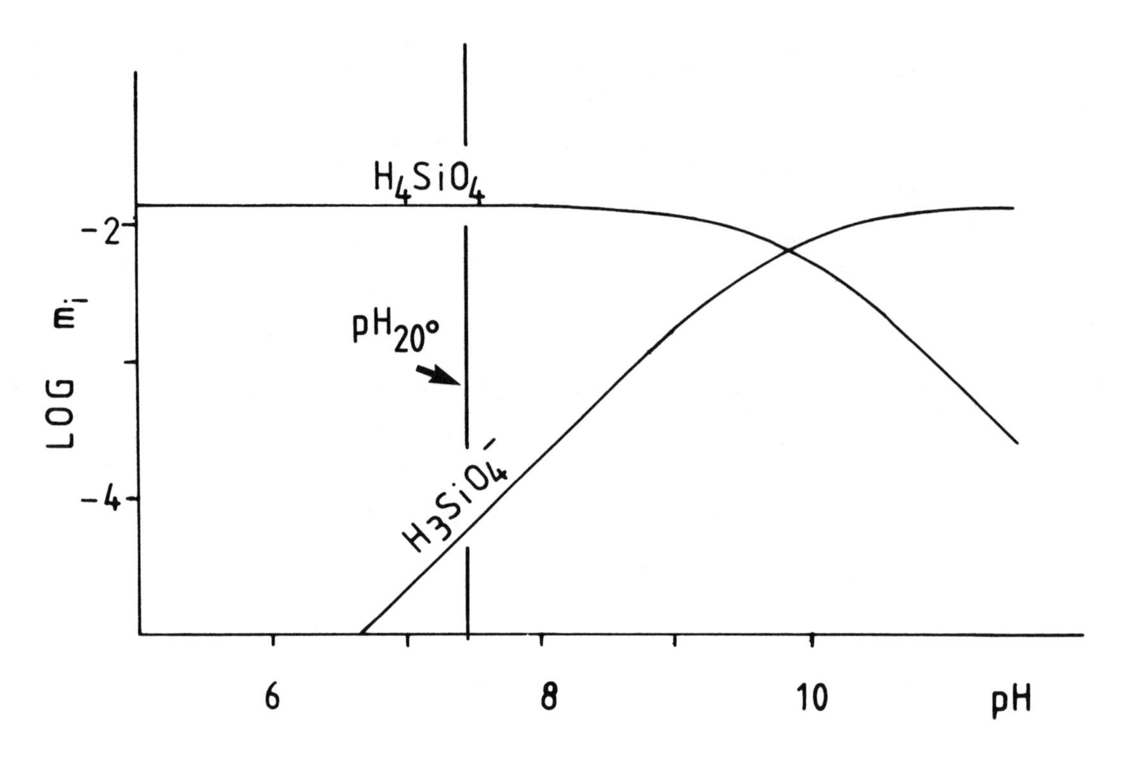

Figure 7.3. Molality-pH diagram for the species in the SiO_2-H_2O system at 25°C and up to pH = 12. The total dissolved silica is that of the BR22 sample used in the example calculation in the text.

> Using the program you have written -- or the program PKN given in Appendix I -- plot the variations of m_{HCO3^-} , $m_{CO3^=}$ and m_{H2CO3} as functions of pH in Figure 7.3 Use the concentration of total bicarbonate given in the BR22 water analysis.

Calculation of pH at elevated temperatures

In discussing the ion balance equation (7) we grouped the principal pH - dependent ions (HCO_3^-, HS^-, $H_3SiO_4^-$, etc.) and suggested that with certain assumptions their sum could be considered a constant with respect to temperature change. We define this constant as Δ (following Glover, 1982). On this basis the unknown pH of a high temperature solution may be calculated from its low temperature analysis, using only data for total bicarbonate, borate, etc. concentrations. (A reasonable estimate of the ionic strength of the solution is also required in order to calculate activity coefficients).

At any temperature

$$\Delta = m_{HCO_3^-} + m_{H_3SiO_4^-} + m_{H_2BO_3^-} + m_{OH^-} \ldots - m_{NH_4^+} - m_{H^+} \qquad (16)$$

and each term is pH dependent.

We can readily calculate species concentrations and hence Δ using the above equations, if we have a pH measurement at laboratory temperature (e.g. 20°C), the concentrations of the weak acids and bases and an estimate of ionic strength (to evaluate γ_i). In the Broadlands sample at laboratory temperature Δ = 3.43 millimoles/kg and this charge sum is due largely to HCO_3^-. At another temperature of interest we can then calculate the species concentrations at any pH_t we choose, by inserting appropriate dissociation constants from Table 7.3 but only one of these pH's will obey equation (16). This is the basis of the iterative calculation of high temperature pH_t where we use a computer program to successively guess pH_t, calculate species concentrations and test equation (16) or some equivalent of it.

> Derive a new log m_i vs pH diagram showing the distribution of species in the BR22 high pressure water sample at the separation temperature (184°C). Your answer may be compared with the example given in Appendix I. The diagram you need may be calculated using program PKN and appropriate dissociation constants from Table 7.2. Alternatively the diagram may be drawn using overlays of the 20°C speciation curves making due allowance for relative changes in pK_1 and pK_2. The individual ion activity coefficients are given in Chapter 6 and may be used to save time.

For most waters separated at the wellhead ($t \simeq$ 100-180°C) the difference between $pH_{20°C}$ and pH_t is small ($\leq$0.2 in the Broadlands example) provided gases have not been separated. However, we have previously seen that removal of CO_2 and H_2S in separated steam significantly increases water sample pH with respect to the reservoir fluid pH.

In order to tackle this problem $\Delta_{20°C}$ is first calculated for the separated water, but must itself be corrected along with the individual components for the effects of steam loss to obtain Δ_t.

$$\text{i.e.} \quad \Delta_t = (1 - y)\, \Delta_{20°C} \qquad (17)$$

Then CO_2, H_2S (and where analyzed, NH_3) from the separated steam analysis are combined with the water analysis to obtain the total discharge composition of the well. Once a new ionic strength for the reservoir has been estimated ($I_t \simeq (1-y)I_{20°C}$), activity coefficients have been calculated and dissociation constants have been obtained for the reservoir temperature, we have all we need to calculate the reservoir pH.

We achieve this by solving equation (16) to find some unique value of a_{H^+} for the given total discharge composition and Δ_t. This can either be done by writing an iteritive program or satisfactory results can be achieved using the geometric method described in the box above. Clare(1979) outlines a Newton-Raphson iteration scheme for use in such a calculator program; a "curve-crawling" technique is used in the illustrative program PH listed in Appendix I.

A steam sample taken at the same time and pressure as the BR22 water sample used above, contained 533 millimoles CO_2/kg, 10.3 millimoles H_2S/kg and 1 millimole NH_3/kg. The total discharge enthalpy of the well at the time of sampling was 1159J/gm.
Complete the following table to obtain the total discharge composition of the well including a value for Δ_t.

Total Discharge Composition (millimoles/kg)

	Cl	B	SiO_2	NH_3	HCO_3	H_2S	t
BR22			9.533				

Speciation curves for this total discharge composition at 265°C, i.e. $t_{enthalpy}$, are shown in Figure 7.4b. Only curves for carbonate-carbon and sulfide are shown because by inspection of pK data and the concentrations of the other weak acids, you can readily see that (in the pH range for aquifer fluids which we have deduced at the beginning of this chapter) the contributions of their dissociation products are negligible in this first-order calculation.

Use the PKN program to solve equation (16) for Δ_t using only the species NH_4^+, HCO_3^-, and HS^- at 265°C, and with pH_t set at 5.6, 5.8, 6.0, 6.2, and 6.4. Plot the Δ_t values against pH to find the pH_t of this well discharge.

In this example $\Delta_{20°C}(1-y)$ = 2.78 millimoles/kg. This Δ_t value is equal to m_{HCO3} at a pH_t of about 6.03. Compare this result with that obtained using the program PH (Appendix I). The value obtained there is 6.07* .

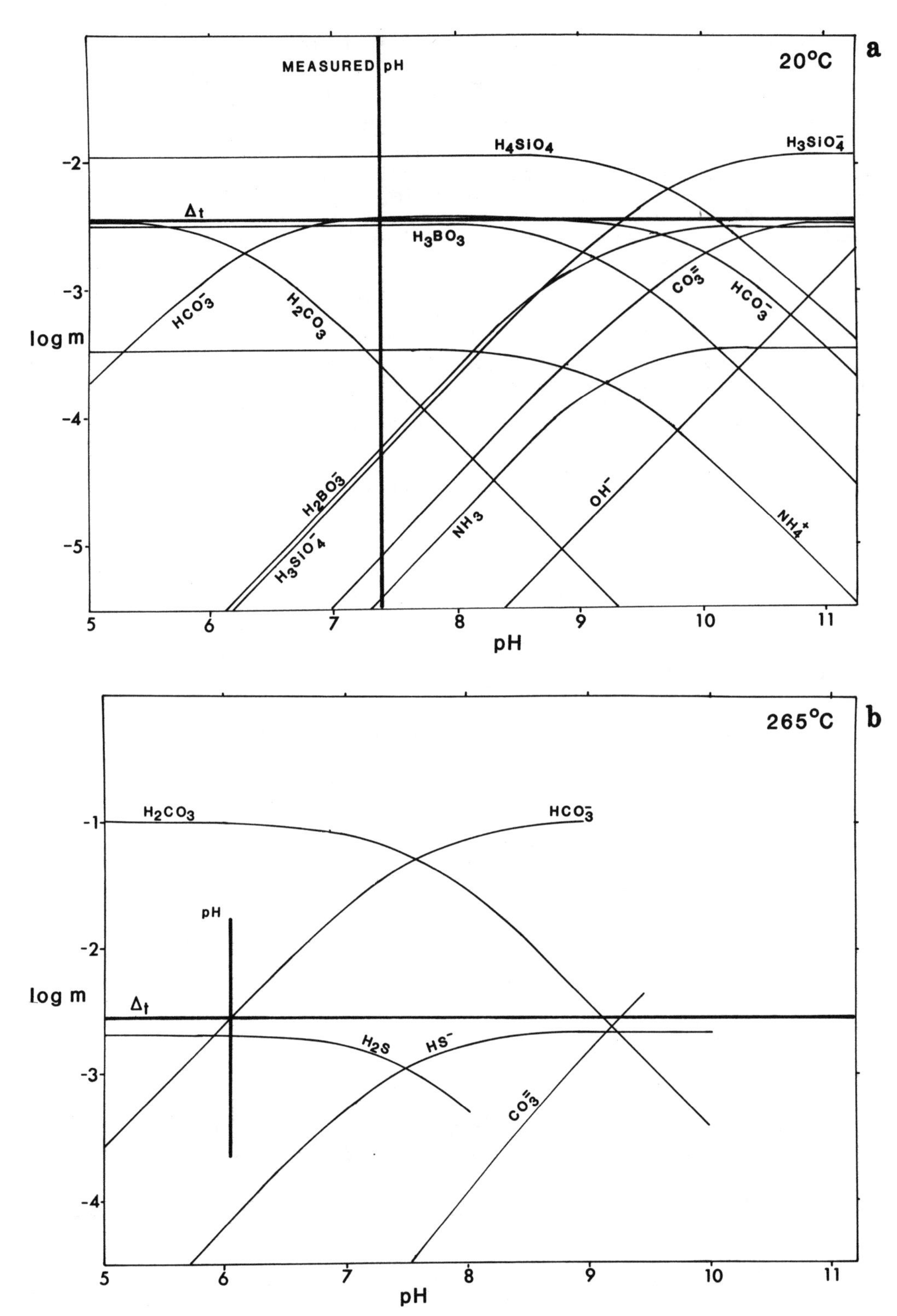

Figure 7.4. Log molality vs pH diagrams for the principal weak acid-base species in water from Broadlands BR22; a) water sample at 20°C, b) reservoir fluid at 265°C.

---

* For comparison, the computer program ENTHALP, which incorporates a much wider range of solution species, gives pH_t = 6.1 for the same BR22 data. This program is outlined later in the chapter.

In the example illustrated by Figure 7.4b, the pH_t calculation is carried out at $T_{enthalpy}$ = 265°C. We could equally well have chosen t_{NaKCa} or a measured downhole temperature depending upon what we know of the well and its discharge characteristics.

Problem

Use the geometric approach, or the program pH (Appendix I), to calculate pH_t for the Hatchobaru fluid whose analysis was given in the previous chapter.

Effect of ion association at high temperatures -- Complexes like $CaHCO_3^+$, introduce a problem because they decrease the effective concentration of dissolved carbonate carbon so that the pH_t estimate is slightly too high. Reliable high temperature data for their formation are not available but some of the larger programs use low temperature data extrapolated up to 350°C. The effect is not significant for the examples considered here where calcium concentrations are low.

Ion pairing reactions like $Na^+ + Cl^- = NaCl$ may become important at high temperatures. Reliable data are not available for these reactions but fortunately they do not affect the calculation except through their effect on ionic strength and activity coefficients. These reactions would clearly be important in calculating the pH of a brine like that in the Salton Sea geothermal system. Sodium sulfate ion pairing may also become significant but again reliable data are not available; in the cases we consider here sulfate ion concentrations are generally negligible.

For many geothermal purposes estimates of pH_t to ± 0.1 units only are required and these are not jeopardized by the assumptions in this simple method.

Earlier, in Chapter 6, we used raw experimental data for the estimation of a_{K^+}/a_{H^+} without correcting for ion association. Charge and mass balance equations must be used for the more rigorous determination of this ratio from laboratory measurements of $m\Sigma K$ and pH.

This procedure was applied by Montoya and Hemley (1975) to experimental data ranging from 300°C to 500°C. The problem in this calculation is that association constants for KCl and HCl are poorly known. For example, Montoya and Hemley used at 300°C log K_{HCl} = -0.8 compared to the value of -1.37 given in Table 7.3. This discrepancy may lead to an over correction for ion association at this temperature. For example, the value they derived for the Kspar - Kmica reaction is 4.4 compared to the raw experimental data shown in Figure 6.2, where the equilibrium ratio is 4.7. It is for this reason we have used the experimental data directly and semi-dependent thermodynamic data in discussing activity diagrams and geothermal fluid pH at temperatures below 300°C. Notice incidentally the chain reaction initiated here; the silicate reaction data demand more accurate assessment of other constants for their interpretation and these in turn demand more refined measurements in other systems.

Alternative Calculation Procedures

Merino (1975, 1979) adopted a similar approach using 'titration alkalinity', $\underline{A}$, both to check for the internal consistency of the analytical data and as the iteration base where,

$$\underline{A} = m_{HCO_3^-} + 2m_{CO_3^=} + m_{H_3SiO_4^-} + m_{H_2BO_3^-} + m_{NH_3} + m_{OH^-} - m_{H^+} \quad (18)$$

An alternative basis is the proton balance - or sum of total ionisable hydrogen ion (HTOT) - as used by Truesdell and Singers (1971,1974), in the computer program ENTHALP.

$$HTOT = 2m_{H_2CO_3} + 2m_{H_4SiO_4} + m_{HCO_3^-} + m_{H_3BO_3} + m_{NH_4^+} + m_{H_2S} + m_{H_3SiO_4^-} + m_{CaHCO_3^-} \ldots\ldots + m_{H^+} \quad (19)$$

(In the temperature and pH range of geothermal interest H_2O may be considered essentially unionizable and therefore is not considered in this sum). This approach has a number of advantages: it limits the number of species which play significant roles in the balance equation and by the same token limits the overall number of species which must be assumed for the solution model.

Appendix I lists an illustrative HP41C program, PH, for the calculation of high temperature pH_t using the HTOT iteration base described above. In this code an ionic strength loop has not been included but satisfactory results are achieved compatible with larger computer programs. It has the advantage of flexibility and ready access compared to many such facilities. A useful comparison of low and high temperature solution programs is provided by Nordstrom et al. (1979).

The procedures we have described for the calculation of pH_t are based on a simple solution model which involves experimentally based dissociation constants, the Debye-Hückel expression for activity coefficients and a charge or proton balance based on some assumptions concerning the principal species present in solution. The results of such calculations are therefore no more than "model pH's" to which we hope we can attach some real significance, but you should always remember the limitations of the model used. With these in mind, as well as the analytical errors on component concentrations it is seldom realistic to calculate a model pH to any more than three significant figures and frequently only two are required for application to mineral alteration problems.

Problems with aquifer boiling and excess steam

In the examples discussed in this and the previous section we assumed that we need to calculate aquifer pH at the temperature given by the quartz geothermometer. As shown in Chapter 11, it is often the case that the quartz temperature only reflects conditions close to the well, and a number of processes may have occurred such that the total discharge composition of the well does not represent the aquifer fluid; the most common problems are due to boiling and gain of enthalpy with or without additional reservoir steam.

Where these processes have occurred, it is unlikely that a pH_t value calculated using methods described above will have any significance and great care must be taken to avoid such problems. In some rare cases boiling in the vicinity of a well may lead to local temperature reduction reflected in the silica temperature due to rapid quartz precipitation, but all the steam and water produced feed the well, and heat may be extracted from wall rocks as discussed in Chapter 11. In some cases, the discharge enthalpy may be close to that of fluid at the alkali geothermometer temperature so that our pH calculation procedures are quite valid provided that either the higher t_{NaK} or t_{NaKCa} is chosen, rather than t_{quartz}. Problems like these are common in interpreting data from exploited geothermal fields but development of programs to routinely handle these contingencies is quite a problem (Singers et al., in preparation).

Coupling Mineral Stability and Fluid Composition

The examples considered in this chapter have focussed on geothermal systems, but the calculation procedures find application elsewhere. In mineral deposit studies much effort has been devoted to computing changes in fluid composition as a result of mineral deposition and dissolution reactions in a closed system. To do this, mineral stability data must be coupled with the fluid speciation routines, an inventory of component mass balance must be kept, and some assumptions made about the kinetics of the mineral-fluid reactions. The procedures are complex and the reader is referred to Helgeson(1970), Wolery(1979) and Read(1982) for discussion and some applications.

Review Problems

Using data from Chapter 6 calculate the pH of fluids with: a) 0.01 m total Cl; and b) 1.0 m total Cl in equilibrium with an assemblage of quartz, K-feldspar, Kmica and albite at

250°C. In the listings below, circle the principal dissolved species you would expect to find in each of these two solutions at 250°C.

	$m_{Cl} = 0.01$	$m_{Cl} = 1.0$
pH ≃		
	H_2CO_3/HCO_3^-	H_2CO_3/HCO_3^-
	H_2S/HS^-	H_2S/HS^-
	$HSO_4^-/SO_4^=$	$HSO_4^-/SO_4^=$
	NH_4^+/NH_3	NH_4^+/NH_3
	HF/F^-	HF/F^-
	$H_3BO_3/H_2BO_3^-$	$H_3BO_3/H_2BO_3^-$
	$H_4SiO_4/H_3SiO_4^-$	$H_4SiO_4/H_3SiO_4^-$

How does the species distribution change as you a) cool the solution conductively to 25°C; or b) allow the solution to boil adiabatically to 100°C?

REFERENCES

(sources of pK data are given below Table 7.3)

Barton, P.B., Bethke, P.M., and Roedder, E., 1977, Environment of Ore Deposition in the Creede Mining District: Economic Geology, v. 72, p. 1-24.

Butler, J.N., 1964, Ionic Equilibrium: A Mathematical Approach: Addison-Wesley, Reading, Mass., 547 p.

Clare, B., 1979, Evaluation of Cation Hydrolysis Schemes with a pocket calculator: Journal of Chemistry Editors, v. 56, p. 784-787.

Giggenbach, W.F., 1981, Geothermal Mineral Equilibria: Geochimica et Cosmochimica Acta, v. 45, p. 393-410.

Glover, R.B., 1982, Calculation of the Chemistry of some Geothermal Environments: New Zealand D.S.I.R. Chemistry Division Report CD 2323.

Goguel, R., 1977, Thermal water transport of some major rock constituents at Wairakei: New Zealand Journal of Science, v. 19, p. 359-368.

Helgeson, H.C., 1970, A chemical and thermodynamic model for ore deposition in hydrothermal system: Mineralogical Society of America, Special Paper No. 3, p. 155-186.

Merino, E., 1975, Diagenesis in Tertiary Sandstones . . . II Interstitial solutions at 100°C: Geochimica et Cosmochimica Acta, v. 39, p. 1629-1645.

Merino, E., 1979, Internal consistency of a water analysis and uncertainty of the calculated distribution of aqueous species at 25°C: Geochimica et Cosmochimica Acta, v. 43, p. 1533-1542.

Montoya, J.W., and Hemley, J.J., 1975, Activity relations and stabilities in alkali feldspar and mica alteration reaction: Economic Geology, v. 70, p. 577-582.

Nordstrom, D.K., Plummer, L.N., Wigley, T.M.L., Wolery, T.J., Ball, J.W., Jenne, E.A., Bassett, R.L., Crerar, D.A., Florence, T.M., Fritz, B., Hoffman, M., Holdren, G.R., Jr., Lafon, G.M., Mattigod, S.V., McDuff, R.E., Morel, F., Reddy, M.M., Sposito, G., and Thrailkill, J., 1979, Comparison of computerized chemical models for equilibrium calculations in aqueous systems in Jenne, E.A. Chemical Modelling in Aqueous Systems. American Chemistry Society Symposium Series No. 93, p. 857-892.

Reed, M.H., 1982, Calculation of multicomponent chemical equilibria and reaction processes in systems involving minerals, gases and an aqueous phase: Geochimica et Cosmochimica Acta, v. 46, p. 513 - 528.

Singers, W.A., Henley, R.W., and Giggenbach, W.F., GEODATA-ENTHALP: REVISIONS AND USERS GUIDE NZ DSIR Chemistry Division Report, in prep.

Truesdell, A.H., and Singers, W.A., 1971, Computer calculation of downhole chemistry in geothermal areas: NZ DSIR Chemistry Division Report CD2136, 145 p.

Truesdell, A.H., and Singers, W.A., 1974, Calculation of aquifer chemistry in hot-water geothermal systems: Journal Research U.S. Geological Survey, v. 2 (3), p. 271-278.

Wolery, T.J., 1979, Calculation of chemical equilibrium between aqueous solution and minerals: the EQ3/6 software package: UCRL-52658, Lawrence Livermore Laboratory.

Chapter 8
REDOX REACTIONS IN HYDROTHERMAL FLUIDS

Many elements participate in oxidation-reduction reactions in the geothermal/epithermal environment. These include C, S, H, O, N, Fe, Mn, U, W, As, Sb, Bi, Cu, Ag, Au, Te, and Sn. The first six or seven elements listed are much more abundant than the rest and they interact to buffer the redox state; the remaining (and to a large extent the most interesting economically) elements are usually much less abundant, and they only respond to the chemical environment imposed by the dominant redox systems. In this chapter we shall investigate methods of determining the oxidation state of a system, either directly by calculations based on the chemistry of geothermal gases and liquids, or indirectly by interpreting the phases and phases assemblages observed in fossil hydrothermal systems.

Redox reactions are important in such diverse areas as the corrosion and scaling of geothermal production pipes, the interaction of organic matter with fluids, the oxidation of H_2S, the precipitation of native metals and pyrite and other sulfides, the destruction of sulfides by oxidation, and the disproportionation of SO_2 into H_2S and $SO_4^=$ on cooling from high temperature.

OXIDATION STATE OF GEOTHERMAL SYSTEMS

Exploration drilling has identified two end-member redox environments in active geothermal systems: a) the relatively reducing environment of the deep, chloride-water systems which are associated with pyrite-rich propylitic alteration and with H_2S as the dominant aqueous sulfur species, and b) the relatively oxidizing environment in the upper part of systems where H_2S oxidation results in acidic, sulfate-dominated waters associated with advanced argillic alteration. A similar, relatively oxidizing environment occurs in the upper part of a volcano where volcanic H_2S and SO_2 are oxidized to H_2SO_4. Chapter 12 deals in some detail with additional aspects of the volcanic gas chemistry. In this chapter we shall begin with the determination of the oxidation state of the same deep system described in previous chapters, well BR22 at Broadlands. The gas analysis is given in Chapter 2. We can calculate that the m_{H2} in the part of the reservoir feeding BR22 is 0.000107 moles per kg (X_{H2} = 0.00000194); the temperature is 260°C and pH 6.1. From Henry's Law we obtain P_{H2} (= f_{H2}) as follows:

$$P_{H_2} = K_H \times 0.00000194 = 21599 \times 0.00000194 = 0.04 \text{ bars} \quad (1)$$

This pressure is small compared to CO_2, but let us consider how it compares to that of O_2 by applying the reaction

$$2\ H_2 + O_2 = 2\ H_2O \quad (2)$$

At 260°C log K is 39.42; log P_{O_2} = -36.6.

$ The Henry's Law constant for O_2 is 21000; calculate the molality of O_2. That's not very many molecules per kg is it?

The O_2 calculation illustrates a recurrent situation in geochemical computations; chemical species having next-to-nil concentrations can be very useful in describing the conditions of equilibrium, because such components are precise measures of chemical potentials. Nevertheless, it may be very poor judgement to assign to such fictive components any role which requires their actual presence, as for example, in dealing with the mechanistic kinetics of a reaction. Also, it points out that dissolved O_2 in a reduced hydrothermal environment is certainly negligible as a buffer. This discussion applies particularly to S_2, sulfide ion ($S^=$), and the electron (e^-) as well as to H^+ and O_2. Even dissolved H_2, which is many, many times more abundant than O_2 in the example above, is a poor bet to be an effective redox buffer. Giggenbach (1980, 1981) has made this point for the New Zealand geothermal fields and shows how several silicate reactions buffer gas chemistry. An example from the ore deposit at Creede, Colorado is given later in this chapter.

Log f_{O2} has a long and honored history as a coordinate for useful diagrams, emanating from metallurgical studies and being adopted widely by geoscientists. We shall continue to use f_{O2} in this text; but inasmuch as the chemical environment of geothermal systems and hydrothermal ore deposits is saturated with water, and because P_{H2} is a readily measurable property, we might well consider using H_2 rather than O_2 as the principal redox parameter.

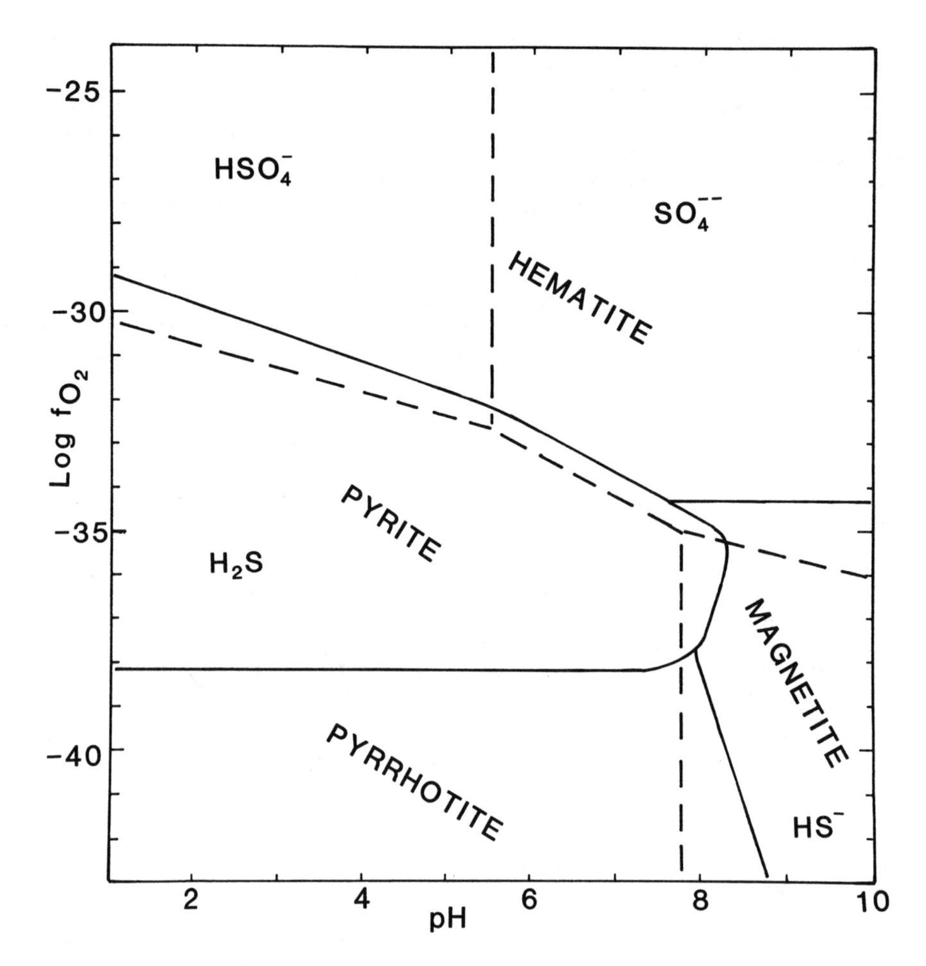

Figure 8.1. Log a_{O2} - pH diagram for well BR22 at Broadlands, N. Z. The temperature is 260°C; composition of the gas is given in Chapter 2, and additional information is in the text of this chapter

Figure 8.1 presents part of the log f_{O2} - pH diagram for BR22 (the pyrrhotite field is calculated assuming that the activity of FeS in pyrrhotite is unity); plot the position of the BR22 sample.

$ Suppose that well BR22 has produced some iron-mineral cuttings from the bottom of the hole; would these be anticipated to be pyrite, magnetite, hematite pyrrhotite, or native iron?

Such drill cuttings are not available from this well, but elsewhere in the Broadlands field pyrite and pyrrhotite coexist (Browne and Ellis, 1970). Pyrrhotite in equilibrium with pyrite at 260°C has an a_{FeS} of about 0.55. Recalculate pyrrhotite field on Figure 8.1. Does this change have any influence on the mineralogy predicted?

No, it doesn't, and for the explanation we need to look at the geothermometry of the well discharge. Using analyses for BR22 from Tables 2.3 and 2.5 and the methods developed in Chapters 3 and 5, calculate temperatures using the following geothermometers:

t_{quartz} = 265°C

t_{NaKCa} = 306°C

$ t_{CH4} =

t_{NH3} =

In Chapter 11 we shall discuss such patterns at greater length, but for now let us note that boiling (either adjacent to the hole or throughout the reservoir) consequent to exploitation results in lowered fluid temperatures. The quartz and the gas geothermometers respond relatively rapidly, but t_{NaKCa} responds only slowly; thus it remembers the initial temperature of the reservoir. The boiling depletes the liquid in CO_2, H_2S, H_2, and other gases. Therefore the gases still remaining in the partially degassed liquid no longer represent the original oxidation state of the reservoir; instead they are more oxidized, as the point you plotted in Figure 8.1 should indicate. Elsewhere in the Broadlands field the H_2 from less disturbed waters fits closely that predicted from the buffering assemblages pyrite + pyrrhotite or pyrite + iron-silicate (Seward, 1974; Giggenbach, 1980).

f_{O2} - pH, AND OTHER ACTIVITY - ACTIVITY DIAGRAMS

Because hydrolysis and redox reactions each have such strong influences on mineral - fluid equilibria, and because each may be represented by a potential that is independent of major element chemistry, the f_{O2}-pH diagrams have become very useful base "maps" for chemical interpretation. [Because the standard state for gases is at 1.0 bar, the activity is numerically equal to the fugacity. Moreover, at these low pressures and moderate temperatures the partial pressure is a reasonably good approximation of the fugacity.] They are examples of activity - activity diagrams which we shall find quite useful in displaying noninterfering, complex relationships. Examples of such diagrams are given in figure 8.2 (from Barton and Skinner, 1979). We shall return to Figure 8.2a and c, but first let us dispose of Figure 8.2b. Topologically, the Eh-pH and f_{O2}-pH diagrams are identical, but we shall abandon the Eh diagrams for the present in favor of the log a_{O2} diagrams which are more straightforward to calculate and have less tendency to crowd interesting features into narrow and sloping parts of the diagram.

Figure 8.2 gives diagrams for the system O-H-S-Fe-Zn-Cu-Pb. The solid lines separate fields of the iron minerals (pyrite, pyrrhotite, magnetite, and hematite). The heavier, more closely spaced dashed lines separate fields of dominance of the various aqueous sulfur species ($SO_4^=$, H_2S, etc.). The lighter dashed lines are contours of sphalerite composition. The dotted lines mark the limits of the chalcopyrite field where it converts to bornite plus an iron mineral. A light, broken line marks the anglesite-galena boundary, and a similar line marks the place where P_{H2} = 1 atm. providing a general frame of reference. Such a system is manageable in these coordinates because solid solutions are minimal (except for sphalerite which has its composition contoured). Some conditions must be specified in order to permit the calculation. Here we have chosen: equilibrium among all species considered, a temperature of 250°C, a pressure of about 40 bars (which puts us on the boiling curve), consideration only of those Cu-Fe-S phases (notably chalcopyrite and bornite) compatible with the simple iron minerals noted above, and a total sulfur concentration of 0.01 moles/kg. One more decision must be made regarding the nature of the termination of the mineral fields against aqueous species (e.g. whether and if so, where, the pyrrhotite field terminates against aqueous Fe species); this would require some decision about the metal concentrations <u>or</u>, as done here, a decision to let all minerals extend throughout the diagram without consideration of solution species for metals.

$ Write a reaction equation for the oxidation of aqueous H_2S to $SO_4^=$.

$ Write the equilibrium constant for the reaction between magnetite and pyrite in the presence of an appropriate aqueous sulfur species.

In fact, there are two pyrite + magnetite reactions, one involving HS^-, the other involving $SO_4^=$; write the other one. Note that the O_2 has changed sides of the reaction relative to pyrite and magnetite.

Note that we may write a great many reactions; all of them are valid if written properly, and all may be evaluated provided that the basic data are available. However, many

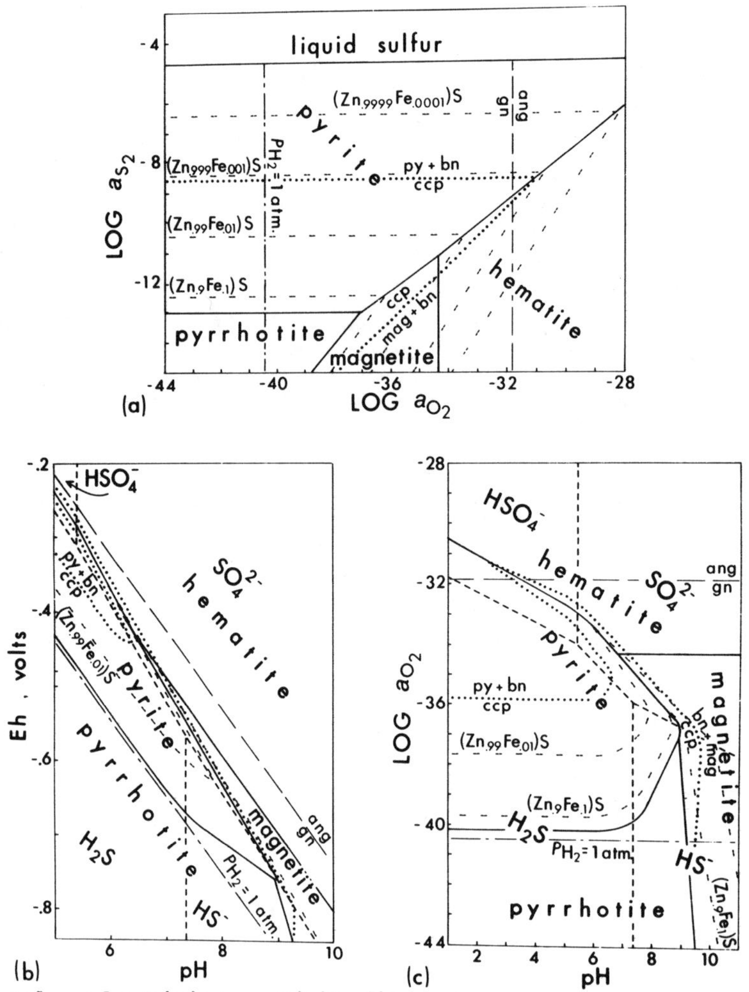

Figure 8.2. Examples of activity - activity diagrams.

All three diagrams have consistent line types representing each type of reaction; the text supplies additional explanation. All are calculated for 250°C, a pressure of 40 bars (which is on the boiling curve), an ionic strength of 1.0, and a total sulfur concentration of 0.02 mol/kg. The field of $FeSO_4$ has been neglected. Abbreviations: py = pyrite; gn = galena; ang = anglesite; ccp = chalcopyrite; bn =bornite; and mag = magnetite. From Barton and Skinner (1979).

Figure 8.2a is the log a_{O_2} - log a_{S_2} diagram; all of the curves shown thereon are independent of pH. Figures 8.3 and 14.4 are similar diagrams on which pH-dependent curves have been plotted.

Figure 8.2b is the Eh - pH diagram.

Figure 8.2c is the log a_{O_2} - pH diagram.

of the reactions may not be representative of equilibrium conditions in the real world in that they may deal with combinations of components that are not compatible when each has unit activity, or they may constitute metastable equilibria wherein some other field has preempted the reaction written. If in answering the questions immediately above you had chosen to write a reaction involving H_2S instead of HS^- and $SO_4^=$, you would have experienced one of these metastable results. When undertaking the calculation of a novel chemical system, some degree of trial and error is to be expected until the stable and metastable curves and parts of curves are sorted out.

The redox reactions involving sulfate in solution constitute a particularly sticky problem in interpretation. A useful, although perhaps overly simplistic explanation is that the sulfur atom in sulfate is shielded from its environment by four tetrahedrally-arranged oxygen atoms; thus, unless some critter has some smart enzymes to open up the tetrahedra, the sulfate is rather inert to the external redox environment. Of course, at high temperatures (>300°C) sulfate redox reactions do proceed at geologically rapid rates. The isotopic fractionation between sulfide and sulfate is large and a strong function of temperature, but samples of apparently coexisting sulfate and sulfide are notoriously erratic as geothermometers in either ore deposits or geothermal systems, presumably because sulfate and sulfide seldom are in equilibrium. Obviously, it is unrealistic to expect that the sulfate-sulfide reaction will be an effective buffer for redox state in Mississippi Valley-type or epithermal ore deposits; and it should be suspect in most submarine massive sulfide environments as well. Ohmoto and Lasaga (1982) have provided a quantitative treatment of sulfate redox kinetics, showing that pH and sulfur concentration are important in addition to temperature; Sakai (1983) has provided an additional analysis of rate constants. Sulfate redox disequilibrium may eventually provide a measure of the duration of a thermal process, but the achievement of that goal lies in the future.

The triple junction between fields of magnetite, pyrite, and hematite in Figures 8.2b and c is angular, whereas the pyrite - pyrrhotite boundary is curved as it crosses the H_2S - HS^- boundary. The reason for the former situation is that the activities of all of the components in the balanced reaction $FeS_2 + Fe_3O_4 + 4\,O_2 + 2H_2O = 2Fe_2O_3 + 2SO_4^= + 4\,H^+$ are either: of constant activity, as for the pyrite, magnetite, hematite, and sulfate; or they are related to the coordinates, as for pH and O_2. In the situation with the pyrite + pyrrhotite boundary the activity of the sulfur species is not constant because either H_2S or HS^- diminishes in anticipation of entering the field of the other, thereby gradually shifting the equilibria toward higher O_2 values.

Table 8.1 -- Equilibrium constants (log K values) for some hydrolytic and redox reactions (data from Helgeson, et al., 1978; Robie et al., 1978; and Fisher and Barnes, 1973), except for reaction (2), which is as shown in earlier section. a and b are linear regression equations for these data: log K = a + b (1/T K) (range: 200-300°C)

	Reaction	25	50	100	150	200	250	300	350°C	a	b
1.	$SO_4^- = H^+ + SO_4^=$	-1.99	-2.31	-2.99	-3.72	-4.48	-5.27	-6.08	-6.90	-13.59	4324.08
2.	$H_2S_{aq} = HS^- + H^+$	-6.98	-7.72	-6.61	-6.81	-7.17	-7.60	-8.05	-	-12.18	2377.5
3.	$HS^- + 2O_2 = SO_4^= + H^+$	132.55	120.43	100.91	85.86	73.82	63.90	55.38	47.28	-31.75	49975.5
4.	$3Fe_2O_3 = 2Fe_3O_4 + 1/2O_2$	-36.15	-32.82	-27.51	-23.45	-20.24	-17.65	-15.51	-13.70	6.9	-12826.9
5.	$3Fe + 2O_2 = Fe_3O_4$	177.40	162.37	138.21	119.78	105.27	93.56	83.91	75.83	-17.13	57893.3
6.	$3FeS + 3H_2O + 1/2O_2 = Fe_3O_4 + 3HS^- + 3H^+$	-6.14	-6.41	-7.34	-8.64	-10.28	-12.08	-14.71	-19.17	-35.25	11905.6
7.	$FeS + H_2S_{(g)} + 1/2O_2 = FeS_2 + H_2O$	46.09	41.54	34.29	28.77	24.40	20.88	17.98	15.55	-12.4	17410.7
8.	$Fe^{+2} + 2H_2S_{(g)} + 1/2O_2 = FeS_2 + 2H^+ + H_2O$	30.24	27.45	23.07	19.80	17.32	15.48	14.37	14.88	.263	8038.5
9.	$Fe^{+2} + 3H_2O = Fe(OH)_3^- + 3H^+$	-29.45	-27.58	-24.64	-22.42	-20.70	-19.33	-18.21	-	-6.42	-6753.56
10.	$H_2O_{(1)} = H_2 + 1/2O_2$	-41.55	-37.68	-31.53	-26.85	-23.18	-20.23	-17.81	-15.78	7.6	-14564.13
11.	$H_2O_{(1)} = H^+ + OH^-$	-13.99	-13.26	-12.23	-11.59	-11.21	-11.08	-11.28	-12.35	-	-
12.	$1/2\ H_2 = H^+ + e^-$	0.0	0.0	0.0	0.0	0.0	0.0	0.0	0.0	-	-
13.	$2H_2S + O_{2(g)} = S_{2(g)} + 2H_2O$	59.44	54.4	46.27	39.99	35.05	30.06	27.21	-	-9.99	21287.7
14.	alunite + qtz + H_2O = Kmica + kaolinite + $2K^+ + 6H^+ + 8HSO_4^-$	-	-	-	-	-48.0	-41.7	-36.5	-	17.90	-31168
15.	$N_2 + 3H_2 = 2NH_3$	5.75	4.48	2.44	0.86	-0.42	-1.46	-2.34	-3.09	-11.41	5201.7
16.	$C + O_2 = CO_2$	69.09	63.79	55.26	48.74	43.60	39.45	36.02	33.13	0.168	20543.6
17.	$C + 2H_2 = CH_4$	8.88	7.86	6.20	4.92	3.89	3.04	2.32	1.71	-5.10	4253.4
18.	$NH_4^+ = NH_3 + H^+$	-9.27	-8.53	-7.42	-6.53	-5.79	-5.14	-4.57	-4.12	-1.19	3303.4

$ Write the expression for the reaction of pyrrhotite + O_2 + sulfide to yield pyrite. Using the data in Table 8.1 calculate the log f_{O2} at 200, 250, and 300°C for a fluid in equilibrium with pyrite and pyrrhotite and possessing an a_{H2S} of 0.2 molal. Is f_{H2} highest, or lowest, at 300°C?

In Figures 8.2b and c why is the triple junction between the three aqueous sulfur species, H_2S + HSO_4^- + $SO_4^=$, angular and not curved?

Note that whether the junctions are rounded or angular is NOT simply a function of whether the buffer is of the fixed point or sliding scale variety. (A discussion of buffers in general is included later in this chapter.)

Now let us return to the activity - activity diagram, Figure 8.2a. The features calculated in Figure 8.2a are all independent of pH, there being no hydrolysis reactants or products among the things plotted. The compositions of sphalerite in equilibrium with the various iron minerals are derived from the simple Henry's Law expression for the solution of FeS in sphalerite (Barton and Toulmin, 1966):

$$XFeS_{sp} = a_{FeS}/2.4 \quad (3)$$

plus a general iron-mineral formula, FeS_mO_n

Thus the a_{FeS} for any of the plotted iron minerals anywhere on the diagram can be represented by the equation:

$$FeS_mO_n = (m-1)/2\ S_2 + (n/2)O_2 + FeS \quad (4)$$

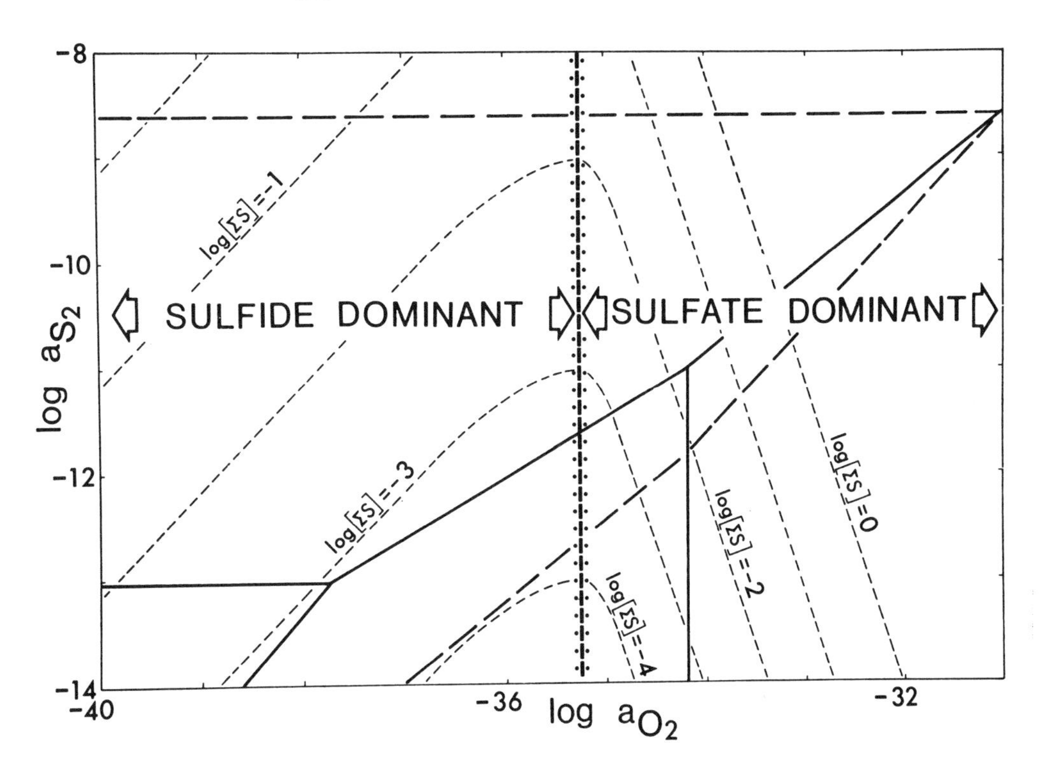

Figure 8.3. Log a_{O2} - log a_{S2} diagram showing the total concentration of sulfur. Most of the diagram is similar to Figure 8.2a, except that the magnetite field has been replaced by a slightly larger field of chlorite, as described in the section titled "Buffer Systems" in this chapter. From Barton et al. (1977).

If species are added which correspond to hydrolysis reactions, e.g., Fe^{++}, HS^-, $SO_4^=$, H^+, they, too, may be placed on the diagram in the way in which total sulfur (summation of H_2S, HS^-, and $NaSO_4^-$) has been made in Figure 8.3. Note that the concentration of sulfur, the pH, the a_{O2}, and a_{S2} cannot all be independent variables; any three fix the fourth. The behavior of sulfate in brines has not been discussed here, and it is a partially known topic of some import which we shall identify, but not resolve, here. Sulfate forms rather stable ion pairs with both Na^+ and K^+ ions so that the dominant oxidized sulfur species may well be $NaSO_4^-$ or KSO_4^- instead of $SO_4^=$ or HSO_4^-. The equilibrium constants are imperfectly known and have been ignored in most discussions here, but for the strong brines associated with many ores, sulfate-alkali ion-pairing will be significant.

NITROGEN AS AN OXIDANT?

At 20°C the solubility of N_2 in water in equilibrium with air is about 20 ppm (about twice that of oxygen from air). When N_2 is reduced (i.e., when it oxidizes something else) the N-bearing component is NH_3 or NH_4^+ which has a gain of 3 electrons per atom compared to a gain of 2 for the reduction of O_2. Thus in air-saturated water, the nitrogen has 3 times the oxidizing capacity of oxygen. Therefore it is appropriate to ask whether nitrogen can act as an oxidant for geologic systems, and if so, under what conditions. There are two parts to this question: thermodynamics and kinetics. We shall deal here only with the former, but we suggest that you look at the gas analyses from geothermal samples (Table 2.5) and decide for yourself about the latter.

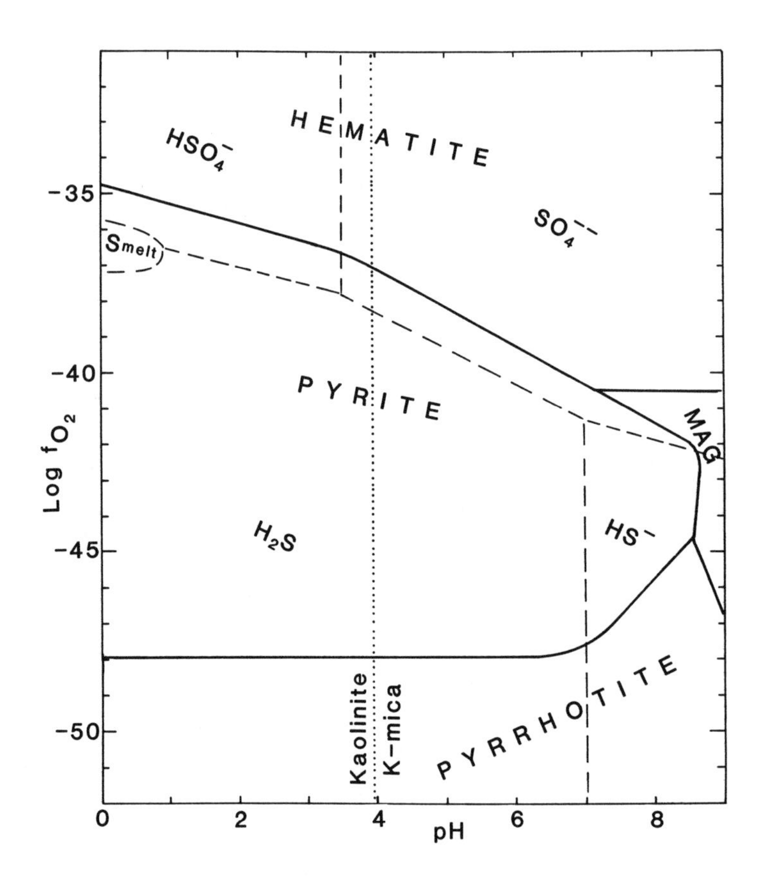

Figure 8.4. Log f_{O2} - pH diagram for 200°C.
This is a base on which to plot nitrogen equilibria, alunite reactions, and the position of the CO_2 - CH_4 buffer, as discussed in the section titled "Nitrogen as an Oxidant?". The Kmica - kaolinite curve is taken from Hemley et al. (1969) so as to make it as consistent as possible with the alunite equilibria (see discussion in Chapter 6). Log a_{K^+} is -1.25.

Table 8.1 lists several mineral and fluid equilibrium constants as functions of temperature. Assuming that NH_3 has the same activity coefficient as H_2S and that NH_4^+ has the same activity coefficient as K^+, plot on Figure 8.4 (pH - log f_{O2} diagram on which we have already plotted some basic information for 200°C) the fields of N_2, NH_3, and NH_4^+ that show if, and if so, under what conditions, nitrogen from air will oxidize pyrrhotite, pyrite, H_2S, HS^-, CH_4, graphite, or magnetite. Note that this calculation is slightly different from that for the sulfur species. The boundary between NH_3 and NH_4^+ is simply one of determining where each has equal concentration, and the location will be independent of total N in solution. In contrast, the N_2 field boundary is between species that have different numbers of atoms per aqueous species; thus the N_2 boundary is not independent of concentration. Use the air-saturated value of 20 ppm total nitrogen.

Hemley et al. (1969) have performed experiments permitting an estimate for the equilibrium constant for the stability of alunite plus quartz relative to Kmica or kaolinite. The reactions are:

$$3Al_2Si_2O_5(OH)_4 + 2K^+ + 4SO_4^= + 6H^+ = 2KAl_3(SO_4)_2(OH)_6 + 6SiO_2 + 3H_2O$$

and

$$KAl_3Si_3O_{10}(OH)_2 + 2SO_4^= + 4H^+ = KAl_3(SO_4)_2(OH)_6 + 6SiO_2$$

The respective log equilibrium constants are 37.72 and 21.54. Alunite may pre-empt part of the diagram; assuming that unit activities are assigned to all solid phases and H_2O, and that total sulfur is .01 mol/kg, calculate and plot the alunite field on Figure 8.4. Substitute the values for a_{K^+} and $a_{SO_4^=}$ or $a_{HSO_4^-}$ as appropriate into the equilibrium constant and solve for pH; then add reactions 2) and 3) from Table 8.1 to the reactions above to get the alunite + quartz reaction yielding kaolinite + H_2S + K^+. By neglecting the KSO_4^- ion pair (which constitutes about 40 percent of the total S^{+6} in the $SO_4^=$ field, and a variably lower percentage in the HSO_4^- field, under the conditions specified for this exercise) we slightly over-estimate the extent of the alunite field.

\$ Will N_2 dissolved in water oxidize reduced sulfur species to sulfate to help produce alunite at 200°C? If so, under what condition?

\$ Ammonium ion, NH_4^+, substitutes for K^+ in many minerals; under what conditions would one expect to find the most ammonium-rich alunite? What about NH_4^+-bearing Kmica or feldspar?

BUFFER SYSTEMS

Buffers are those reacting systems which modify or control the magnitude of change of any intensive variable (pressure, temperature, a_{O2}, a_{H2S}, a_{SiO2}, a_{K+}, pH, etc.). They are commonly used in the laboratory to assure that variables remain within some desired limits. We shall discuss buffers from the point of view of the phase rule:

$$\phi = c - \nu + 2 \qquad (5)$$

where ϕ is the number of stable phases, c is the number of independent components, ν is the number of degrees of freedom. [If exotic conditions of state are considered (i.e., a magnetic, gravitational, or electrostatic field) the "2" must be increased by 1 for each condition so added.] The factor of special interest here is ν, the number of degrees of freedom, for this is the number of independent intensive parameters that we must specify or measure in order to define uniquely the state of the system.

Let us consider the 1-component system H_2O as an example: let us have 2 phases (liquid and vapor), thus ν equals 1 and only one additional parameter may be specified arbitrarily. The liquid + vapor places us along the boiling curve; if pressure is specified there are no longer any variables that may be independently specified: the temperature is fixed, and we thereby have a thermometer.

Why do we care? All geothermometers, geobarometers, and methods used to estimate pH, a_{O2}, and so on, rely on constraining the system sufficiently to have a single "unknown", namely temperature, pressure, pH, a_{O2} and so on. In effect, we are asking the mineral-fluid assemblage (including the compositions of minerals and fluids) to indicate the conditions under which it is reacting. Mineral buffers are performing exactly similar roles, except that instead of indicating the answer to the question, they are specifying it. Now, if via the phase rule we establish that there is only 1 undefined degree of freedom, and if we have not yet specified temperature, the system is potentially a geothermometer.

In much of the discussion in this text we have neglected the effect of pressure, recognizing that within the range of pressures extant in geothermal/epithermal systems, pressure does not have a significant effect. (This is comparable to substituting 1 for the 2 in equation (5)).

We have two kinds of buffers, which we shall designate as "fixed point" and "sliding scale". In the former, we have "saturated" the phase rule with components having immutable activities (e.g., phases having fixed compositions) and/or by specifying the conditions of state (pressure and temperature). Thus each intensive variable has one, and only one, possible value. An example of a fixed point buffer would be the assemblage pyrite + pyrrhotite at a specified temperature and pressure; for this condition the a_{S2} (or for that matter, any other conceivable component in the Fe-S system) has a unique value. On the other hand, a sliding scale buffer exists when the number of phases is less than the number of components ($\phi < c$) and some continuously variable intensive parameter, such as the composition of a phase, is specified in order to reduce the degrees to 0. In this situation the position of the buffer is not unique, for it may slide as the composition (and therefore the activities of all components within the solution) of the phase slides. An example of a sliding scale buffer would be the pH of a solution in equilibrium with K-feldspar + Kmica + quartz at a predetermined temperature and pressure; the pH value can slide in response to changes in the a_{K^+} in the liquid.

There is yet another aspect to the buffer concept; in this instance the aspect is geological rather than chemical. This is the question of whether the buffer in question has sufficient capacity to control the reaction being considered, in which case it is a true buffer; or whether other reactions provide the real control and the would-be buffer merely goes along for the ride. We term the passive sort of buffer role an "indicator" since remnants of the reaction only record that it happened and did not control the course of the overall chemistry appreciably. An example would be the sulfidation of native silver to argentite; in this example there simply is not enough silver to function effectively as a buffer for the whole hydrothermal system, and the preservation of silver + argentite merely records passage from one side of the buffer to another. Similarly the HCO_3^- - CO_2 buffer in a geothermal fluid has a smaller buffer capacity than mineral pH buffers so it plays the role of an "indicator". Indicators are probably far more common than buffers in geologic situations, and their interpretation is an essential part of the geochemistry of geothermal/epithermal systems.

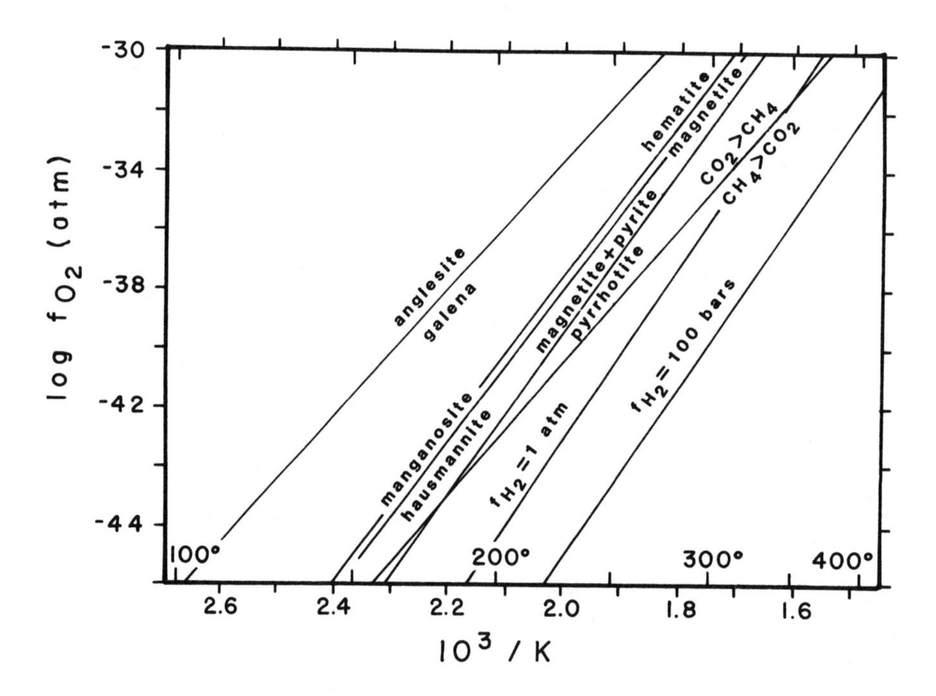

Figure 8.5. Log f_{O2} - 1/t diagram showing typical oxidation reactions among minerals. The figures inside the lower margin are °C.

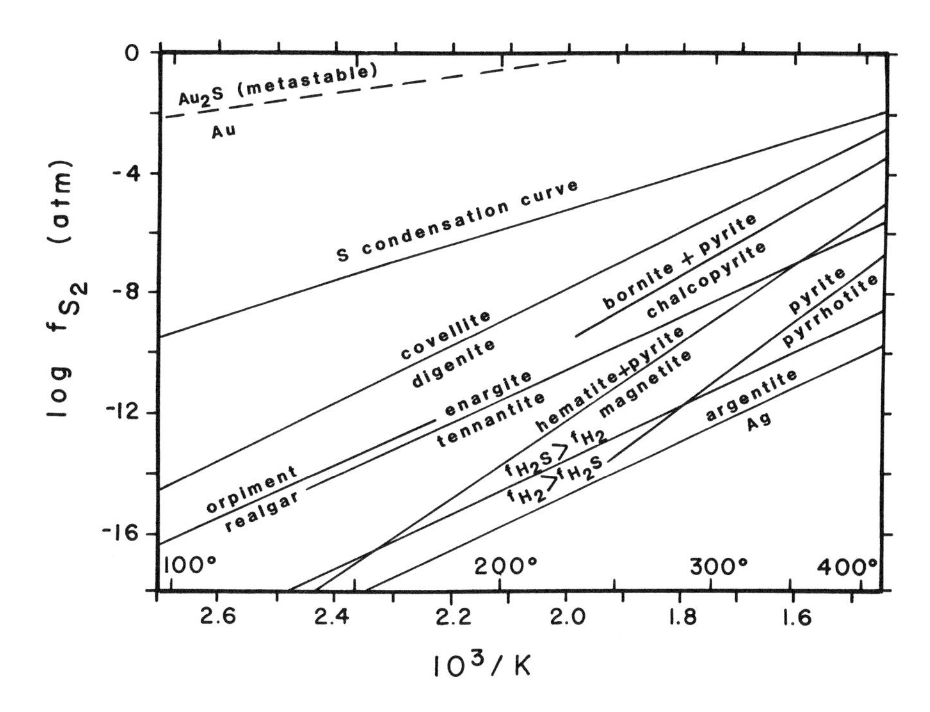

Figure 8.6. Log f_{S2} - 1/t diagram showing sulfidation reactions among minerals and emphasizing those equilibria typical of high sulfidation states. The figures inside the lower margin are °C.

The variation of oxidation or sulfidation state for many buffer and indicator systems (represented by the fugacity of O_2 or S_2) as a function of temperature is easily depicted in f_{O2} - t and f_{S2} - t diagrams, examples of which are given in Figures 8.5 and 8.6 respectively. Note that the log f is plotted against 1/K; this choice is made to produce diagrams in which the oxidation and sulfidation curves are sensibly linear. These diagrams are useful references for the equilibrium constants of redox reactions (written in terms of 1.0 mole of O_2 or S_2) and also serve as visual comparisons of the redox state of various reactions. An extensive compilation of such information for sulfur-bearing systems is available in Barton and Skinner (1979).

In their interpretation of the geochemistry of the OH vein at Creede, Colorado, Barton et al. (1977) constructed a diagram similar to Figure 8.2a; but although the deposit contains abundant pyrite and hematite and sphalerite (with a wide range of iron content), there is no magnetite at all. Instead, there is an abundance of iron-rich chlorite. A log f_{S2} - log f_{O2} diagram was calculated in which a hypothetical, iron end-member chlorite ($H_{12.73}Fe_{12}Si_8O_{36}$) was assumed, in agreement with the evidence given by the mineral assemblage, to just eliminate the magnetite field along the hematite + pyrite boundary. The sulfidation of this chlorite (perhaps we should call it a "partial molal chlorite") to pyrite + quartz and the oxidation of the chlorite to hematite + quartz provided a pair of new bounding curves for the pyrite and hematite field that neatly pre-empted the magnetite field and identified the geologic buffer system that evidently controlled the covariation of f_{S2} and f_{O2} during the formation of the OH vein. Figure 8.3 shows such a chlorite field. Giggenbach (1980, 1981) has shown that analyses of gases from geothermal fields indicate that redox conditions generally are controlled by reactions involving pyrite, quartz, and iron-aluminum silicates (especially chlorite and epidote).

The volume change for the balanced magnetite + hematite reaction is -3.548 cm^3/mole O_2 (about 2% $\Delta \underline{V}$, which is relatively large). The change in the equilibrium constant with pressure is given by the expression:

$$d \ln K/dP = - \Delta \underline{V}/RT \qquad (6)$$

$ Calculate the shift in the log f_{O2} for the hematite - magnetite reaction at 300°C caused by a pressure change of 1.0 kilobar (a pressure considerably greater than that extant for geothermal or epithermal systems).

It should be clear that a pressure change such as that made above is relatively minor, although by no means always negligible; it does not have nearly the influence on stability fields as does pH or a_{O2}.

EFFECT OF BOILING ON REDOX STATE

Minerals and solution species (especially the dissolved gases) buffer the redox state of a hydrothermal system. If boiling occurs the dissolved gases (e.g., CO_2, H_2S, H_2, N_2, NH_3) are selectively transferred into the vapor thereby disturbing the existing equilibria. For example, a system containing pyrrhotite + magnetite + pyrite may react to boiling and consequent hydrogen loss through the reaction:

$$6FeS + 4H_2O = Fe_3O_4 + 3FeS_2 + 4H_2$$

Thus boiling would favor the oxidation of pyrrhotite to pyrite + magnetite. Drummond (1981) has calculated the changes in dissolved gases caused by single-step boiling processes; in Chapter 4 we have presented additional considerations concerning boiling.

As we have seen earlier in this chapter in the interpretation of the data from BR22, boiling in the aquifer in the vicinity of the drill hole resulted in the selective loss of reduced gases (especially H_2), locally lowered the temperature, and shifted the chemistry from that of the original fluid at 300°C (similar to drillhole BR25 see Table 11.1) into one that was in the pyrrhotite-absent field. The rates of re-equilibration of gas reactions of this type are poorly established, but it is known that they are slow (see Figure 14.1 and the discussion of BR22 earlier in this chapter).

CONVENTIONS: HALF CELLS AND GAS FUGACITIES

As we have seen throughout the previous discussions, scientists dealing with high temperature reactions usually use the fugacity (or activity) of O_2, H_2, S_2, and so on to define the position of equilibrium for a reaction. In low-temperature aqueous systems another common approach is based on electrochemistry. The approaches are equivalent, as the following example shows; either may offer advantages or disadvantages.

Consider the reaction of magnetite oxidizing to hematite:

$$4\ Fe_3O_4 + O_2 = 6\ Fe_2O_3 \qquad (7)$$

for which the equilibrium constant has the form

$$K = a^6_{Fe_2O_3}/(a^4_{Fe_3O_4} \times f_{O_2}) = 1/f_{O_2} \qquad (8)$$

Alternatively, one may algebraically sum two half cells:

$$6\ Fe° + 9\ H_2O = 3\ Fe_2O_3 + 18\ H^+ + 18\ e^-$$

$$18\ E° = 18(-0.051)\ \text{volts} \qquad (9)$$

$$2\ Fe_3O_4 + 16\ H^+ + 16\ e^- = 6\ Fe° + 8\ H_2O$$

$$16\ E° = 16(+0.085)\ \text{volts} \qquad (10)$$

summing (and dividing the E by 2):

$$2\ Fe_3O_4 + H_2O = 3\ Fe_2O_3 + 2\ H^+ + 2\ e^- \qquad E° = 0.221\ \text{volts}\ (11)$$

Note that adding half-cell equations, as above, requires that volt-equivalents be summed; that is, each half-cell potential is multiplied by its respective stoichiometric coefficients (the e^- coefficients) yielding a volt-equivalent value. The summation is then divided by the number of electrons exchanged in the final reaction to yield the cell (or half cell) potential. In the iron oxide example above

$$E°(8.11) = [(18 \times -0.051) + (16 \times 0.085)] \times 1/2 = 0.221 \text{ volts}$$

If the final equation represents a complete cell (rather than another half-cell as given in the example) the coefficients for the e^-'s cancel out and the simple summation of the half-cell potentials represents the reaction.

The superscript "°" denotes that all components are in their standard states. The potentials are for 25°C and are from Pourbaix (1966) which is an excellent reference for this type of data. By convention, the half-cell potential for the reaction $1/2\ H_2 = H^+ + e^-$ is 0.0; that is, the standard free energy of formation of the hydrogen ion is 0.0. A similar convention is followed by assigning 0.0 as the standard free energy of formation of an electron, e^-. Half-cell reactions written with the electrons on the right are oxidations; negative values for oxidation reactions mean that the half cell is more reducing than H_2 (care regarding the polarity of half cells is warranted because most older American texts, e.g. Latimer (1952) or Lewis and Randall (1961), use an opposite convention.)

Reaction (10) gives the oxidation potential for the hematite + magnetite pair. To get the fugacity of O_2 one must apply the water reconstitution half-cell reaction to the oxidation potential.

$$2\ e^- + 1/2\ O_2 + 2\ H^+ = H_2O \qquad \text{for which} \quad E° = -1.229 \text{ volts} \qquad (12)$$

Thus the E° for $2\ Fe_3O_4 + 1/2\ O_2 = 3\ Fe_2O_3$ is -1.008 volts; that is, 1.008 volts below O_2 = 1 atm. The standard cell potential, E°, is related to the standard free energy change, ΔG_R°, and the equilibrium constant through the relations,

$$\Delta G_R^\circ = neE°F = -RT\ln K \qquad (13)$$

where ne is the number of electrons exchanged and F is the Faraday constant, 96487 joules/volt equivalent. The oxidation potential is directly measurable at low temperatures using a platinum electrode in combination with a reference such as the calomel electrode and in principle can be measured at higher temperatures. The oxidation state for most geothermal and hydrothermal systems is such that the hydrogen concentration or hydrogen partial pressure can be measured, and oxygen can only be calculated (see comment in the box at the beginning of this chapter). A few years ago when the American convention for the sign of electrochemical cells was the opposite of that now in acceptance, the term Eh (which then was defined as equal to -E relative to the hydrogen electrode) was introduced. Now the American sign convention has been abandoned and all potentials are recorded relative to the hydrogen electrode; so Eh = E.

Complete the following exercises:

$ 1. How much difference is there between the fugacity of O_2 values calculated for hematite + magnetite assemblage for 25°C from the emf relationship above and reaction 4 in Table 8.1? The level of discrepancy is greater than we should like, but it is typical of data from different sources for equilibria that are difficult to measure.

$ 2. Wustite is nominally FeO in composition. The half-cell potential for the reaction $Fe° + H_2O = FeO + 2\ H^+ + 2\ e^-$ is E° = -0.047. What is the f_{O2} compatible with the wustite + iron assemblage at 25°C?

$ 3. Calculate the potential for the wustite + magnetite pair at 25°C.

$ 4. Is wustite stable at 25°C?

One additional variant of cell potential is in occasional use, pE. The pE is defined as the negative log of the activity of the electron, (see, for example Stumm and Morgan, 1981; Thorstenson, 1984). The pE concept leads one into some interesting logical corners when one attempts to understand what sort of activity coefficient an electron might have, or to speculate about the maintenance of charge balance through a redox reaction; but it has the advantage of removing the Faraday constant and particularly the temperature from redox reactions.

An example of how metastability might be studied through the emf method is provided by considering the disturbance of the regular crystal structure of placer gold by the mechanical battering it receives in the stream bed. Visualize an experiment wherein a simple electrochemical cell is set up having: an annealed gold reference electrode, an electrolye of dilute $AuCl_3$, and a second electrode of unannealed, battered gold. Write the half-cell reaction for each electrode. The standard half-cell potential is 1.50 volts. Both half-cells have the same reaction, but in summing them to give the complete cell one is written as the reverse of the other, the result being

$$Au_{battered} = Au_{annealed}$$

Let us suppose that a potential of 0.015 volts is measured. What will be the "standard" $ potential of the battered electrode? (Note, the logic used to decide the polarity of the change is exactly the same as that used to deduce metastable-extention rule used in phase-equilibria studies.)

$ Now, using equation (13), what is the "standard" free energy of the gold in the battered gold electrode?

$ What is the equilibrium constant for the formation of the battered electrode?

$ And now the point of this exercise: how much more soluble (expressed as percent supersaturation) is the battered gold than the annealed gold?

Finally, the numerical aspect of this calculation is only an exercise. John Hass tried this experiment several years ago and could not get stable potentials because the Au^{+++} ions oxidized the Cl^- ions to Cl_2 gas.

REFERENCES

Barton, P. B., and Skinner, B. J., 1979, Sulfide mineral stabilities; in Geochemistry of Hydrothermal Ore Deposits: 2nd edition, edited by H. L. Barnes, Wiley-Interscience, New York, p. 278-403

Barton, P. B., Jr., and Toulmin, P., III, 1966, Phase relations involving sphalerite in the system Fe-Zn-S: Economic Geology, v. 61, p. 815-849.

Drummond, S. E. Jr., 1981, Boiling and mixing of hydrothermal fluids: chemical effects on mineral precipitation: PhD Thesis, Pennsylvania State University, 380 p.

Fisher, J. R., and Barnes, H. L., 1972, The ion-product constant of water to 350°: Journal of Physical Chemistry, v. 76, p. 90-99.

Giggenbach, W. F., 1980, Geothermal gas equilibria: Geochimica Cosmochimica Acta, v. 44, p. 2021-2032.

Giggenbach, W. F., 1981, Geothermal mineral equilibria: Geochimica et Cosmochimica Acta, v. 45, p. 393-410.

Helgeson, H. C., Delany, J. M., Nesbitt, H. W., and Bird, D. K., 1978, Summary and critique of the thermodynamic properties of rock-forming minerals: American Journal of Science, v. 278-A, p. 1-229.

Hemley, J. J., Hostetler, P. B., Gude, A. J., and Mountjoy, W. T., 1969, Some stability relations of alunite: Economic Geology, v. 64, p. 599-612.

Latimer, W. M., 1952, The Oxidation States of the Elements and their Potentials in Aqueous Solutions: 2nd edition, Prentice-Hall, New York, 392 p.

Lewis, G. N., and Randall, M., 1961, Thermodynamics: 2nd edition, revised by K. S. Pitzer and L. Brewer, McGraw-Hill, New York, 723 p.

Ohmoto, H., and Lasaga, A. C., 1982, Kinetics of reactions between aqueous sulfates and sulfides in hydrothermal systems: Geochmica et Cosmochimica Acta, v. 46, p. 1727-1746.

Pourbaix, M., 1966, Atlas of Electrochemical Equilibria in Aqueous Solutions: Pergamon Press, Oxford, 644 p.

Robie, R. A., Hemingway, B. S., and Fisher, J. R., 1978, Thermodynamic Properties of Minerals and Related Substances at 298.15°K and 1 Bar (105 Pascals) and at Higher Temperatures: U. S. Geological Survey Bulletin 1452, 456 p.

Sakai, H., 1983, Sulfur isotope exchange rate between sulfate and sulfide and its application: Geothermics, v. 12, p. 111-117.

Seward, T. M., 1974, Equilibrium oxidation potential in geothermal waters at Broadlands, New Zealand: American Journal Science, v. 274, p. 190-192.

Stumm, W., and Morgan, J. J., 1981, Aquatic Chemistry: Wiley, New York, 780 p.

Thorstenson, D. C., 1984, The concept of electron activity and its relation to redox potentials in aqueous geochemical systems: U.S. Geological Survey, open-file report, 84-072, 67 p.

Chapter 9
METALS IN HYDROTHERMAL FLUIDS

The recognition that some present day geothermal systems may be active analogues of metal depositing hydrothermal systems of the past has promoted a great deal of interest in metal transport and deposition in present day systems (White, 1981; Weissberg, et al., 1979; Henley and Ellis, 1983). Relatively few thermodynamic data are available from which to calculate the high temperature solubilities for metals and metal sulphides. The calculations discussed below focus on recent experimental data up to about 350°C and on metal transport in geothermal and analogous epithermal environments. In Chapter 14 similar calculation procedures are used to examine metal deposition in some other ore-forming environments. Studies of high temperature metal complexing may ultimately become significant in geothermal corrosion and scale control, mineral recovery from brines and development of chemical processes for the control of toxic metals.

METALS IN ACTIVE GEOTHERMAL SYSTEMS

Table 9.1 shows metal and major component concentrations for fluids in the Broadlands and Imperial Valley Geothermal systems.

Table 9.1 — Concentrations of metals and major components in aquifer fluids at Broadlands and at Niland, California. (mg/kg)

	t°C	pH_t	Cl	CO_2	H_2S	Fe	Pb	Zn	Cu	Au	Ag	As	Sb
Broadlands, BR2*	260°	6.1	1140	5000	136	0.4	$1x10^{-3}$	$1x10^{-3}$	$9x10^{-4}$	$4x10^{-5}$	$7x10^{-4}$	5.7	0.2
Imperial Valley, Magmamax 1**	260°	$\sim$5.4†	82000	13500	10-30	175	53	154-247	0.7	0.07	0.34	0.14	2.7

* Weissberg et al., 1979
** Austin et al., 1977
† McKibben, M. A., and Elders, W. A., in press

In the Salton Sea IID wells a copper rich sulfide (together with calcite and amorphous silica) scale deposits from the flashed brine in surface equipment and well casing (Skinner, et al., 1967) and its presence has been a constraint on exploitation. Only minor amounts of sphalerite have been observed in drillcore. At Broadlands, galena and sphalerite occur in vugs and fissures in the deep aquifer and also as a minor component in calcite scale deposited within discharging wells. Recently chalcopyrite and electrum have been found coating the inside of a separator at Broadlands well 22 (Kevin Brown, personal communication). In the nearby Ohaaki Pool, an amorphous Sb-As-Hg-Tl-sulfide precipitate enriched to ore grade in gold and silver coated the silica sinter (Table 9.2) Here, and at Waiotapu, New Zealand, the geologic setting of the metaliferous precipitates, and the occurrence of base metal sulfides at deeper levels strongly resemble features in a number of epithermal type ore deposits (e.g. Round Mountain and Tonopah, Nevada, McLaughlin, California).

Table 9.2 -- Metal Content of Precipitates at Broadlands and at Niland, Imperial Valley, California.

mg/kg or % where indicated

	t°C	As	Sb	Au	Ag	Hg	Tl	Cu	Pb	Zn
Broadlands										
Ohaaki Pool*	98	400	10%	85	500	2000	630	-	25	70
BR2 Inside Silencer	98	50	1000	50	2000	600	150	2.5%	400	50
Outside Silencer	98	250	8%	55	200	200	1000	-	-	-
Imperial Valley										
Magmamax No. 1 Wellhead**	220	na	trace	na	3400	na	na	0.02%	76%	700
Separator**	200-220	1500	190	na	80	na	na	1.0%	1.3%	1100

*Weissberg, 1969;** Maimoni (1982). na = not analysed

In both the Broadlands and Magmamax discharges strong zonation of metals occurs as a result of temperature decrease and pH and ΣH_2S change due to flashing. At Magmamax No. 1, a silver-rich galena (PbS) precipitate occurs with iron-rich silica near the wellhead, while the copper enrichment occurs in the first stage separator as lead decreases. Silica and carbonate increase downstream (Austin, et al., 1977).

We shall tackle two questions in this chapter using lead and gold as case histories;

1. What are the constraints on the metal content of geothermal fluids?

2. What processes may lead to metal deposition in the reservoir or during discharge?

METAL COMPLEXING IN HYDROTHERMAL SOLUTIONS

Case History I: Lead

The solubility of a metal may be written in a generalized form as follows:

$$M + zH^+ = M^{z+} + z/2\ H_2 \qquad (1)$$

where z is the valency of the metal. Notice that we have written a redox reaction involving pH so that immediately we see the need for the earlier studies of mineral-fluid chemistry (Chapters 6, 7 and 8), in constraining the f_{O_2}-pH conditions under which we consider metal transport in geothermal systems.

Metal ions, such as lead, form complex ions by reaction with available ligands such as chloride or bisulfide ions. e.g.

$$Pb^{++} + Cl^- = PbCl^+ \qquad (2)$$

$$PbCl^+ + Cl^- = PbCl_2 \qquad (3)$$

$$PbCl_2 + Cl^- = PbCl_3^- \qquad (4)$$

$$PbCl_3^- + Cl^- = PbCl_4^= \qquad (5)$$

and so on.

Equilibrium constants for these reactions are called stepwise formation constants.

e.g.

$$K_{PbCl^+} = \frac{a_{PbCl^+}}{a_{Pb^{++}}\, a_{Cl^-}} \qquad (6)$$

$$K_{PbCl_2} = \frac{a_{PbCl_2}}{a_{PbCl^+}\, a_{Cl^-}} \qquad (7)$$

These last two expressions can be combined to give

$$(K_{PbCl^+})(K_{PbCl_2}) = \frac{a_{PbCl_2}}{(a_{Pb^{++}})\, a^2_{Cl^-}} = K'_{PbCl_2} \qquad (8)$$

The new constant K' is the cumulative formation constant for the $PbCl_2$ complex ion and is the equilibrium constant for the reaction,

$$Pb^{++} + 2\,Cl^- = PbCl_2$$

Constants such as this are assigned the symbol β_n, where n is the ligand number for the complex.

$ Find an expression for the cumulative formation constant, β_4, of the $PbCl_4^=$ ion in terms of the stepwise formation constants of complexes with ligand numbers 0-4.

We may write similar expressions for any ligand or metal, but it is essential to have good experimental data before we can sensibly evaluate the relative importance of each of the possible complex species in any chemical environment.

$ Write a set of reactions describing the complexing of lead with bisulfide (HS^-) ions.

A great deal of controversy has attended many studies of metal transport in hydrothermal solutions as a result of ad hoc assumptions about the principal ligand involved in metal complex formation. Bisulfide and chloride are usually the most important, but H_2S, NH_3, OH^-, $CO_3^=$ and organic ligands may also be important depending on their concentration, the temperature regime and the stability of the metal-ligand bond. Useful discussions of these factors are to be found in Seward (1984) and Barnes (1979) as well as in other recent literature.

Over the last twenty years or so experimental techniques have gradually evolved to allow determination of some complex and solubility equilibria up to about 300°C. Experiments are much more difficult at higher temperatures, so that there is a paucity of reliable data through which to tackle ore metal transport problems in the higher temperature magmatic or metamorphic environments. You should always carefully examine the format of solubility data reported in the literature--check whether individual ion or mean ion activity coefficients have been used in interpretation of experimental data and recalculate the original data where necessary.

Simple inspection of formation constant data reveals which are the dominant -- that is most stable -- complexes. Table 9.3 shows cumulative formation constants for lead-chloro complexes (from Seward, 1984).

Table 9.3

t°C	cumulative formation constants				stepwise formation constants			
	$\log \beta_1$	$\log \beta_2$	$\log \beta_3$	$\log \beta_4$	$\log K_1$	$\log K_2$	$\log K_3$	$\log K_4$
25	1.41	1.97	1.66	1.46	1.41	0.56	-0.31	-0.20
100	1.67	2.62	2.21	1.93				
150	2.09	3.18	2.84	-				-
200	2.55	4.00	3.81	-				-
250	3.18	4.98	5.03	-				-
300	3.89	6.26	6.76	-				-

1. Draw a graph showing these data as a function of temperature.
2. Complete Table 9.3, using equations similar to equation (8) to derive stepwise formation constants for the different species at 300°C.

The total concentration of a metal in solution is the sum, Σ , of the concentrations of each of its complexes in solution. For lead

$$\Sigma Pb = m_{Pb^{++}} + m_{PbCl^+} + m_{PbCl_2} + m_{PbCl_3^-} + m_{PbCl_4^=} + m_{Pb(HS)^+} + m_{Pb(HS)_2} + \text{etc.} \quad (9)$$

If conditions are such that bisulfide complexes of lead may be discounted, as in the boxed example below and later problems, a simplified expression may be written

$$\text{e.g.,} \quad \Sigma Pb \simeq m_{Pb^{++}} + m_{PbCl^+} + m_{PbCl_2} + m_{PbCl_3^-} + m_{PbCl_4^=} \quad (10)$$

Weissberg, et al. (1979) showed, using experimental data, that thiocomplexes of lead could only account for about 0.5% of the lead in solution at Broadlands (250-300°C) so that we can assume that chloride dominates in our first calculation below.

The programmable calculator allows us to be a little more rigorous in routine calculations. First we substitute the cumulative formation constants (β_n) into equations (9) or (10), at the same time substituting activities by molalities and appropriate activity coefficients (see box below and Chapter 1);

$$\Sigma Pb = m_{Pb^{++}} \left\{1 + \frac{\beta_1 m_{Cl^-} \gamma_{Pb^{++}} \gamma_{Cl^-}}{\gamma_{PbCl^+}} + \frac{\beta_2 (m_{Cl^-} \gamma_{Cl^-})^2 \gamma_{Pb^{++}}}{\gamma_{PbCl}} + \ldots\right\} \quad (11)$$

Notice how each of the terms in equation (11) provides the relative contribution of each complex ion. i.e. the percentage of all the lead in solution which is complexed as $PbCl_2$ is given by

$$\%PbCl_2 = 100 \left\{ \frac{\beta_2 m^2_{Cl^-} \gamma^2_{Cl^-} \gamma_{Pb^{++}}}{\gamma_{PbCl}} \right\} / \left\{ 1 + \frac{\beta_1 m_{Cl^-} \gamma_{Pb^{++}} \gamma_{Cl^-}}{\gamma_{PbCl^+}} + . . \right\} \quad (12)$$

What are the principal factors controlling the relative contributions of the different chloro-complexes?

Calculate the relative contributions of each of the lead chloro-complexes to the total lead in solution at 250°C in a) a 0.14 m chloride solution and b) a 2.0 m chloride solution. To solve the equations you will need values of the activity coefficients for the lead complexes and chloride ion. For help here see the next paragraph. Programs like ION and PBS in Appendix I may be used for this problem.

Which complex dominates the solution chemistry of lead in these chloride solutions at this temperature?

Activity Coefficients -- The individual ion activity coefficients for each of the species are obtained from the extended Debye-Hückel expression (Chapters 1 and 7) but some estimate must be made for the ion size parameter, $\mathring{a}$. $\mathring{a}$ should not be considered as representing the real size of an ion in aqueous solution as its role in the extended Debye-Hückel equation is to accommodate ion-ion and ion-solvent interactions. Kielland (1937) calculated $\mathring{a}$ values for about 130 common ionic species. For most ions use of $\mathring{a}$ = 3 to 5 is usually satisfactory. For lead complexes Seward (1983) used values of 5, 4, 4 and 5 for Pb^{++}, $PbCl^+$, $PbCl_3^-$ and $PbCl_4^=$ respectively. For neutrally charged complexes like $PbCl_2$ we may take $\gamma = 1$ for all but the most saline solutions.

Solubility of Metal Sulfides -- The solubility of a sulfide, like galena, may be written

$$PbS + 2H^+ = Pb^{++} + H_2S \quad (13)$$

for chemical conditions where H_2S is the dominant sulfur species.

The solubility product of PbS ($PbS = Pb^{++} + S^=$) has not been determined in chloride solutions up to high temperatures, but Helgeson (1969) has extrapolated low temperature data up to 300°C. These data are shown in Table 9.4. The $S^=$ ion does not occur in the pH range of interest here so that additional expressions are necessary to obtain an equilibrium constant for reaction (13). These are available in Table 8.1 where they were used for redox equilibrium calculations.

Table 9.4

	log K					
	25°C	100°C	150°C	200°C	250°C	300°C
$PbS = Pb^{++} + S^=$	-28.57	-23.96	-21.93	-20.36	-19.14	-18.31
$S^= + H^+ = HS^-$	13.90	11.78	10.62	9.57	8.61	7.72
$HS^- + H^+ = H_2S$	6.98	6.61	6.81	7.17	7.60	
$PbS + 2H^+ = Pb^{++} + H_2S$	-7.69	-5.57	-4.50	-3.62		

COMPLETE THE TABLE AND ADD THESE DATA TO YOUR EARLIER GRAPH

$$K_S = \frac{a_{Pb^{++}} \; a_{H_2S}}{a^2_{H^+}} \qquad \text{so that}$$

$$m_{Pb^{++}} = \frac{K_S \; a^2_{H^+}}{\gamma_{Pb^{++}} \; m_{H_2S}} \qquad (14)$$

Equation (14) now substitutes nicely into equation (11) to give an expression for the solubility of galena as a function of temperature, pH and the concentrations of H_2S and chloride. This is readily programmed into a hand calculator (Appendix I). Some of these parameters may not be well defined, and in some cases a simple calculation based on the dominant complex species may suffice.

> $ Calculate the solubility of galena (mg_{Pb}/kg) as $PbCl_2$ at 300°C in a solution containing 2.0 moles/kg free Cl^- and 4 x 10^{-4} moles/kg H_2S at pH = 5.4.
>
> Compare your calculated lead content with the analytical data for the Magmamax No. 1 well in Table 9.1.

Deposition of Galena -- If $PbCl_2$ is the dominant lead complex in hydrothermal solutions above about 200°C, then we may write

$$PbS + 2H^+ + 2Cl^- = PbCl_2 + H_2S \qquad (15)$$

From the reaction equation we can see that processes such as dilution or addition of H_2S (how?) would be able to deposit galena from an initially saturated solution. Increases in pH, due to boiling, will also be effective and cooling, of course, reduces solubility. The relative significance of each of these processes may be assessed quantitatively using the PBS program or semi-quantitatively as illustrated below.

Consider a solution initially at 300°C which may undergo one of three possible processes to drop in temperature to 280°C: conductive cooling, dilution, and boiling.

The solubility constant for reaction (15) is given by

$$K_{PbS} = \frac{a_{PbCl_2} \; a_{H_2S}}{a^2_{Cl^-} \; a^2_{H^+}}$$

so that

$$m_{PbCl_2} = \frac{K_{PbS} \; m^2_{Cl^-} \; a^2_{H^+} \; \gamma^2_{Cl^-}}{m_{H_2S}}$$

Values of log K_{PbS} can be obtained from our earlier data:

$$\log K_{PbS,300} = 3.72, \quad \log K_{PbS,280} = 2.95.$$

For conductive cooling from 300 to 280°C, m_{Cl} and m_{H2S} are constant and changes in pH and activity coefficients are negligible

$$m_{PbCl2,280}/m_{PbCl2,300} \simeq K_{PbS,280}/K_{PbS,300} = 0.17$$

i.e. the solubility at 280°C is less than one fifth that at 300°C.

For the dilution process, $m_{Cl,280} \simeq m_{Cl,300} \times 280/300$*, $m_{H2S} \simeq m_{H2S} \times 280/300$ and again the pH change is negligible,

$$\frac{m_{PbCl2,280}}{m_{PbCl2,300}} \simeq 0.17 \left\{\frac{280}{300}\right\}^2 \left\{\frac{300}{280}\right\} \simeq 0.16$$

The solubility change due to dilution is dominated by the decrease in temperature. Notice that for a solution initially saturated with respect to galena, dilution decreases the total lead in solution so that the supersaturation at 280°C is 5.8.

For adiabatic boiling, $m_{Cl} \simeq m_{Cl,300} \times 300/280$, $m_{H2S,280} \simeq m_{H2S,300} \times .2$ ** and, since $m_{CO2,280} \simeq m_{CO2,300} \times .1$, the pH change is about + 1 unit***

$$\frac{m_{PbCl_{2,280}}}{m_{PbCl_{2,300}}} \simeq 0.17 \left\{\frac{300}{280}\right\}^2 \times \left\{\frac{10^{-2}}{0.2}\right\} \simeq 0.01$$

---

* For ease of computation in these "thought experiments", and since the temperature interval is small, we equate water temperature with water enthalpy; you may wish to compute the error introduced by this assumption using data from the Steam Tables. Hence, the factor 280/300 approximates the dilution factor responsible for decreasing temperature from 300 to 280°C by mixing hot chloride and cold fresh water. Similarly the factor 300/280 approximates the concentration increase due to adiabatic boiling through this temperature interval.

** See Chapter 4.

*** If the solution does not undergo reaction with host rocks to buffer pH (e.g. in a discharging well or silica-lined fissure) a tenfold change of m_{CO2} results in a pH increase of about 1 unit through charge balance and the reaction $H_2CO_3 = HCO_3^- + H^+$.

---

i.e. , the solubility at 280°C after boiling is one hundred times less than that of the original 300°C solution. For a solution initially saturated with galena, boiling increases the total lead in solution by only x 1.07 but the supersaturation attained due largely to the pH change is about 107.

Although all three processes may deposit galena from an initially saturated solution, clearly boiling is the most efficient process. Dilution changes galena solubility by 6 times and boiling by about 100 times. The degrees of supersaturation are 6 and 107, respectively, so that the boiling process results in much more favorable kinetics for sulfide deposition. Also, if the original solution was undersaturated by a factor of say 5 x, dilution only just leads to saturation but boiling leads to extreme sulfide supersaturation.

Here we have considered the response of an isolated body of fluid to three processes. In ore veins you might expect silicate reactions in the wall-rocks to buffer the pH of the vein fluid during dilution or boiling. Evidence from exploited geothermal systems suggests that these mineral reactions respond more slowly to changing conditions than do most of the solutes. As an exercise you can calculate the effect of silicate buffering on the three processes considered above; the easiest method by which to obtain pH's for this exercise is to follow the procedure used in calculating Table 7.2.

Saturation states may also be calculated for non-metallic phases -- gangue -- as functions of boiling, cooling, or mixing. (Calcite and silica are discussed in Chapter 11.) Solubility data for these minerals are covered in Chapter 6.

1. Account for the deposition of galena scale in the Imperial Valley wells and in the aquifer.

$ 2. What mineralogical changes would you expect to accompany galena deposition from an ore forming fluid undergoing (a) boiling, (b) dilution?

$ 3. Discuss the effects of wall-rock buffering (Kmica-K-feldspar equilibrium) on these processes, and the solubility consequences of changing silica concentration from quartz to cristobalite saturation (see Chapter 6).

Silver, copper, and perhaps zinc are probably transported predominantly as chloride complexes. Similar calculations to those outlined above may be made using data from Barnes (1979) and Seward (1976).

Table 9.5 -- High temperature association constants (log B_1, log B_2) for silver chloride species are now available (Seward, 1976) and summarized in the Table below, together with log dissociation constants for sulfide species and the solubility product of acanthite (Ag_2S).

Reaction	200	250	300°C
$Ag^+ + Cl^- = AgCl°$	2.89	3.16	3.6
$Ag^+ + 2Cl^- = AgCl_2$	4.59	4.94	5.6
$Ag_2S = 2Ag^+ + S^=$	-31.71	-28.74	-26.36
$S^= + H^+ = HS^-$	9.57	8.61	7.72
$HS^- + H^+ = H_2S$	7.17	7.60	8.05
$Ag_2S + 2H^+ = 2Ag^+ + H_2S$	-14.97	-12.53	∿-10.6

$ Calculate the relative changes of solubility of acanthite as a consequence of boiling, dilution with cold water (Cl = 0 mg/kg) or conductive cooling over the temperature range 250 to 200°C.

Case History II: gold

Largely through its romantic history and continuing importance in the international economy, gold has been the focus of a great deal of speculation, but rather less experimental study concerning its mode of transport in a wide range of hydrothermal environments. Recently Seward (1973, 1982) has shown that thio-complexes of gold are stable to at least 300°C and dominate transport of the metal in geothermal fluids. At higher temperatures, chloride complexes may become more important (Henley, 1973). Studies of thio-complexing for other metals are few so that here we shall carefully examine only the transport of gold. Arsenic, antimony and mercury may also form stable thio-complexes as suggested by their close geologic association and chemical affinities.

Seward (1973) showed that up to 300°C, two gold-thio-complexing reactions were dominant in the pH range in which H_2S dominates sulfide speciation. Notice that we write the complexing reaction in terms of the stronger HS^- ligand rather than the weak dipolar H_2S.

$$Au^+ + HS^- = Au(HS) \quad (16)$$

$$Au^+ + 2HS^- = Au(HS)_2^- \quad (17)$$

In the HS^- field another complex, $[Au_2(HS)_2S^=]$, begins to dominate but this has little significance for most epithermal ore depositing environments in which pH remains in the H2S field.

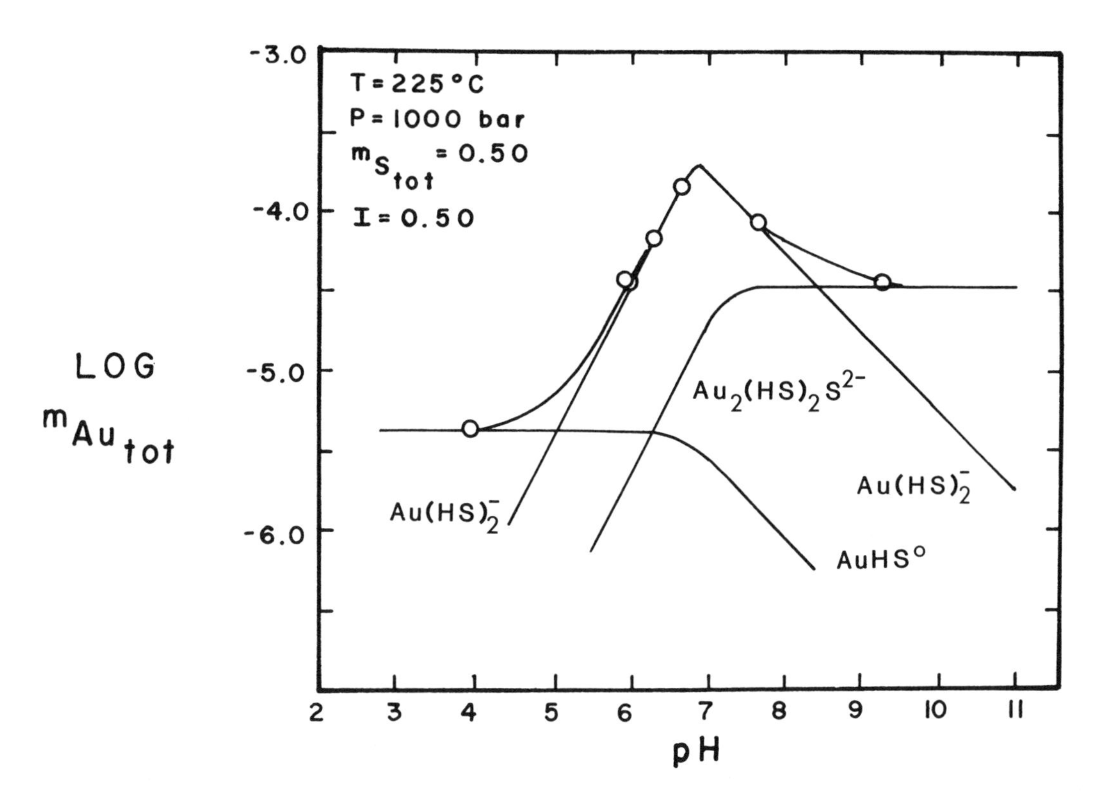

Figure 9.1. Calculated solubility curves for three thio-gold complexes compared with experimental data at 225°C and 1000 bars pressure (redrawn from Seward, 1973).

The solubility curves for these reactions are shown in Figure 9.1; the heavy line is the measured concentration of gold in the specified solution. Notice especially the solubility maximum close to the HS^- - H_2S boundary ($pK_{H2S,\ 225°C} \simeq 7.3$).

Complete the diagram by converting the log molality units to mg/kg units for total dissolved gold (the atomic weight of gold is 198).

The other reaction we require is for the formation of the Au+ ion

$$Au + H^+ = Au^+ + 1/2H_2 \qquad (18)$$

Now complete and balance the following reaction equations by inserting the appropriate principal gold complex. Assume that the pH refers to a solution at 225°C.

pH = 6.5 $\quad Au + HS^- + H_2S = Au \quad + H_2 \qquad (19)$

pH = 3 $\quad Au + HS^- + H_2S = Au \quad + H_2 \qquad (20)$

Table 9.6 gives experimental data for reaction (19)

Table 9.6 -- Solubility constant data for the complex, $Au(HS)_2^-$ (from Seward, 1973)

t°C	pK	
175	+1.29	
200	+1.28	
225	+1.22	
250	+1.19	
300°	______	Find this datum by extrapolation

Write a program for the calculation of gold solubility as a function of $m_{\Sigma S}$, pH, f_{H2} and temperature (Appendix I gives a similar program, AU, involving f_{O2} instead of f_{H2}).

$ Using data from Table 9.6, calculate the solubility of gold at 250°C when pH = 6.1, f_{H2} =.33 bars and the total sulfur concentration = 3×10^{-3} (assume H_2S and HS^- are the major sulfide species; pK for H_2S is given in Table 7.2, and that $\gamma_{HS^-} = \gamma_{Au(HS)_2^-} = 0.7$)

Compare your calculated solubility with that in the deep Broadlands waters (Table 9.1). Is this fluid supersaturated or undersaturated with respect to gold? (For the K-feldspar-Kmica pH use a_{K^+} = 85 mg/kg)

For the Broadlands fluids, Seward (1973) showed, using available low temperature chloride complexing data, that of the 4×10^{-5} mg/kg Au only 6×10^{-9} mg/kg could be carried as gold chloride complexes. (It is possible that chloride complexing may be more significant at high temperatures than these data suggest, but if so it might be expected that gold should deposit along with lead as a result of deep boiling, rather than in a nearer surface hot spring environment).

Figure 9.2 is an f_{O2}-pH diagram for gold and a number of mineral-fluid equilibria at 250°C. The conditions are those calculated for the Waiotapu geothermal system in New Zealand - where a gold-silver-arsenic-antimony-rich precipitate occurs at surface and sphalerite and galena are disseminated at depths of 200 to 700 m.

Complete the diagram by inserting a) f_{H2} values on the right hand axis b) gold concentrations on the isosolubility curves. [m_{H2S} = 0.003. Assume that negligible chloride complexing occurs in the f_{O2} - pH field of the diagram]

Consider a solution at 250°C initially undersaturated with gold at the pyrite-pyrrhotite phase boundary at pH = 6.0. What are the expected effects of

(a) mixing with fresh water to reach a temperature of 200°C?

(b) boiling with (1) continuous and (2) single step steam loss to 200°C?

(c) rapid oxidation of H_2S at the surface of a boiling hot spring derived from the original deep water?

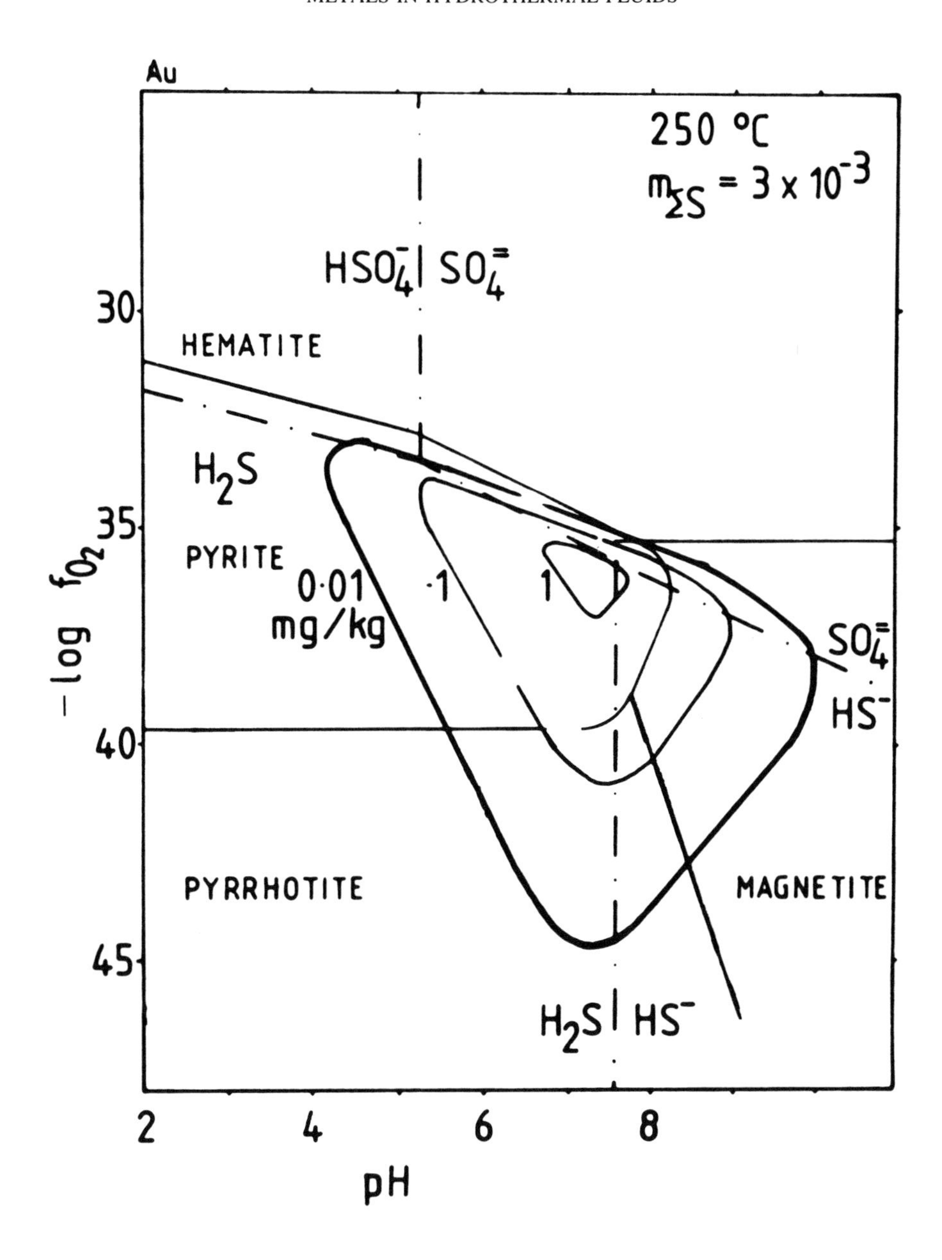

Figure 9.2. f_{O2} - pH diagram at 250°C showing the stability fields of the principal sulfur species and solubility contours for gold in mg/kg as $Au(HS)_2^-$, see text

SUMMARY PROBLEMS

1. Barnes (1979) provides data for the solubility of galena as bisulfide complexes (Table 9.7). The relative importance of these and the chloride complexes may be examined as follows:

$ a) Determine the relative stabilities of the thiosulfide complexes at 250°C by plotting curves of equal activity (e.g. $a_{Pb(HS)2}/a_{Pb(HS)\bar{3}} = 1$ or $a_{Pb(HS)2H2S}/a_{Pb(HS)2} = 1$) as a function of pH and total dissolved sulfur. You will need to make your own decisions about the range of $m_{\Sigma S}$ and ionic strength that you consider.

Indicate on your figure the general pH-$m_{\Sigma S}$ range for geothermal and analogous ore depositing systems.

$ b) Draw another diagram showing iso-pH solubility curves for PbS as chloride and thiosulfide complexes at 250°C in the pH range 5 to 7 with chloride = 1000 mg/kg. On the y-axis plot log $m_{\Sigma Pb}$ from -10 to +2 and on the x-axis plot log $m_{\Sigma S}$ = log(m_{H2S}+m_{HS^-}) from -9 to +1. Plot the relation $m_{Pb(HS)2} = m_{PbCl2}$ and the relation Σ Pb mg/kg = m_S. Plot the analytical data for the Broadlands well, BR2 (Table 9.1) on your diagram and comment on their relation to the solubility curves. (Cl = 1000mg/kg, $f_{fO2} = 10^{-36.6}$, pH = 6.1 and $m_{H2S} = 3 \times 10^{-3}$)

2. Pyrite is probably the most common sulfide present in geothermal and ore depositing systems. Using the data for ferrous (Fe^{2+}) chloride complexing summarized in Table 9.5, calculate the solubility of pyrite at 250°C in the Broadlands well fluid considered above.

3. Draw solubility curves for galena and pyrite onto Figure 9.2. The chloride concentration is 1000 mg/kg. What are the anticipated effects of the above three processes?

Table 9.7 -- Equilibrium constants (log K) for ferrous chloride and lead bisulfide complexes (from Barnes, 1979).

t°C:	25	200	250	300	n
$Fe^{2+} + Cl^- = FeCl^+$	-0.51	1.58	2.34	2.33	1
$Fe^{2+} + 2Cl^- = FeCl_{2(aq)}$	---	---	---	2.57	2
$FeS_2 + 2H^+ + n\,Cl^- + H_2O_{(1)}$	---	-24.77	-19.47	-14.88	1
$= Fe\,Cl_n^{2-n} + 2H_2S_{(aq)} + 1/2O_{2(g)}$	---	---	---	-14.64	2
$FeS + 2H^+ + nCl^- =$	---	1.04	2.56	4.18	1
$= Fe\,Cl_n^{2-n} + H_2S_{(aq)}$				4.42	2
$PbS + H_2S_{(aq)} + HS^- = Pb(HS)_3^-$	-5.62	-5.49	---	---	
$PbS + H_2S_{(aq)} = Pb(HS)_{2(aq)}$	-7.6	-4.97	-4.87	-4.78	
$PbS + 2H_2S_{(aq)} = PbS(H_2S)_{2,(aq)}$	---	-4.88	---	-4.4	

REFERENCES

Austin, A.L., Lundberg, A.W., Owen, L.B., and Tardiff, G.E., 1977, The LLL Geothermal Energy Program Status Report; January 1976-January 1977: Lawrence Livermore Laboratory, UCRL 50046-76.

Barnes, H.L., 1979, Solubilities of Ore Minerals; in Barnes, H.L. (ed), Geochemistry of Hydrothermal Ore Deposits (2nd Edition): Wiley Interscience, p. 404-508.

Helgeson, H.C., 1969, Thermodynamics of hydrothermal systems at elevated temperatures and pressures: American Journal of Science, v. 267, p. 729-804.

Henley, R.W., 1973, Solubility of gold in hydrothermal chloride solutions: Chemical Geology, v. 11, p. 73-87.

Henley, R.W., and Ellis, A.J., 1983, Geothermal systems Ancient and Modern: Earth Science Reviews, v. 19, p. 1-50.

Maimoni, A., 1982, U. California, Lawrence Livermore Laboratory Report UCRL 53252.

Seward, T.M., 1973, Thiocomplexes of gold and the transport of gold in hydrothermal ore solutions: Geochimica et Cosmochimica Acta, v. 37, p. 379-399.

Seward, T.M., 1976, Stability of chloride complexes of silver in hydrothermal solutions up to 350°C: Geochimica et Cosmochimica Acta, v. 40, p. 1329-1341.

Seward, T.M., 1982, Transport and Deposition of Gold in hydrothermal systems: Proceedings of Conference on Geology, Geochemistry, and Origin of Gold Deposits, Zimbabwe. 1982.

Seward, T.M., 1984, The formation of lead (II) chloride complexes to 300°C: Spectrophotometric study: Geochimica et Cosmochimica Acta, v. 48, p. 121-134.

Skinner, B.J., White, D.E., Rose, H.J., and Mays, R.E., 1967, Sulfides associated with the Salton Sea geothermal brine: Economic Geology, v. 62, p. 316-330.

Weissberg, B.G., 1969, Gold-silver ore grade precipitates from New Zealand thermal waters: Economic Geology, v. 64, p. 95-108.

Weissberg, B.G., Browne, P.R.L., and Seward, T.M., 1979, Ore Metals in Active Geothermal Systems; in Barnes, H.L. (ed) Geochemistry of Hydrothermal Ore Deposits (2nd Edition): Wiley Interscience, p. 738-780.

White, D.E., 1981, Active geothermal systems and hydrothermal ore deposits: Economic Geology, v. 75th Anniversary, p. 392-423.

Chapter 10
STABLE ISOTOPES IN HYDROTHERMAL SYSTEMS

Isotopes are forms of an element with the same number of electrons and protons but a different number of neutrons and therefore different masses. Although chemical behavior differs very little between isotopes of the same element, the mass differences between isotopes do produce small chemical differences and equilibrium constants (fractionation factors) for isotopic exchanges are typically close to one. The relative abundances of isotopes commonly analyzed in geothermal fluids were given by Panichi and Gonfiantini (1976).

Isotope	Abundance, %	Isotope	Abundance, %
^{1}H	99.985	^{16}O	99.76
^{2}H (= D)	0.015	^{17}O	0.04
^{3}H (= T)	10^{-15} to 10^{-12}*	^{18}O	0.20
^{12}C	98.89	^{32}S	95.0
^{13}C	1.11	^{33}S	0.76
^{14}C	1.2×10^{-10}**	^{34}S	4.22
		^{36}S	0.016

* approximate range for pre-1952 to 1963 precipitation in the northern hemisphere.
** pre-1952 modern carbon.

Results of the application of isotopes to geothermal systems have been reviewed by Craig (1963), Panichi and Gonfiantini (1976), Truesdell and Hulston (1980), and Giggenbach et al. (1984).

> Calculate the abundance ratios of $^{1}H^{1}H^{16}O$ to $HD^{16}O$, $^{1}H^{1}H^{18}O$, and $HD^{18}O$.

COLLECTION AND ANALYSIS

Samples for water isotope analysis (oxygen-18, deuterium, tritium) should be collected with care because the analyses cost between $25 and $400 each. Containers should be glass with vapour-tight lids. Although 20 ml is sufficient for ^{18}O and D, about 100 ml should be collected to allow for chloride analysis, and repeat analyses if required. For tritium, larger samples (500 ml to 1 litre) are required. Comprehensive field data are necessary in the interpretation of isotope analyses.

Oxygen-18 analysis

The water sample is degassed and equilibrated with CO_2 at 25°C. The CO_2 is drawn off and analysed in a mass spectrometer.

> What mass numbers are analysed?

Since dissolved CO_2 equilibrates rapidly with water through the symmetrical $CO_3^{=}$ complex, half a day with shaking or 2 days without shaking are sufficient for equilibrium.

Deuterium analysis

A small quantity (5-10 mg) of water is reacted with hot metal (U, Zn) to produce hydrogen gas. The hydrogen gas is introduced into the mass spectrometer.

> The range of D/H ratios in natural waters is nearly 10x greater than that of $^{18}O/^{16}O$. Why?

Tritium analysis

About 500 ml of water are decomposed by electrolysis to leave 5 ml of tritium-enriched water. This process takes up to 3 months. The residual water is either decomposed to hydrogen and counted in a gas counter usually at a pressure above 1 atmosphere (why?), or mixed with a scintillation material ("cocktail") and counted in a scintillation counter.

> Why would it be advantageous to be able to measure tritium directly in a mass spectrometer?

NOTATION AND FRACTIONATION

To describe the variation in nature of water isotopes we must understand their concentration units. A ratio mass spectrometer simultaneously collects molecules containing isotopes of different masses and measures the ratio of their abundance. Because of the many chemical, physical and electronic processes involved, this ratio is not absolutely accurate but can be compared accurately to a similar mass ratio of a standard gas measured alternately with the unknown through a double inlet system and gas switching valves.

Thus the mass ratio R_{unk}(= $^{18}O/^{16}O$ or D/H) is not known accurately but may be accurately compared to a standard as

$$\frac{R_{unk} - R_{std}}{R_{std}}$$

This ratio is used in the del notation as parts per thousand deviation (permil)

$$\delta \equiv 10^3 \, [(R_{unk}/R_{std}) - 1] \qquad (1)$$

This notation arises from the method of measurement but is also convenient in expressing fractionations. Consider the fractionation of oxygen isotopes between water vapor and liquid.

$$H_2{}^{18}O, vapor + H_2{}^{16}O, liquid = H_2{}^{16}O, vapor + H_2{}^{18}O, liquid \qquad (2)$$

For this reaction we can write an equilibrium constant which, if only one atom in each molecule is exchanged, is identical to the fractionation constant,

$$\alpha \equiv \frac{H_2{}^{16}O_v \; H_2{}^{18}O_l}{H_2{}^{18}O_v \; H_2{}^{16}O_l} = \frac{R_l}{R_v} \qquad (3)$$

From our definition of δ,

$$10^3 + \delta = 10^3 \, (R_{unk}/R_{std}), \qquad (4)$$

$$\alpha = \frac{10^3 (R_l/R_{std})}{10^3 (R_v/R_{std})} = \frac{10^3 + \delta_l}{10^3 + \delta_v} \qquad (5)$$

These are exact relations; some approximations are useful as well. Almost all isotope fractionation factors are close to unity and therefore

since $\ln \alpha \simeq \alpha - 1$ for $\alpha \simeq 1$,

$$\ln \alpha \simeq \frac{10^3 + \delta_l}{10^3 + \delta_v} - 1 = \frac{\delta_l - \delta_v}{10^3 + \delta_v} \qquad (6)$$

and since $(10^3 + \delta_v) \simeq 10^3$ for $\delta_v << 10^3$,

$$10^3 \ln \alpha \simeq \delta_l - \delta_v . \qquad (7)$$

Thus we may expect the differences of δ values to be nearly constant at constant temperature.

> Write some pairs of δ values that may be involved in isotope exchange in geothermal systems. Use notation like "$\delta^{18}O$ (vapor) - $\delta^{18}O$ (liquid)".

APPLICATIONS OF WATER ISOTOPES

The most important applications of water isotopes in geothermal studies are as natural tracers. These are applied in two ways, as tracers of water origins and as tracers of reservoir processes. This chapter focuses on active geothermal systems, but stable isotopes are equally important in ore deposit studies. In these studies, isotope analyses of minerals are used to first obtain temperatures of formation, and then to estimate the isotope composition of ore-forming fluids in order to trace water origin and chemical processes. In some cases, the fluids trapped in inclusions may be analyzed directly.

Water Origins

Most early workers in geothermal and ore deposit studies thought that all water came from juvenile or magmatic sources. Craig (1963) and others analysed rain and snow and found that samples from higher latitudes and elevations or further inland were progressively lighter (lower values of δ) and that δD and $\delta^{18}O$ were approximately related by the meteoric water line.

$$\delta D = 8\ \delta^{18}O + 10 \qquad (8)$$

The depletion in heavy isotopes (^{18}O, D) with distance from tropical seas (Figures 10.1 and 10.3) results from separation of these isotopes into condensed phases (rain and snow) and enrichment of light isotopes in the remaining vapour. This can be described as a Raleigh (open system) process in which the products are continuously lost and the reaction proceeds to completion.

> Why do reactions in closed systems not proceed to completion?

Craig (1963) also found that the $\delta^{18}O$ values of geothermal waters were higher (more positive) than those of local meteoric waters, but δD values were the same. The change in $\delta^{18}O$ might result from mixture of meteoric with magmatic waters, but then all the δD values from a given geothermal area would not necessarily match local meteoric water (why?). The "oxygen isotope shift" was interpreted by Craig to result from isotopic equilibration of water with rock minerals rich in oxygen with high $\delta^{18}O$ values. Why are hydrogen isotopes little affected? Figure 10.1 shows observed oxygen isotope shifts for geothermal systems with various local meteoric water isotopic compositions. Figure 10.2 shows the isotopic fractionation ($10^3\ln \alpha$) between quartz-water, calcite-water, and feldspar-water as a function of temperature (the water is always isotopically lighter). Figure 10.3 shows meteoric water $\delta^{18}O$ and δD compositions and rock $\delta^{18}O$ compositions. Using these figures answer the questions in the box.

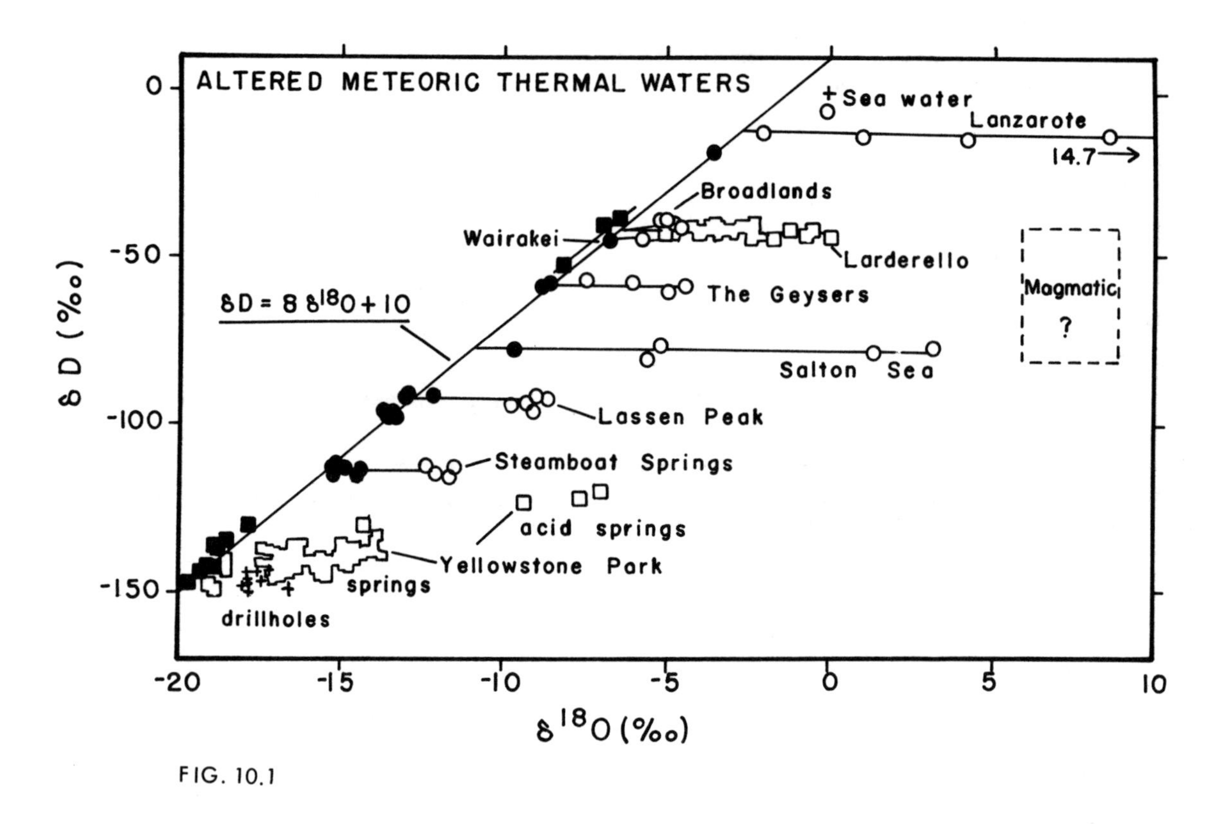

Figure 10. 1. Oxygen-18 and Deuterium compositions of hot spring, fumarole, and drill hole thermal fluids derived from meteoric waters (o) and of meteoric waters local to each system (o). From Truesdell and Hulston (1980).

The "oxygen isotope shift" of geothermal waters is related to (1) the original $\delta^{18}O$ of the water and rock, (2) the magnitude of the fractionation (temperature, mineralogy), and (3) the water/rock ratio. Describe localities, rocks, temperatures and water:rock ratios that would give minimum or zero shifts, maximum shifts, or reverse shifts. all. Describe shifts in rock compositions. Indicate localities that would be best for searching for magmatic water contributions to hydrothermal systems.

A more quantitative approach to the oxygen isotope shift was made by Taylor (1977) and others, who found that isotope exchange between cooling granitic plutons and meteoric waters was a general phenomenon. The change in isotopic compositions of water and rock in a closed system follows from the isotope balance and fractionation equations,

$$W(\delta_{f,H_2O} - \delta_{i,H_2O}) = -R(\delta_{f,rock} - \delta_{i,rock}) \qquad (9)$$

$$\Delta = \delta_{rock} - \delta_{water} \qquad (10)$$

Therefore

$$W/R = \frac{(\delta_{f,rock} - \delta_{i,rock})}{\delta_{i,H_2O} - (\delta_{f,rock} - \Delta)} \qquad (11)$$

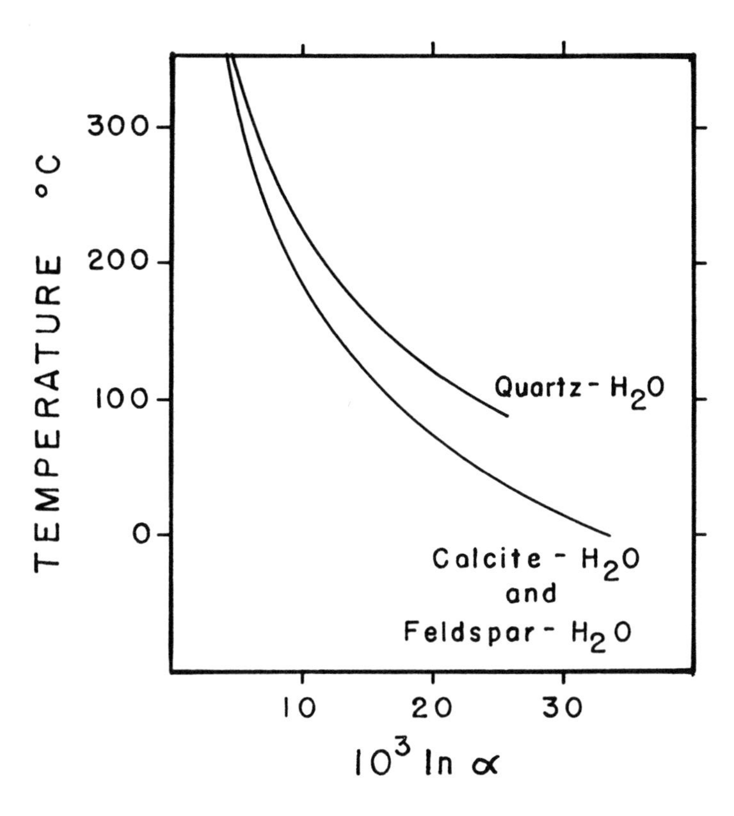

Figure 10.2. Isotopic fractionation factors, $10^3 \ln \alpha$, between quartz and water, calcite and water, and feldspar and water as a function of temperature. The mineral is isotopically heavier in each case.

where W and R are the exchangeable molar amounts of oxygen or hydrogen in water and rock, and i and f stand for initial and final states.

> Write an equation similar to (11) for the final composition of water in terms of the inital water and rock compositions and the W/R ratio.
>
> Plot a $\delta^{18}O$ (range -20 to +10) vs δD (range -160 to -60) graph of the trajectories of water compositions and rock compositions as a function of molar W/R ratios from 0.001 to 0.99 for the interaction of a granite (initial composition 50 mole % feldspar with $\delta^{18}O$ = +9 and 10 mole % biotite with δD = -65) with meteoric water (initial composition $\delta^{18}O$ = -16 and δD = -120). Assume that only the feldspar (with 61 mole % O) exchanges $\delta^{18}O$ and only the biotite (with 9 mole % H) exchanges D. Use fractionation factors of $\Delta^{18}O$ feldspar-water = +2 and ΔD biotite-water = -35.

We can use the natural variation in isotopes of rain and snow to indicate the source of recharge for a geothermal system, but we must be careful to allow for the effects of boiling discussed below. Surface evaporation also produces isotopic changes, and we can trace infiltration of lake and seawater from their distinct isotopic compositions.

Reservoir Processes

Subsurface geothermal waters at temperatures above surface boiling must cool as they rise to the surface. This can occur by three processes: conduction, mixing, and boiling. Each of

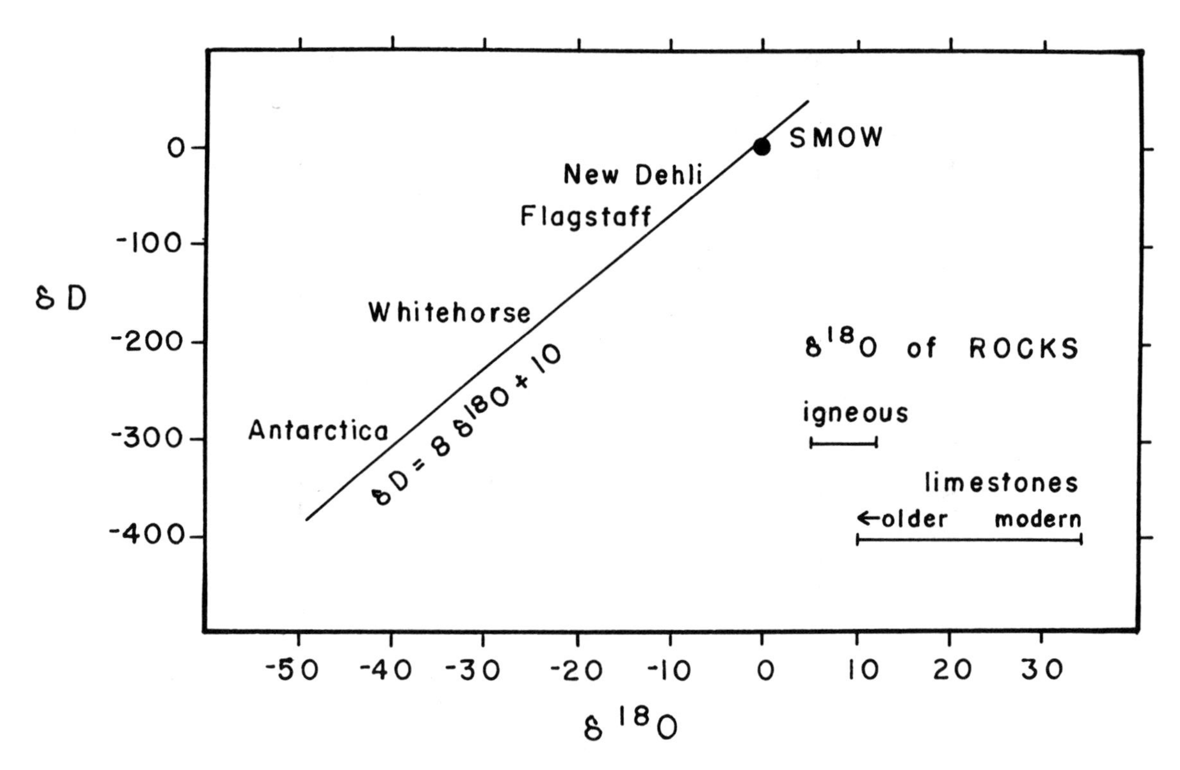

Figure 10.3 Isotopic compositions of meteoric waters (^{18}O and D) and of some common rocks (^{18}O only).

these processes can be traced isotopically (and chemically) and in the case of boiling, the mechanism can be studied through isotopic changes.

Conductive heat loss -- During ascent with conductive heat loss no changes are expected in deuterium because liquid water is the only phase containing significant hydrogen and exchange is impossible. Oxygen-18 may exhange with rock oxygen, but this proceeds very slowly at temperatures below 200°C. Conductive heat loss has been calculated for vertical pipe flow (Truesdell et al., 1977) and shown to reduce temperatures by half for waters flowing at 0.4 ℓ /sec from a 1 km deep reservoir, with proportional heat losses for flows from deeper or shallower reservoirs (0.2 ℓ /sec from 500m, etc.) Conductive cooling should be assumed to be important for isolated springs flowing at less than 1 ℓ/sec. (For groups of springs the aggregate flow should be taken.)

Mixing with cold water -- Mixing with cold water is very common in hot spring systems. Cold waters can be identified by isotope compositions falling on the meteoric water line (i.e. without an oxygen isotope shift) and are usually local to the spring area while deep recharge can come from a long distance. Mixing may occur before or after boiling and both possibilities should be considered in the interpretation. Variations of chloride and $\delta^{18}O$ greater than 40% usually indicate mixing (Why?).

Boiling and steam loss -- Boiling and steam loss are the most important cooling processes in high temperature geothermal systems with large surface dischanges. This is because the amount of heat is greater than can be lost by conduction, and because mineral deposition seals upflow paths from invading cold waters. We will use our knowledge of water-steam isotope fractionation to calculate the isotopic changes in hot spring waters that lose heat by boiling during ascent from a reservoir at 260°C to a 100°C hot spring flowing at 5 ℓ /sec. The boiling point to depth curve can be approximated (Russell James, unpublished report: NZDSIR) as

$$t(°C) = 69.56\ h^{(1/4.7962)} \qquad (h = \text{meters}) \qquad (12)$$

If we assume boiling throughout, the reservoir depth is 550m and our spring (5 ℓ/sec) cannot be much cooled by conduction (about how much ?). The following data for α water-steam separation are from Truesdell et al. (1977).

Table 10.1 -- Oxygen 18 and Deuterium fractionation factors, $10^3 \ln\alpha$, and values for Integrals, I, for Oxygen 18 and Deuterium, and, I_n, for the water fraction in Continuous Steam Separation (for definitions of I see equations (24) and (27) and Truesdell et al., 1977).

	Oxygen 18		Deuterium		
t°C	$10^3 \ln\alpha$	I	$10^3 \ln\alpha$	I	I_n
0	11.20	0.000	106.0	0.000	0.000
20	9.54	0.349	81.5	3.021	0.034
40	8.17	0.651	61.3	5.377	0.068
60	7.03	0.916	46.4	7.197	0.103
80	6.07	1.150	36.1	8.640	0.139
100	5.24	1.357	27.8	9.792	0.176
120	4.53	1.542	21.5	10.710	0.214
140	3.91	1.707	16.3	11.443	0.253
160	3.37	1.855	11.7	12.008	0.294
180	2.90	1.989	7.4	12.414	0.337
200	2.48	2.110	3.5	12.654	0.382
220	2.10	2.219	0.1	12.739	0.430
240	1.77	2.319	-2.2	12.680	0.482
260	1.46	2.410	-3.6	12.509	0.538
280	1.19	2.494	-4.0	12.261	0.602
300	0.94	2.571	-3.4	11.986	0.675
320	0.70	2.644	-2.2	11.735	0.764
340	0.45	2.710	-1.3	11.536	0.881
360	0.19	2.769	-0.5	11.374	1.073
374	0.00	2.805	0.0	11.279	2.018

Boiling and steam loss (as you learned from considering gas depletions in Chapter 4) have different effects according to the mechanism of steam loss. If steam stays with ascending water and separates at one temperature (usually at the surface), maximum isotope effects are found. This is single stage separation. At the other extreme, the process is continuous steam separation where steam separates continuously as formed. With this process, minimum isotope effects are found. The effects of multistage separation are intermediate.

Let us first calculate the effects of single stage separation at the spring temperature (100°C). We will assume that the isotope composition of our 260°C reservoir water is $\delta^{18}O$ = -5 and δD = -40. From the tables

$$10^3 \ln\alpha \ (100°C) = 5.24 \ (^{18}O) \text{ and } 27.8 \ (D)$$

Write enthalpy and isotope balances

$$h_l 260 = y\, h_v 100 + (1 - y)\, h_l 100 \qquad (13)$$

$$\delta_l 260 = y\, \delta_v 100 + (1 - y)\, \delta_l 100 \qquad (14)$$

Use the steam tables to calculate y.

Combine the isotope balance equation with the equation derived earlier,

$$10^3 \ln\alpha \simeq \delta_l - \delta_v \qquad (15)$$

to obtain

$$\delta_l 100 \simeq \delta_l 260 + y\, 10^3 \ln\alpha \qquad (16)$$

From the values of y and $10^3 \ln \alpha$ we have

$$\delta^{18}O_l 100 = -3.33 \quad \text{and} \quad \delta D_l 100 = -31.3$$

Calculate the composition of the escaping steam. Can you fill in the isotope balance equation?

Now calculate the effects of multistage steam separation using 40°C stages.

$ Set up a table -

Temp.	y	$10^3 \ln \alpha$		water composition	
		^{18}O	D	$\delta^{18}O$	δD
260	-	-	-	-5	-40
220					
180					
140					
100					

At each stage we can use the single stage equation and the water composition from the previous stage. Fill in the table. From this calculation we find that the hot spring water has $\delta^{18}O = -3.81$ and $\delta D = -35.9$.

An equation for continuous steam separation was given by Truesdell et al. (1977). When a small amount $\Delta \underline{m}$ of steam is produced from water of mass $\underline{m}$, we can write equations for mass and isotope balances.

$$\underline{m} H_l(\underline{m}) = (\underline{m} - \Delta \underline{m}) H_l(\underline{m} - \Delta \underline{m}) + \Delta \underline{m} H_v(\underline{m} - \Delta \underline{m}) \quad (17)$$

$$\underline{m} \ \delta_l(\underline{m}) = (\underline{m} - \Delta \underline{m}) \ \delta_l(\underline{m} - \Delta \underline{m}) + \Delta \underline{m} \ \ \delta_v(\underline{m} - \Delta \underline{m}) \quad (18)$$

Rearranging and taking the limit as $\Delta \underline{m}$ goes to zero,

$$\underline{m}(dH_l/d\underline{m}) = H_l - H_v \quad (19)$$

$$\underline{m}(d\delta_l/d\underline{m}) = \delta_l - \delta_v \quad (20)$$

combining these equations

$$(d\delta_l/dH_l) = (\delta_l - \delta_v)/(H_l - H_v) \quad (21)$$

and using equation (5)

$$(10^3 + \delta_l)/(10^3 + \delta_v) = \alpha$$

to substitute for δ_v, we can integrate from initial conditions (i) to final conditions (f),

$$\ln \frac{10^3 + \delta_l f}{10^3 + \delta_l i} = \int_{H_l i}^{H_l f} \frac{1/\alpha - 1}{H_v - H_l} dH_l \quad (22)$$

Defining I as the integral from a reference state (r) to the specified state (s),

$$I = \int_{H_l r}^{H_l s} \frac{10^3(1 - 1/\alpha)dH_l}{(H_v - H_l)} \quad (23)$$

then

$$(10^3 + \delta_l f)/(10^3 + \delta_l i) = \exp[(I_i - I_f)/10^3] \quad (24)$$

This integral has been solved numerically and is given in Table 10.1.

For our example

$$\frac{10^3 - \delta_l 100}{10^3 + \delta_l 260} = \exp[(I_{260} - I_{100})/10^3] \quad (25)$$

which can be approximated by

$$\delta_l 100 - \delta_l 260 \simeq I_{260} - I_{100} \quad (26)$$

Values of I are given in the data table:

$$I_{260} = 2.41 \ (^{18}O) \text{ and } 12.51 \ (D)$$

$$I_{100} = 1.36 \ (^{18}O) \text{ and } 9.79 \ (D)$$

and we calculate that with continuous steam separation the hot spring water has $\delta^{18}O = -3.95$ and $\delta D = -37.3$.

If we had used 20°C steps in our multistage calculation, the resulting composition would have been $\delta^{18}O = -3.88$ and $\delta D = -36.6$.

Chloride concentration changes from boiling with continuous steam separation can be calculated using water fractions, x, from

$$x\ (t_1 \rightarrow t_2) = \exp\ (In,t_f - In,t_i) \quad (27)$$

Values for the integral In are given in Table 10.1.

For boiling from 200 to 100°C single stage and continuous steam separation result in Cl concentration increases of 1.238 times and 1.229 times respectively.

> Would the multistage calculation give the same results as the continuous calculation if the steps were vanishingly small?
> Why do multistage and continuous steam separations give smaller isotopic and chemical changes than single stage processes?

Steam-heated waters -- Steam-heated waters are common in the surficial zone of geothermal systems where boiling occurs in the reservoir. They are characterized by high sulfate to chloride ratios and local advanced argillic alteration where the pH is low. These waters originate by the condensation by groundwater of steam rising from the underlying reservoir. Since steam may be adsorbed at temperatures above surface boiling by deep groundwater which then rises toward the surface, the resultant steam-heated water may itself boil and lose mass relative to that of the reservoir steam and original groundwater. A steady state model can be developed to calculate the isotope composition of the steam-heated water.

For example, consider a shallow aquifer through which cold, 10°C, groundwater flows at 1 kg/unit time. At some point the groundwater encounters an upflow of 240°C steam whose flux is ψ kg/unit time. Interaction of the two phases produces a steam-heated water which boils near the water table at 110°C to give a steam flux ω kg/unit time and a stable isotope enriched steam-heated water. Heat and mass balances require that

$$(H_l 110 - H_l 10) = \psi (H_v 240 - H_l 110) + \omega (H_l 110 - H_v 110) \quad (28)$$

If ψ = 0.95 kg/unit time, calculate the mass fluxes of the 110°C water using equation (28).

The isotope balance equations for deuterium and oxygen are similar;

$$(\delta_l 110 - \delta_l 10) = \psi (\delta_v 240 - \delta_l 110) + \omega (\delta_l 110 - \delta_v 110) \quad (29)$$

$\delta_l 10$ is obtained by analysing local groundwaters and $\delta_v 240$ may be calculated from the δ value of the reservoir fluid. ($\delta_l 110 - \delta_v 110$) = $10^3 \ln \alpha_{110}$ which is obtained from tabulated data (Table 10.1).

Rearrange equation (29) and obtain $\delta_l 110$ by substituting the ω and ψ values obtained above and the following values;

	$\delta^{18}O$	δD
$\delta_l 10°$	- 7.4	- 46
$\delta_v 240°$	- 6.8	- 43

$ Is the steam-heated water enriched or depleted in ^{18}O and D with respect to the fresh groundwater?

What isotopic trend would you expect to see if fresh groundwater dilutes the steam-heated water derived from the above process?

$ What would be the effect of increasing the flux of steam from the reservoir?

This heat and mass balance model was used by Henley and Stewart (1983) to help in evaluating changes in the shallow parts of the Tauhara geothermal system due to the drop in pressure resulting from the exploitation of the adjacent Wairakei field. Tauhara was the system whose original hydrology you investigated in Chapter 2. At Tauhara, as at Karapiti (see Plate 2), a pressure drop in the reservoir led to an increased steam flux to the surface and this in turn to measurable increases in surface heat flow. In this case a heat flow of 340 MW_H in steaming ground areas could be related, by simultaneously solving equations (28) and (29), to heat and mass flows of reservoir steam of 414 MW_H and 148 kg/sec, respectively.

Similar isotope enrichments are found in the steam condensate formed in and around fumarolic vents. Giggenbach and Stewart(1982) have suggested some ways in which their isotope composition may be used prior to exploration drilling to estimate the isotope composition of the underlying reservoir. Lake waters are also isotopically enriched but in their case the heat responsible for the evaporation process is mainly solar.

$ Graphing reservoir processes -- It is very useful in the study of reservoir processes to graph the data and indicate on the graph the direction of expected processes. As an exercise in what you have learned about the behavior of isotopes and dissolved salts in rock-water systems, please show the following on $\delta^{18}O$ - δD graphs (ranges of 0 to -10 in $\delta^{18}O$ and -20 to -70 in δD will be required).

1. A typical 10°C central North Island rainwater with δD = -40 falling on the meteoric water line -- equation (8).

2. A geothermal aquifer water at 300°C originating from (1) but oxygen isotope shifted + 4 per mil.

3. The geothermal aquifer water cooled conductively to 100°C.

4. The composition of this aquifer water after boiling to 200 and 100°C with steam loss at each of these temperatures (2 stage steam separation).

5. The composition of the steam boiling off the water at each of these temperatures.

6. The composition of 80°C steam heated hot spring waters produced by mixing the 10°C cold water of (1) with the 200°C steam.

7. The compositions of 100°C hot spring waters produced by continuous and single stage steam loss from (a) the 300°C aquifer water; (b) a 2 to 1 mixture of the 300°C aquifer water and the 10°C cold water; (c) a 1 to 2 mixture.

On a separate chloride-enthalpy plot, show the above processes. Assume the cold water has 10 ppm Cl and the geothermal aquifer water has 1000 ppm Cl.

ISOTOPIC GEOTHERMOMETERS

As we have seen for oxygen and hydrogen in water and steam, isotope fractionations are generally temperature dependent and, because their rates of equilibration differ widely, provide geothermometers which can indicate temperatures in various parts of a geothermal system. Several conditions should be met in order to use isotope compositions as geothermometers.

1) An element with variable isotope composition (H, O, C, S) must exist in two different compounds in fluid or solid phases of the geothermal reservoir.

2) These different compounds must be or have been in isotope equilibrium and the isotopic fractionation factor must show a measurable change over the expected range of temperatures.

3) The rate of isotope exchange must be sufficiently slow that the compounds do not re-equilibrate significantly between the establishment of equilibrium and the time of analysis; alternately, the amount of re-equilibration should be evaluated.

4) Other processes which could affect the isotopic compositions should not occur or their effects should be evaluated.

In practice the rates of most isotope fractionations are not experimentally known and evaluation of useful reactions is based on field data. Thus temperatures based on carbon isotope distribution between CO_2 and CH_4 are usually 50 - 150°C higher than reservoir temperatures. This is usually interpreted to indicate equilibrium at greater depths and higher temperatures with very slow re-equilibration. Some workers, however, believe that exchange does not take place and that the indicated temperatures are fortuitous. In contrast, oxygen isotope fractionation between CO_2 and liquid water is known to be very rapid and indicates only collection temperatures. Exchange of CO_2 with water vapor is much slower and has been used on samples from steam wells to indicate reservoir temperatures. Examples and discussions of isotope geothermometry have been given by Panichi and Gonfiantini (1976), Giggenbach et al. (1983), and Truesdell and Hulston (1980). Equilibrium fractionation data from these sources for possible fluid geothermometers are given in Figure 10.4.

From both experimental and field studies the best characterized and most useful isotope geothermometer for hot-water systems appears to be the fractionation of oxygen isotopes between dissolved sulfate and water. The application of this geothermometer was discussed in Chapter 3.

The fractionation factor of this exchange varies with temperature (Mizutani and Rafter, 1969) as

$$1000 \ln \alpha_{(SO_4-H_2O)} = 2.88 \times 10^6/T^2 - 4.1 \qquad (30)$$

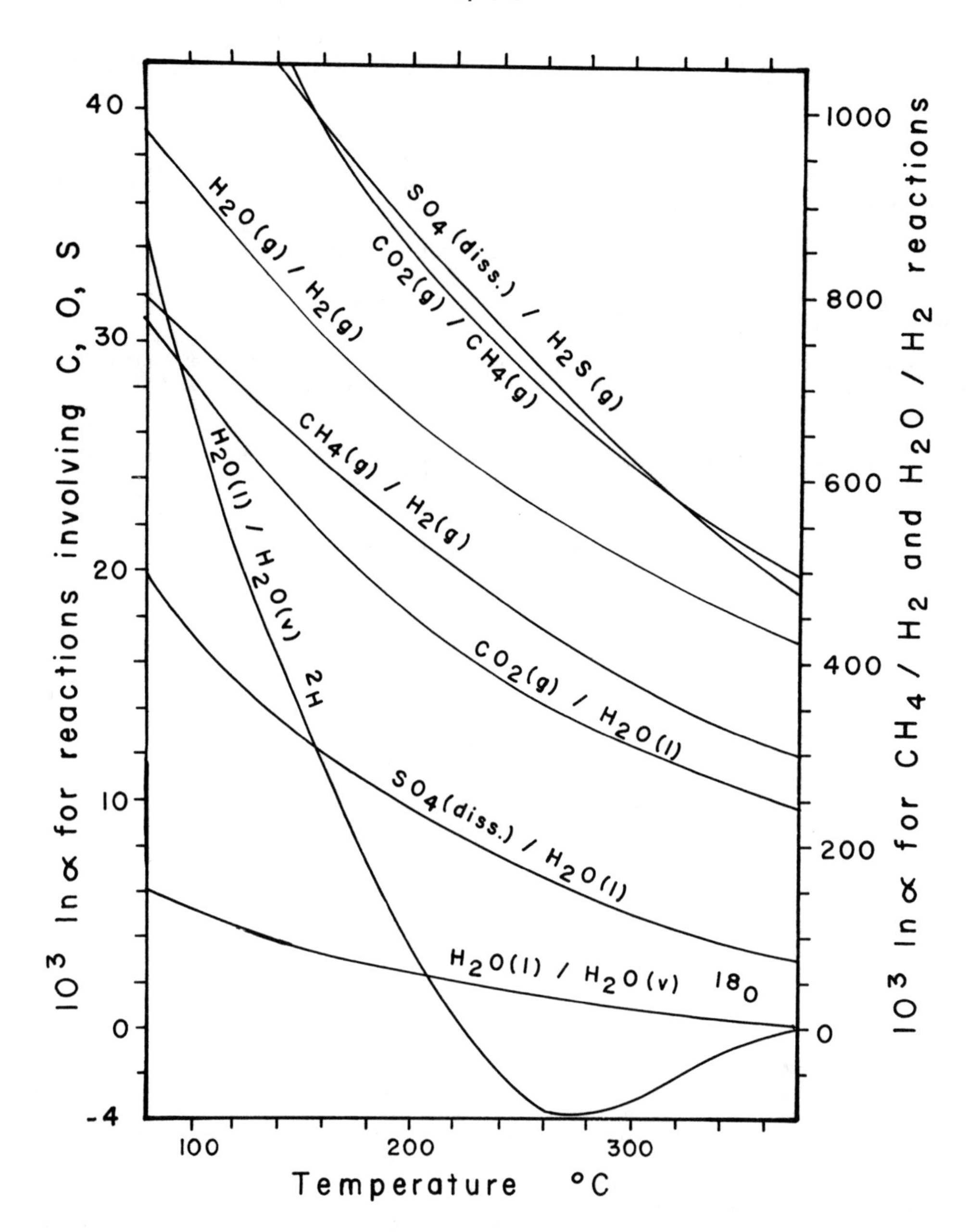

Figure 10.4. Equilibrium fractionation factors for some possible fluid geothermometers. fter Giggenbach et al. (1983) and from data in Friedman and O'Neil (1977).

with T in °K (Figure 10.4). The half time of equilibration varies with the temperature and pH (Lloyd, 1968) as

$$\log t_{1/2} = 2.54\ (10^3/T) + b \tag{31}$$

with $t_{1/2}$ in hours, T in °K and b equal to 0.28 at pH 9, -1.17 at pH 7, and -2.07 at pH 3.8. In time t the fraction equilibrated FE is related to the half time of equilibration by

$$FE = 1 - 0.5^{(t/t_{1/2})} \tag{32}$$

How many half times are required for 97% reaction? How much time is required for 97% reaction in a pH 7 solution at 250°C; at 25°C? Calculate the isotopic composition of 50 mg/kg of dissolved sulfate which equilibrates isotopically with water of $\delta^{18}O$ = -5 at 250°C. What changes in sulfate and water $\delta^{18}O$ would occur if the following processes occurred during ascent to the surface?

$ 1) Without boiling, the pH 7 water entered a shallow reservoir at 175°C and remained for 5 years.

$ 2) The fluid, which was at the boiling point at 250°C, ascends rapidly in a geothermal well and loses steam only at the surface (100°C).

$ 3) After boiling, the water of 2) contains 5 ppm of H_2S. After collection, the H_2S reacts with air and oxidizes to sulfate with $\delta^{18}O$ = +10.

$ 4) What effect would processes described in 3) have on the calculated temperature?

Mineral Isotope Geothermometers

In ore deposit and geothermal studies it is useful to know the geochemical history of a system. To this end fluids may be unavailable, available in too small quantities, or have re-equilibrated to present conditions. Earlier we discussed mineral equilibria and the calculation of fluid compositions from mineral associations. Much the same thing can be done for isotope compositions of fluid and for isotope temperatures. The same conditions must be met as for fluid isotope geothermometers with the additions:

1) The mineral phases must have been deposited from the same fluid and under the same conditions (which usually means at the same time).

2) The compound(s) in the fluid containing the isotopic element should be known and the fractionations with mineral phases included in calculations of the source of the isotopes.

The problem of re-equilibration during cooling is not as severe with solid phases because it apparently requires recrystallization. The interrelation of solution chemistry and isotope fractionation in the determination of isotope sources has been discussed by Sakai (1968) and Ohmoto and Rye (1979).

REFERENCES

Craig, H., 1963, The isotopic geochemistry of water and carbon in geothermal areas; in Tongiorgi, E. (ed.), Nuclear Geology in Geothermal Areas: Pisa, Italy, Consiglio Nazionale delle Ricerche, Laboratorio de Geologia Nucleare, p. 17-53.

Giggenbach, W. F., Gonfiantini, R., and Panichi, C., 1983, Guidebook on Nuclear Techniques in Hydrology: International Atomic Energy Agency, Technical Report 91.

Giggenbach, W.F., and Stewart, M.K., 1982, Processes controlling the isotopic composition of steam and water discharges from steam vents and steam-heated pools in geothermal areas: Geothermics, v. 11, p. 71-80.

Henley, R.W., and Stewart, M.K., 1983, Chemical and isotopic changes in the hydrology of the Tauhara geothermal field due to exploitation at Wairakei: Journal of Volcanology and Geothermal Research, v. 15, p. 285-314.

Lloyd, R. M., 1968, Oxygen isotope behavior in the sulfate-water system: Journal of Geophysical Research, v. 73, p. 6099-6110.

Mizutani, Y., and Rafter, T. A., 1969, Oxygen isotope composition of sulfates - 3. Oxygen isotope fractionation in the bisulfate ion-water system: New Zealand Journal of Science, v. 12, p. 54-59.

Ohmoto, H., and Rye, R. O., 1979, Isotopes of sulfur and carbon: in Barnes, H. L., ed., Geochemistry of hydrothermal ore deposits: New York, Wiley and Sons, p. 509-567.

Panichi, C., and Gonfiantini, R., 1976, Environmental isotopes in geothermal studies: Proc. of Symp. on Inter. sobre Euergia Geotermia en America Latina Guatawala City, Oct. 1976, p. 29-70.

Sakai, H., 1968, Isotopic properties of sulfur compounds in hydrothermal processes: Geochemical Journal, v. 2, p. 29-49.

Truesdell, A. H., and Hulston, J. R., 1980, Isotopic evidence on environments of geothermal systems; Chapter 5 in Handbook of Environmental Isotope Geochemistry, v. 1, The terrestrial environment: P. Fritz and J. Ch. Fontes, eds., Elsevier, Amsterdam, p. 179-226.

Truesdell, A. H., Nathenson, M., and Rye, R. O., 1977, The effects of subsurface boiling and dilution on the isotopic compositions of Yellowstone thermal waters: Journal of Geophysical Research, v. 82, p. 3694-3704.

Chapter 11
AQUIFER BOILING AND EXCESS ENTHALPY WELLS

Analytical data presented in earlier chapters were used to calculate the single phase composition of aquifer fluid through a knowledge of the discharge enthalpy of the well. The assumptions made in this calculation are a) that a single phase is indeed present in the feed zone to the well and b) that no phase separation occurs underground to modify the original fluid composition. While valid for many wells, these assumptions are not universally true. In this chapter we shall examine the effects of underground boiling on the composition of well discharges in relation to original aquifer fluid. These effects are frequently seen in exploration well discharges but become common in production wells as a consequence of exploitation.

NORMAL AND EXCESS ENTHALPY WELLS

The following examples provide a comparison between discharges in which the above assumption may be either true or false.

Table 11.1 — Water and steam compositions for Tauhara, Mokai, and Broadlands well discharges.

Well	Water sample SP	H	Na	K	Ca	SiO_2	Cl	Steam sample SP	CO_2	H_2S
	b.a.	J/gm			mg/kg			b.a.	mmoles/100 moles steam	
TH1	1	1120	1275	223	14	726	2222	9.8	112.3	7.7
BR25	1	1390	934	187	0.8	1069	1355	38.4	2704	25.9
MK3	1	1684	1590	399	12	972	3067	5.9	270	11.4

SP = Sample separation pressure; H = Discharge Enthalpy

1. Calculate the total discharge compositions of the three wells represented in Table 11.1.

2. Calculate the quartz and NaCaK geothermometer temperatures for the wells.

For the Tauhara well (TH1) the silica and NaKCa temperatures* are close to each other. At these estimated temperatures the enthalpy of steam saturated water is also close to the measured enthalpy of the discharge. In this case it is quite reasonable to assume that the well draws on unmodified aquifer fluid. Downhole temperatures, if measured, are also close to the geothermometer temperatures.

* In discussing the relation of fluid enthalpy to measured downhole and geothermometer temperatures, it is useful to define the enthalpy temperature, t_H, as the temperature of liquid water with the observed enthalpy and H_{NaKCa}, etc., as the enthalpy of liquid water with the indicated or observed temperature. Geothermometer temperatures are here abbreviated t_{NaKCa} and t_{quartz}. Discharge enthalpies are similarly abbreviated with the subscripts NaKCa, quartz referring to the enthalpy of steam saturated water at the respective geothermometer temperature. The subscript TD refers to the measured total discharge enthalpy of the well.

For the Broadlands well BR25, $t_{NaKCa} > t_{quartz}$ and $H_{TD} \simeq H_{NaKCa} > H_{quartz}$. In this case no additional heat or steam appears to have been incorporated into the well discharge even though some underground boiling has resulted in the drop in t_{quartz} relative to the reservoir temperature given by t_{NaKCa}.

For the Mokai well, $t_{NaKCa} >> t_{qz}$ and $H_{TD} > H_{NaKCa} > H_{quartz}$. These conditions are characteristic of 'excess' or 'high' enthalpy wells and result from reservoir boiling with preferential steam flow to the well and, in some cases, the gain of heat from reservoir rocks.

> In some discharges
>
> $$t_{qz} < t_{NaKCa} \text{ but } H_{TD} \simeq H_{quartz} < H_{NaKCa}$$
>
> These would be termed 'low enthalpy' discharges and the well itself, a low enthalpy well. Such discharges are not common, but may occur where multiple feed zones intersect the well or where exploitation has led to inflow of relatively cold diluting water. To complete the terminology, well-behaved discharges like TH1 may be designated 'normal' enthalpy discharges.

ORIGIN OF EXCESS ENTHALPY DISCHARGES

Interpretation of excess enthalpy well discharges is difficult, especially for exploration wells in new fields like Mokai, New Zealand (e.g. well MK3, Table 11.1) where no other data are available for undisturbed reservoir compositions. In order to examine some of the interpretation problems we shall develop a synthetic example.

Table 11.2 -- Total Discharge composition from a typical well at Wairakei.

Enthalpy J/gm	Na	K	mg/kg Ca	SiO_2	Cl	mmoles/100 moles CO_2	H_2S
1085	880	140	10	472	1558	5.5	.26

First check geothermometer temperatures and discharge enthalpy to characterize the well as a high, normal, or low enthalpy well.

If this composition represents a single phase reservoir fluid, what happens to the total discharge composition if boiling occurs in the reservoir accompanied by phase separation underground? This might occur if residual liquid flows, due to gravity, to other wells drilled to a level beneath the well-feed.

To answer this question we shall proceed through a series of calculations to simulate reservoir boiling and phase separation (or mixing), ignoring initially the effects of heat transfer from reservoir rocks.

Calculate the composition of the liquid and steam following boiling with 1) 0% 2) 5% and 3) 10% steam separation (assume a single step). The resulting data are provided in Table 11.3.

Table 11.3

Steam Separation	liquid phase Enthalpy J/gm	Cl (mg/kg)	SiO_2	vapor phase Enthalpy J/gm	CO_2	H_2S (mmoles/100 moles)	CO_2/H_2S
1) 0	1085	1558	472	2802	---	---	21.2 (liquid)
2) 5	995	1640	497	2804	99.7	3.93	25.3
3) 10	895	1731	524	2796	53.4		

What are the fluid temperatures resulting from these boiling processes at 1) t = 250° 2) t = 231° 3) t = ?

Complete Table 11.3 by calculating the remaining gas data for case 3)

Calculate the composition of a well discharge made up of a) 90% of the liquid and 10% of the steam derived from process 2) above, and then b) 80% liquid plus 20% steam from process 2). Note that this mixing involves phases both at 231°C.

The total discharge compositions resulting from a) and b) are given in Table 11.4.

Table 11.4 Total Discharge

	% liquid	% steam	Discharge enthalpy J/gm	mg/kg Cl	mg/kg SiO_2	millimoles/100 moles CO_2	millimoles/100 moles H_2S	CO_2/H_2S
	100	0	1085	1558	472	----	----	----
a)	90	10	1176	1476	447	9.97	.393	25.3
b)	80	20	1357	1312	398	19.94	.786	25.3

3. Calculate the composition of water and (assuming all gas is separated into the vapor) of the steam separated from these discharges at 1 b.a. You should then complete Table 11.5.

Table 11.5

	Discharge enthalpy J/gm	y	liquid Cl (mg/kg)	liquid SiO_2 (mg/kg)	vapor CO_2 (mmoles/100 moles)	vapor H_2S (mmoles/100 moles)	CO_2/H_2S
original aquifer fluid	1085	.295	2210	670	18.64	.88	21.2
a) + 10% excess steam	1176	.335	2221	673	29.76	1.12	25.3
b) + 20% excess steam	1357	.416	2245	681	47.93		

The data in Table 11.5 show that the presence of the additional steam leads to slightly higher solute concentrations in the weirbox and much higher gas contents in the steam samples.

4. Next complete the following geothermometry table (Table 11.6).

Table 11.6

	$t_{qz,ad}$ °C	t_{NaKCa}* °C	t_H °C	$t_{"meas."}$ downhole, °C
Original aquifer fluid	251	251	250	250
a) Calc. from weirbox (10% excess steam)	252		268	231
b) " " " (20% excess steam)	253		302	231

*Na, K, and Ca in these fluids will be proportional to Cl.

The compositions and discharge enthalpies that we have calculated fulfill our earlier conditions for the recognition of an excess enthalpy discharge, $H_{TD} > H_{NaKCa} > H_{quartz}$, whereas the fluid we started with was characterised by $H_{TD} \simeq H_{quartz} \simeq H_{NaKCa}$.

5. Finally consider a) the response of the geothermometers to changing local reservoir temperature, and b) the buffering effects of heat stored in wallrocks of the feedzone.

a) The silica geothermometer responds very rapidly to changing temperature while the NaKCa thermometer responds much more slowly. For our synthetic example t_{QA} would trend toward 231°C where the liquid phase silica content is 375 mg/kg (Chapter 3). The silica content of the total discharge is then

$$SiO_{2,TD} = 375 \text{ times } x \qquad (1)$$

where the water fraction, x, is 1.0, 0.9, 0.8 for the cases under consideration. (Note that x here refers to aquifer conditions, not to surface separation.

The silica content of the weirbox water derived from this discharge by flashing at atmospheric pressure (1 bar abs) and the quartz-adiabatic geothermometer are given in Table 11.7.

Table 11.7

	Total Discharge mg/kg	SiO_2 mg/kg (weirbox)	t_{QA}
a)	472	670	251
b)	338	508	230
c)	300	514	231

Notice that although t_{QA} is the local boiling temperature in the reservoir there is a small, but detectable apparent temperature increase as the excess enthalpy increases. If the deep steam component was 40%, the weirbox silica content would become 531 mg/kg and t_{QA} = 233°C.

If the NaKCa temperature is unchanged, then for the new discharge, we have $H_{TD} > H_{NaKCa} > H_{quartz}$. Since extra aquifer steam is present in the discharge its gas content is high relative to that of the original reservoir fluid.

b) Fluid which boils in an aquifer and migrates towards a discharging well may receive additional heat by conduction from the (higher temperature) host rocks. Where fissure flow dominates this effect is short lived, since heat is rapidly removed from the immediate wall rock, but where flow is more intimately related to the host rock permeability as at Cerro Prieto (Truesdell et al., 1982), the heat gain may be large until a stabilized boiling front is established with the major part of the excess heat already mined from the zone closest to the well. Figures 11.1 to 11.3 show the geothermometer and enthalpy changes observed at Cerro Prieto. In such cases the same pattern ($t_{TD} > t_{NaKCa} > t_{quartz}$) is observed but by contrast gas contents remain close to that of the original reservoir fluid.

Consider the unmodified Wairakei reservoir composition given in Table 11.2. If no phase separation occurs despite underground boiling, calculate a) the weirbox composition of the discharge if heat gained from rocks leads to a discharge enthalpy of 1800 J/gm; b) the gas content of the total discharge; and c) the gas content of the steam separated at 1 b.a.

INTERPRETING EXCESS ENTHALPY DISCHARGES

Figure 11.4 shows the results of the above calculations. Where some information about reservoir chemistry is available, the cause of an excess enthalpy discharge may be assessed directly from analytical data. For the case of exotic steam addition due to widespread reservoir boiling, the weirbox chloride and total discharge gas contents of excess enthalpy discharges are respectively apparently normal and high relative to normal enthalpy wells. By contrast, where boiling is local to the well and additional heat has been absorbed from reservoir rocks, weirbox chloride is anomaloously high and separated steam is gas poor, but total discharge chloride and gas are normal. Problems arise, of course, where the heat excess is relatively small. In these cases, the behavior of the gas geothermometers may be helpful as discussed below. For discharges affected by general aquifer boiling, but where steam loss has occured, $H_{TD} \approx H_{quartz}$, H_{TD} is less than H_{NaKCa}, and gas contents are low compared to that of the original reservoir fluid.

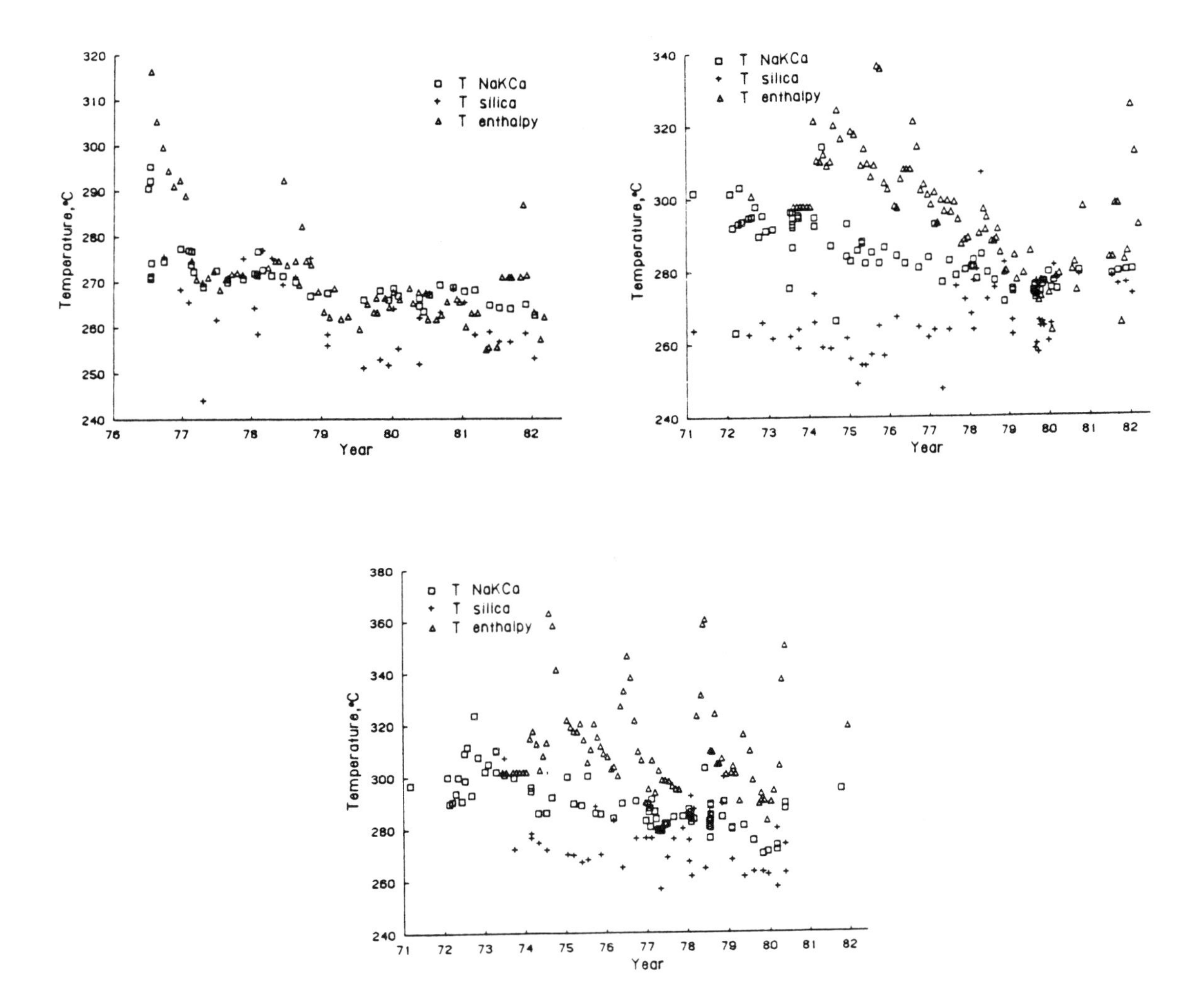

Figure 11.1. Changes with time in NaKCa, silica, and enthalpy temperatures for well CPM14. This well has had a predominantly liquid feed with little near well boiling.

Figure 11.2. Changes with time for fluid temperatures for well CPM31. This well shows near-well boiling with an expanding boiling front which stabilized in 1980.

Figure 11.3. Changes with time for fluid temperatures for well CPM8. The boiling front of this well expanded rapidly when nearby well CPM21A was opened (mid 1974) and again in mid 1976 when well CPM-27 was opened. The opening of CPM46 and diversion of its flow into the CPM8 separator caused similar effects in 1977 as did an earthquake(?) in 1980.

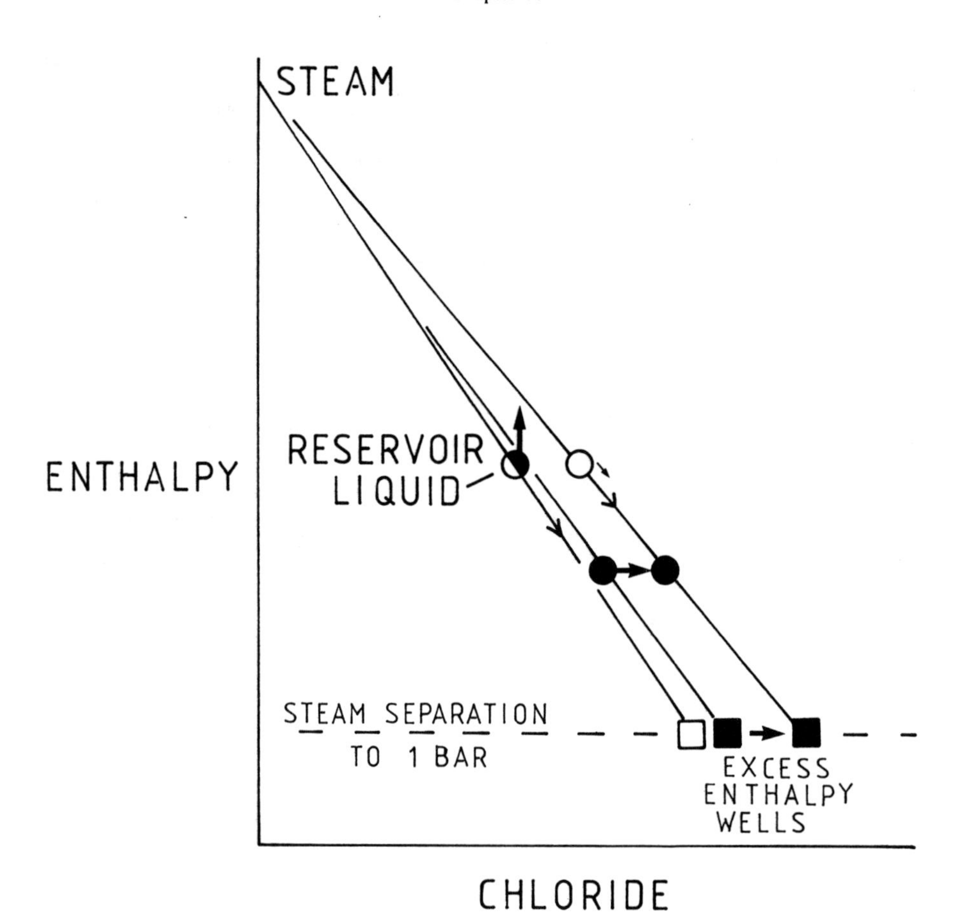

Figure 11.4. Effects of gain of excess enthalpy on the composition of water samples derived from geothermal wells.

Symbols: • - chloride concentration calculated at t_{quartz}

o - chloride concentration calculated through t_{NaKCa}. Arrows indicate effects of reservoir boiling and heat gain from wall rocks. Ultimately the NaKCa geothermometer temperature decreases in response to the temperature of the boiled aquifer fluid. In the later stages of exploitation a) the reverse trends occur as heat is mined from host rocks and b) as dilution catches up with pressure drop. Sketch these trends on the diagram.

> Since boiling in the reservoir is a consequence of pressure drop, cool near surface waters may also flow toward the discharge zone. Sketch the effect of this dilution process on Figure 11.4.
>
> If an initial reservoir fluid is close to saturation with respect to calcite, what effects might you anticipate on the NaKCa geothermometer due to underground boiling?

Clearly, a complete interpretation of reservoir fluid composition from isolated excess enthalpy discharges is impossible, but where reservoir composition is known from earlier discharge data from neighboring exploration or production wells, the processes responsible for the excess enthalpy condition may be identified. In an early exploration stage, as at Mokai (Table 11.1), no such reference reservoir composition is known but some information can be obtained from changes in discharge composition through time provided that environmental or engineering limitations on the allowable period of discharge are not too severe.

A single stage boiling model can be adopted to interpret excess enthalpy discharges using the silica temperature as the reference temperature for calculation; but in assessing such model data it is essential to realize that the calculated temperatures and compositions are artifacts of the model. You may see trends in the apparent aquifer compositions, e.g. from a to c in Figure 11.4, but later reverting to condition a, b or c.

Except where the condition is due only to rock heat, the gas contents of well discharges are most sensitive to excess enthalpy effects (Table 11.4). A correlation of gas content with enthalpy for the early discharge of some wells at Broadlands is shown in Figure 11.5. Can you interpret these trends? Notice that after an initial increase, the well usually trends toward the original reservoir enthalpy - unless some dilution of the reservoir has occured in the meantime.

Glover et al. (1980) have shown that in some excess enthalpy wells a simple relationship exists between total discharge chloride and CO_2 contents, but this is not common.

The excess enthalpy condition may be a long or short term transient effect in the history of a given well or geothermal field. The Cerro Prieto data (Figures 11.1-11.3) show the trend to a steady state discharge condition as the boiling zone stabilizes and heat addition from the rock mass stops. At Broadlands, where excess steam with gas enters the discharge, enthalpy-gas relations are sympathetic (Fig. 11.5). This behavior may relate to general boiling due to production induced pressure drop in the feedzone wallrocks. Steam formed early in this process will be relatively high in gas compared to that formed later. Why?

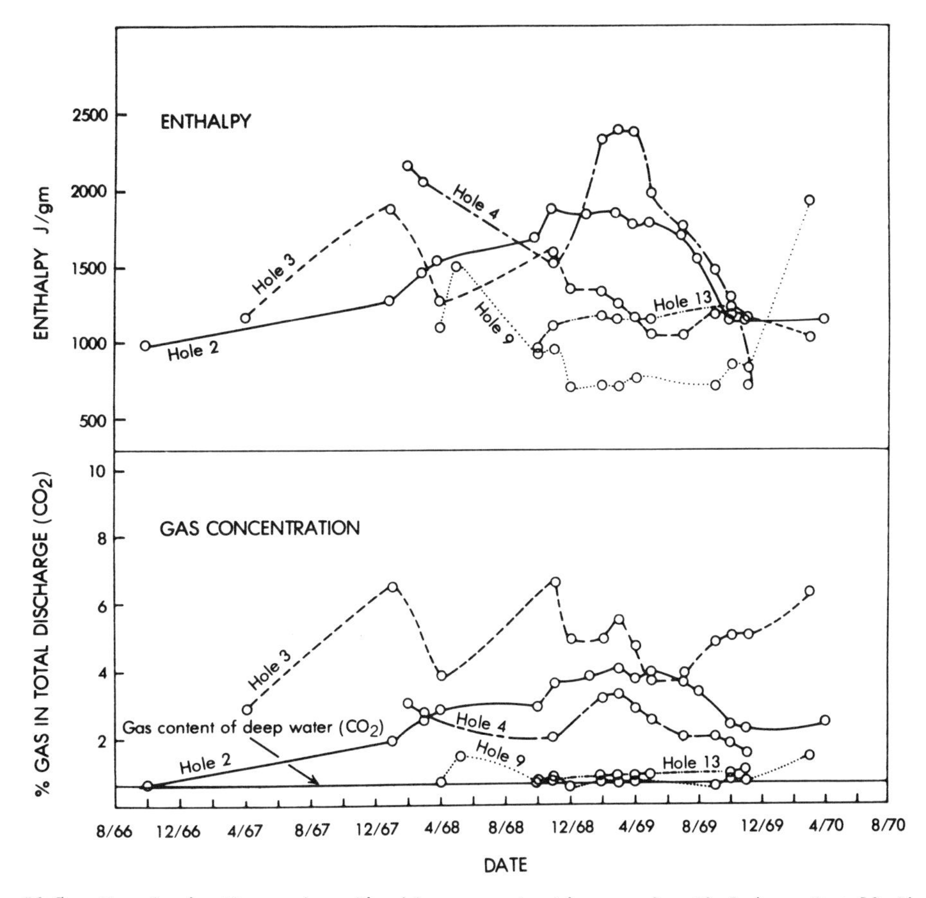

Figure 11.5. Trends in the carbon dioxide concentrations and enthalpies of well discharges at Broadlands between 1966 and 1970. (Modified from Mahon and Finlayson, 1972)

Multiple Feed Zone Wells

The assumption is made in the above discussion that wells are fed from a single feed point, or feed zone, but since many wells produce from a long interval of open hole or slotted liner, this is obviously not always true. In the majority of cases the assumption does lead to useful results for reservoir modeling, suggesting that a single feed zone does dominate the discharge. In some wells cycling may occur between two or more feed zones (Grant et al., 1980). Other effects occur as a result of extensive field exploitation. At Wairakei during early exploitation, shallow wells tended to evolve rapidly to dry steam production before ceasing to discharge, while in some deeper wells, upper feed zones drew on steam generated by widespread reservoir boiling and their lower zones on liquid water. All these effects can only be recognized by careful interpretation of well data, but they provide essential information on the response of a field to exploitation which cannot be obtained from other physical measurements.

Gas ratios and geothermometry

As we discussed in Chapter 5, the ratios of common geothermal gases may be controlled by reactions such as

$$2NH_3 = N_2 + 3H_2 \quad (2)$$

$$CH_4 + 2H_2O = CO_2 + 4H_2 \quad (3)$$

through their equilibrium constants e.g.

$$K''_C = \frac{X_{d,CO_2} X_{d,H_2}}{X_{d,CH_4}} \quad \text{and} \quad K''_N = \frac{X_{d,N_2} X_{d,H_2}}{X^2_{d,NH_3}} \quad (4, 5)$$

where $X_{d,i}$ are the mole fractions in the total discharge, corrected for incomplete partitioning of the gases between water and steam at the sampling temperature. When discussing these equilibria earlier, values of $x_{d,i}$ calculated from the discharge enthalpy and gas content of separated steam were assumed to represent the original reservoir gas contents. Where wells receive extra steam and gas from general reservoir boiling or conversely where the water tapped by the well has already boiled, this assumption is incorrect. Giggenbach (1980) showed that these effects could be assessed by correcting gas contents using distribution coefficients for the apparent 'reservoir' temperature, t_R, and a single step model. This is clearly difficult unless t_R is obtained by other methods (t_{qz}, T_{NaKCa}).

In Figure 11.6, gas concentration quotients K''_C and K''_N are plotted against t_{quartz} and t_{NaKCa} for a set of geothermal wells. The diagrams are contoured to show the affects on K" or steam gain ($+ y_i$) or loss ($- y_i$) relative to single phase fluid reaching the well. The curve for $y_i = 0$ is an old friend - the gas geothermometry relation which you derived earlier. Surprisingly gas thermometer temperatures seem to fit t_{quartz} more closely than t_{NaKCa} presumably due to relatively fast re-equilibration rates, but in a number of cases steam excess or deficit effects are clearly recognizable e.g. Wairakei 25 and 81.

> Using gas distribution coefficients calculate the total discharge CO_2 and H_2S content of BR25 from the high pressure steam sample analysis (Table 11.1).
>
> Find the BR25 gas quotients in Figure 11.6.

Evaluation of expressions such as K''_C and K''_N based on mole fractions requires a knowledge of the total vapor pressure of the deep system. However, Giggenbach (1980) showed that a combination geothermometer, K_{CN}, avoided the problem of guessing P_{gas} in the deep system where both t and x of the gas were unable to be independently estimated. Figure 11.7 shows a calibration curve ($y_1 = 0$) for this thermometer which is based directly on steam analyses, but with gas concentrations corrected for incomplete partitioning at the sampling temperature.

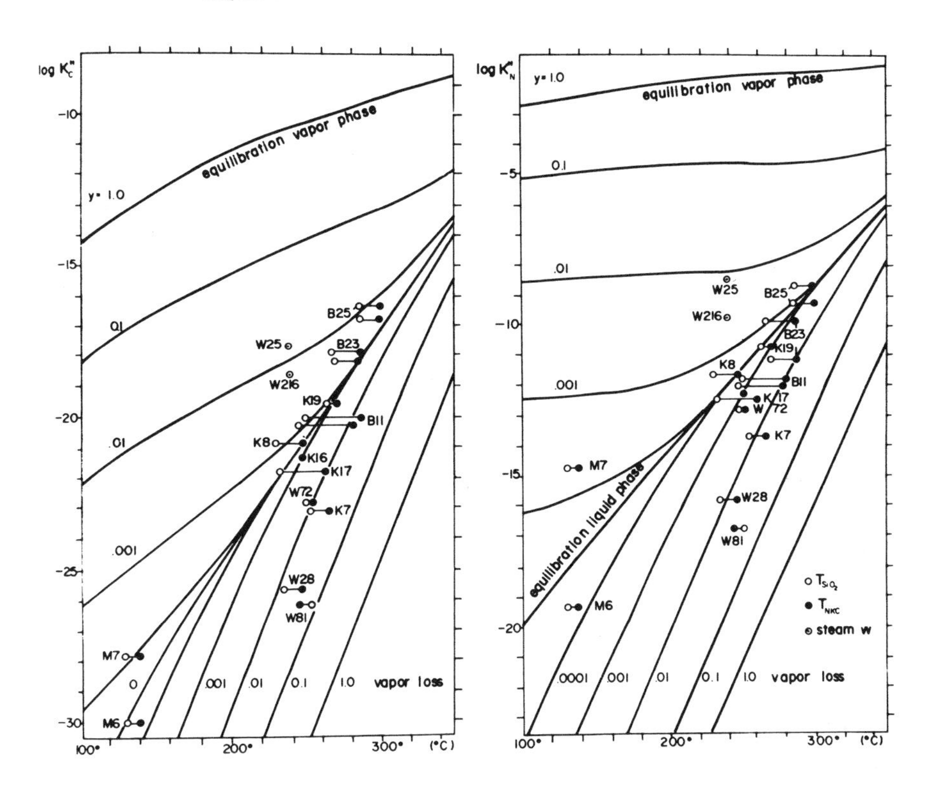

Figure 11.6. Variation of the methane and ammonia concentration quotients, K''_C and K''_N (see text), in relation to the gain or loss of steam underground; y = steam fraction. W = Wairakei, B = Broadlands, K = Kawerau. (From Giggenbach, 1980) The CH_4 and NH_3 geothermometer curves correspond to y = 0.

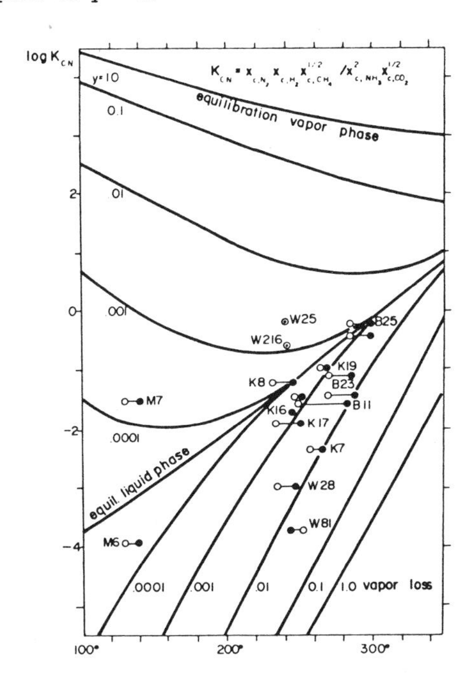

Figure 11.7. The CH_4-NH_3 combination gas geothermometer, K_{CN} (see text) to vapor loss or gain underground. (From Giggenbach, 1980)

Vapor Saturation in Steam Wells in Vapor-dominated Fields

D'Amore and Celati (1983) and D'Amore et al. (1983) have modified the approach of Giggenbach (1980) to use well head steam compositions from vapor-dominated fields to determine the proportions of original reservoir steam to steam produced by vaporization of reservoir liquid caused by exploitation. The application of equilibria for H_2O breakdown to H_2 and O_2 and the CH_4 breakdown reaction described earlier along with mass balance and gas distribution equations (Chapter 1, equation (12); Chapter 4, equations (1) and (2) and Table 4.1) allow the calculation of the original vapor fraction in the reservoir. The basic equations are the equilibrium expressions for the reactions

$$H_2O = H_2 + 1/2\ O_2 \quad , \quad (6)$$

$$CH_4 + 2H_2O = CO_2 + 4H_2 \quad , \quad (7)$$

a set of gas distribution equations, e.g.

$$B_{H_2} = C_{H2,vapor}/C_{H2,liquid} \quad , \quad (8)$$

and mass balance equations, e.g.

$$C_{CO2,wellhead\ steam} = y\ C_{CO2,vapor} + (1-y)\ C_{CO2,liquid} \quad , \quad (9)$$

where y is the original vapor fraction in the reservoir and the concentrations in the vapor and liquid refer to reservoir fluids. The final equation is

$$y = [(\%H_2/100)\ x_g\ (\%CO_2/\%CH_4)^{1/4} P_{H_2O}{}^{1/2}/K_C{}^{1/4} - 1/B,H_2]/(1 - 1/B,H_2) \quad (10)$$

in which the gas mole % refers to the dry gas (without H_2O) steam analyses, x_g is the mole ratio of total gas to steam (H_2O), P_{H2O} is the pressure of steam (see equations below), K_C is the equilibrium constant of equation (7) and B,H_2 is the distribution constant for hydrogen (below and Table 4.1). The derivation of this equation is given by D'Amore and Celati (1983). A similar equation can be written using H_2S concentrations if the action controlling sulfur fugacity is known. D'Amore and Celati (1983) assume that sulfur fugacities may be calculated approximately from the pyrite-magnetite equilibrium and the equilibrium expression for reaction (6).

Temperature functions for the required constants in equation (10) are (from Chapter 4 and 5)

$$\log P_{H2O} = 5.51 - 2048/T\ (°K)$$

$$\log K_C = 10.278 - 9082/T$$

$$\log B,H_2 = 6.2283 - 0.01403\ t\ (°C)$$

Calculator programs (STEAM) by Jill Robinson that solve these equations are given in Appendix I. The constants used in these programs are from D'Amore and Celati (1983) and D'Amore et al. (1982) and differ slightly from those given above.

Three wells at The Geysers have the following steam compositions

Table 11.8

	gas/steam mole (xg)	mole % CO_2	H_2S	H_2	N_2	CH_4	NH_3	Ar
Well A	0.00455	61.0	7.5	14.7	1.0	9.1	6.7	0.006
Well B	0.00727	64.5	4.6	12.8	1.6	12.7	3.8	0.004
Well C	0.00162	54.5	7.4	13.3	0.9	6.6	17.3	0.004

$ Assuming that the reservoir temperature is 250°C, calculate y for each well using equation (10). If the steam reserves are related to the amount of liquid water vaporized in the feed zone, which well has the largest reserve? Using the programs in Appendix I calculate the y values and temperatures using the steam analyses alone.

The values of y you have calculated are the effective steam saturation (steam fraction) in the reservoir or the fraction of the well head steam that was vapor in the reservoir. This number is obviously very useful in estimating reserves since the major part of the reservoir fluid must be water (which is 80 times more dense than steam at 240°C). Do not confuse this y with the steam fraction from boiling processes. Note that these equations (and programs) work equally well for liquid-dominated systems in which y is interpreted as the fraction of reservoir steam in the total discharge.

1. Consider the initial discharge data for well MK3 in Table 11.1 and attempt to model the reservoir composition of this field. Compare your result with the reservoir composition at Wairakei. (The maximum downhole temperature measured at MK3 was 303°C and this was measured several weeks after the initial discharge was terminated).

2. Compare the high enthalpy discharge composition from BR25 (Table 11.1) with other Broadlands well data (Chapter 1, Tables 2 and 3, Chapter 7, Table 1).

REFERENCES

D'Amore, F., and Celati, C., 1983, Methodology for calculating steam quality in geothermal reservoirs: Geothermics, v. 12, no. 2/3, p. 129-140.

D'Amore, F., Celati, R., and Calore, C., 1982, Fluid geochemistry applications in reservoir engineering (vapor-dominated systems): Proceedings 8th Workshop Geothermal Reservoir Engineering, Stanford, Dec. 14-16, 1982, p. 295-308.

Giggenbach, W.F., 1980, Geothermal gas equilibria: Geochimica et Cosmochimica Acta, v. 44, p. 2021-2032.

Glover, R.B., Lovelock, B., and Ruaya, J.R., 1981, A novel way of using gas and enthalpy data: New Zealand Geothermal Workshop, Univ. Auckland, p. 163-170.

Grant, M.A., Bixley, P.F., and Syms, M.C., 1979, Instability in well performance: Geothermal Resources Council, v. 3, p. 275-278.

Mahon, W.A.J., and Finlayson, J.B. 1972, The chemistry of the Broadlands geothermal area, New Zealand: American Journal Science, v. 272, p. 48-68.

Truesdell, A.H., Nehring, N.L., Thompson, J.M., and Janik, C.J., 1982, A Review of Progress in Understanding the Fluid Geochemistry of the Cerro Prieto Geothermal System: 4th Symposium on the Cerro Prieto Geothermal Field, Guadalajara, 1982.

Plate 4. The surface expression of magmatic gases is well illustrated by the acid lakes and fumaroles within the new crater of El Chichon, Mexico. The diameter of the main crater is about 1 km and of the foreground lake about 250 m.

Prior to the 1982 eruptions fumaroles and thermal activity were concentrated in a circular zone about 1.5km in diameter and were associated with acid sulfate(advanced argillic) alteration of host rocks.

The volcano erupted violently three times in March and April, 1982 sending eruption columns of gas and ash into the stratosphere. The SO_2 in the cloud produced a global haze of sulfuric acid droplets which may affect world weather patterns. The eruption produced a new crater within which a lake evolved by condensation of steam and collection of rain water eventually submerging the fumaroles shown in this plate. In January 1983, the lake water was 55°C, acid (pH = 0.56) and high in chloride, sulfate and boron derived from the volcanic gas. Alteration products in the suspended sediment include sulfur, kaolinite, anhydrite and alunite. The erupted material itself contained abundant anhydrite and some sulfides suggesting the presence of an underlying porphyry-type hydrothermal system.

Photo, dated June 2nd, 1982, courtesy of R.I. Tilling, U.S. Geological Survey

Chapter 12
VOLATILES IN MAGMATIC SYSTEMS

The physical and chemical setting of geothermal systems is dominated by waters of surficial origin; nevertheless, the heat sources are believed to be magmas, and there is also a high probability that magmatic fluids contribute heat and some dissolved components (H_2S, SO_2, CO_2, ...) to the modern hydrothermal systems that are tapped for energy. By the same token, epithermal and other fossil hydrothermal deposits may well have received contributions of metals from magmas. Because subsurface zones of magma influence are never directly observed during magmatic activity, the magmatic story is an after-the-fact interpretation of mineral assemblages; and it is an emerging story with many chapters still unwritten. This chapter will develop the basis necessary to deal with the magmatic equilibria responsible for some of the magmatic gases.

For the most part we shall deal with silicic magmas because they are much more commonly associated with both geothermal activity and hydrothermal ore deposits than are mafic ones, but the general calculations described here are relatively insensitive to rock type. It is the minor minerals -- especially the titanium-iron oxides and pyrrhotite -- that are most critical in this discussion.

At magmatic temperatures many species are associated as neutral complexes rather than ionic compounds. So, for example, HCl, NaCl, and KCl are not ionic, but neutral molecules at temperatures above about 500°C or so (Franck, 1956; Quist and Marshall, 1968). Similarly, the sulfurous gases are present as H_2S and SO_2 rather than ionized species. As a consequence, reactions at magmatic temperatures are largely between neutral species. The concentration of a metal, such as iron, is controlled by the relative strength of its complexes with the available anions. Sulfur, for example, is not free to react to form sulfides, unless the reaction involving H_2S or SO_2 is favorable. As temperatures decrease, however, the story changes drastically as species begin to dissociate. H^+, for example, is formed from the dissociation of HCl, and the pH of a chloride solution tends to decrease as temperature falls. Other species such as SO_2 or CO become unstable and will disproportionate. Any of the species may be out of equilibrium with their host rocks and may react, changing both the rock and the fluid.

CALCULATION OF OXYGEN AND SULFUR FUGACITIES

Silicic volcanic rocks contain a number of phenocrystic phases which may be useful in defining the activity of gaseous species. Many of these volcanics contain coexisting magnetite, ilmenite, and pyrrhotite accompanying quartz, feldspar, and various mafic minerals.

Pyrrhotite, $Fe_{1-x}S$, possesses a wide range of solid solution in the Fe-S system. The one-phase field has been calibrated by studies relating the composition to temperature and sulfur fugacity (Toulmin and Barton, 1964; Rau, 1976), and the Gibbs-Duhem equation has been applied to solve for the activity of FeS. Thus simply by determining the composition of a quenched pyrrhotite, and knowing the temperature, we may obtain an accurate measure of the f_{S2} (and thereby any other component in the Fe-S system - e.g., a_{Fe}, a_{FeS}, or a_{FeS2}) in the magma.

The following equation from Toulmin and Barton (1964) gives the activity of S_2 in pyrrhotite (Program FS in Appendix).

$$\log f_S = (70.03 - 85.83\ X)\ (1000/T - 1) + 39.3\ (1 - .9981\ X)^{0.5} - 11.91 \quad (1)$$

$ X is defined as the mole fraction of FeS in the system FeS - S_2. If a pyrrhotite analysis shows 61.5 wt % Fe and 38.5 wt % S, what is X? Calculate log f_{S2} for 750°C.

$ The computation is very sensitive to X; if the analysis for Fe and S were in error by as much as 1 % to yield 61.0 wt % Fe and 39.0 wt % S, what would be the apparent log f_{S2} for 750°C?

This apparently simple procedure has several problems in application to real rocks. First of all, pyrrhotite is present only in very small amounts in most rocks, and separation of enough material that is sufficiently clean for analysis is difficult. Many rocks have deuteric or later oxidation products added to, or altering from, primary pyrrhotite, further complicating the problem of obtaining a valid sample. Natural pyrrhotite also is often altered to a composition other than its initial one, as by sulfidation, by exsolution of chalcopyrite, or by the inversion of the high-temperature pyrrhotite to a complex of low-temperature phases (note the Fe-S phase diagram in Barton and Skinner, 1979). These processes so modify the character of the pyrrhotite that even the method using the x-ray spacing (Toulmin and Barton, 1964) is usually inappliciable. Fortunately, there is another computation strategy based on the Fe-Ti oxides and the f_{O2} - f_{S2} diagram (some aspects of which were discussed in Chapter 8) as developed below.

THE Fe-O-S-SiO_2 SYSTEM

For the purpose of understanding the nature of the magmatic gas sulfur chemistry the f_{O2} - f_{S2} diagram is very informative. We have already shown (in Chapter 8) how to calculate such diagrams. Four isothermic sections for the Fe-O-S-SiO_2 system are shown in Figures 12.1 and 12.2. All of the applications we shall discuss here are related to the magnetite + pyrrhotite assemblage and involve the redox reaction

$$6\ FeS + 4\ O_2 = 2\ Fe_3O_4 + 3\ S_2 \quad (2)$$

$$\log K = 3 \log f_{S_2} + 2 \log a_{Fe_3O_4} - 6 \log a_{FeS} - 4 \log f_{O_2} \quad (3)$$

Now, if some independent measure of f_{O2} could be obtained (provided that temperature can be estimated), one might read the f_{S2} from either Figure 12.1 or from the equilibrium constant given by equation (3). Fortunately, the calibration (by Buddington and Lindsley, 1964) of tie-lines between the magnetite-ulvospinel (Fe_3O_4-Fe_2TiO_4) and ilmenite-hematite ($FeTiO_3$-Fe_2O_3) solid solutions as functions of temperature and f_{O2} provides just the information we need; and the recasting of the data by Spencer and Lindsley (1981) shown in Figure 12.3 enables us to calculate both temperature and f_{O2} from analyses of coexisting magnetite and ilmenite. Stormer (1983) has given a useful discussion concerning the recalculation of analyses to permit one to work around the "impurities" such as Mn, Cr, Mg, Al, or Si in the oxides. Program SL in the appendix maybe used for this calculation.

$ From Figure 12.3 determine the temperature and f_{O2} for a magnetite ($Fe_{27}Ti_3O_{40}$) quenched from an equilibrium with ilmenite ($Fe_{11.1}Ti_{8.9}O_{30}$).

Is fayalite stable under this condition?

$ What is the f_{S2} if pyrrhotite coexists with the magnetite + ilmenite?

This calculation of f_{S2} via f_{O2} is straightforward and avoids many of the difficulties inherent in trying to establish the composition of magmatic pyrrhotite (provided that you have a microprobe).

We may write additional reactions, such as those between S_2 and O_2 to yield SO_2 and SO_3; the former is contoured in Figure 12.2 . It is also desirable to estimate f_{H2S}. The following equilibrium constants of formation from the elements (interpolated from Robie et al., 1979) are needed:

	600°C	700°	800°	900°
Log $K_{f(H_2O)}$	11.933	10.413	9.173	8.141
Log $K_{f(SO_3)}$	18.837	16.008	13.710	11.806
Log $K_{f(SO_2)}$	17.842	15.614	13.803	12.301
Log $K_{f(H_2S)}$	2.844	2.290	1.838	1.462

Figure 12.2 is calculated for a f_{H2O} of 2000 bars. If the f_{H2O} were only 500 bars, how much would the H_2S contours be shifted, and which way would they move? What about f_{SO2}?

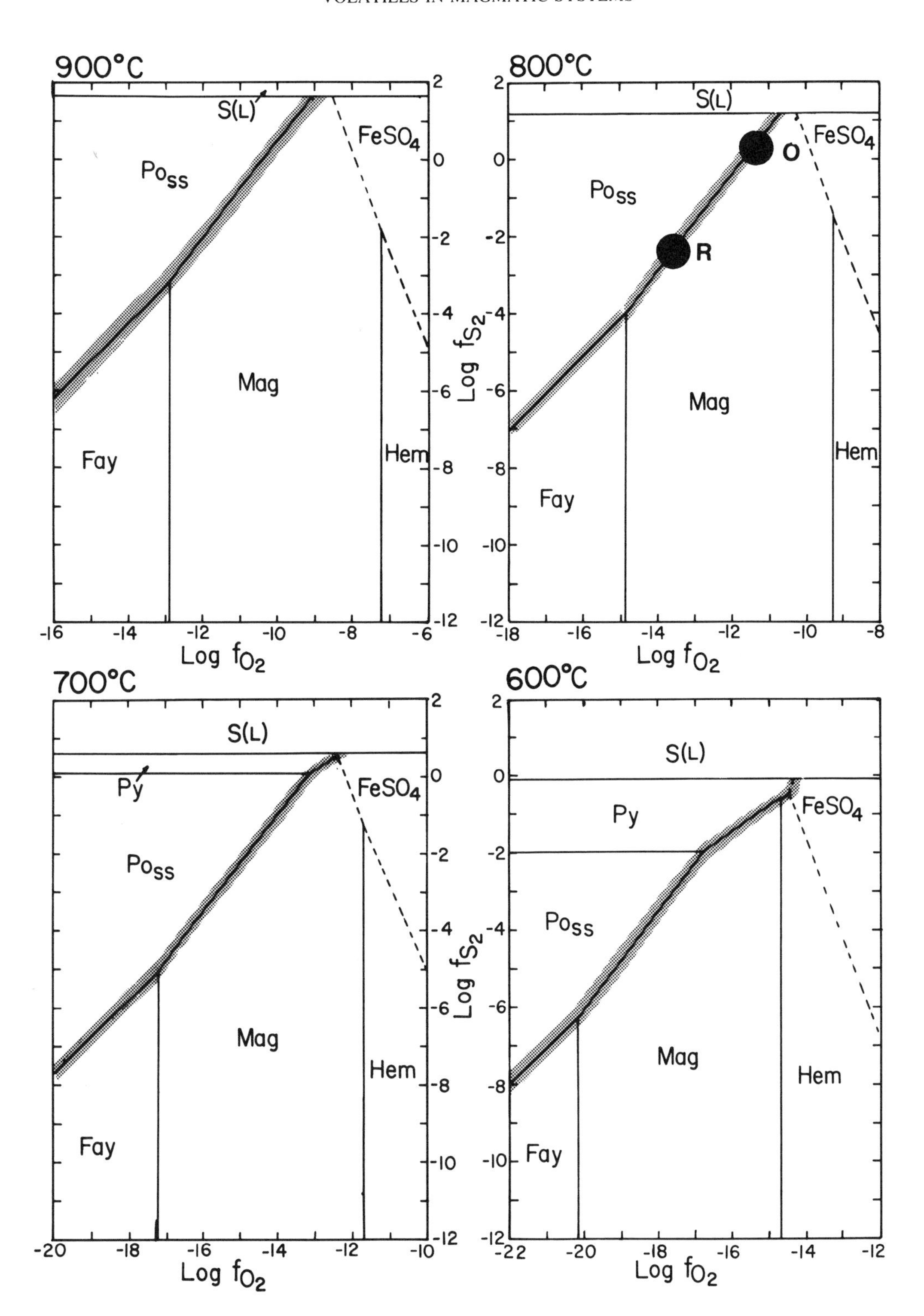

Figure 12.1. Isothermal, quartz saturated, log f_{O_2} diagrams for the system Fe-O_2-S_2-SiO_2. The sulfide saturation surface is shown in gray. S(L) = sulfur liquid, Poss = pyrrhotite solid solution, Fay = fayalite, Mag = magnetite, Hem = hematite, Py = pyrite.

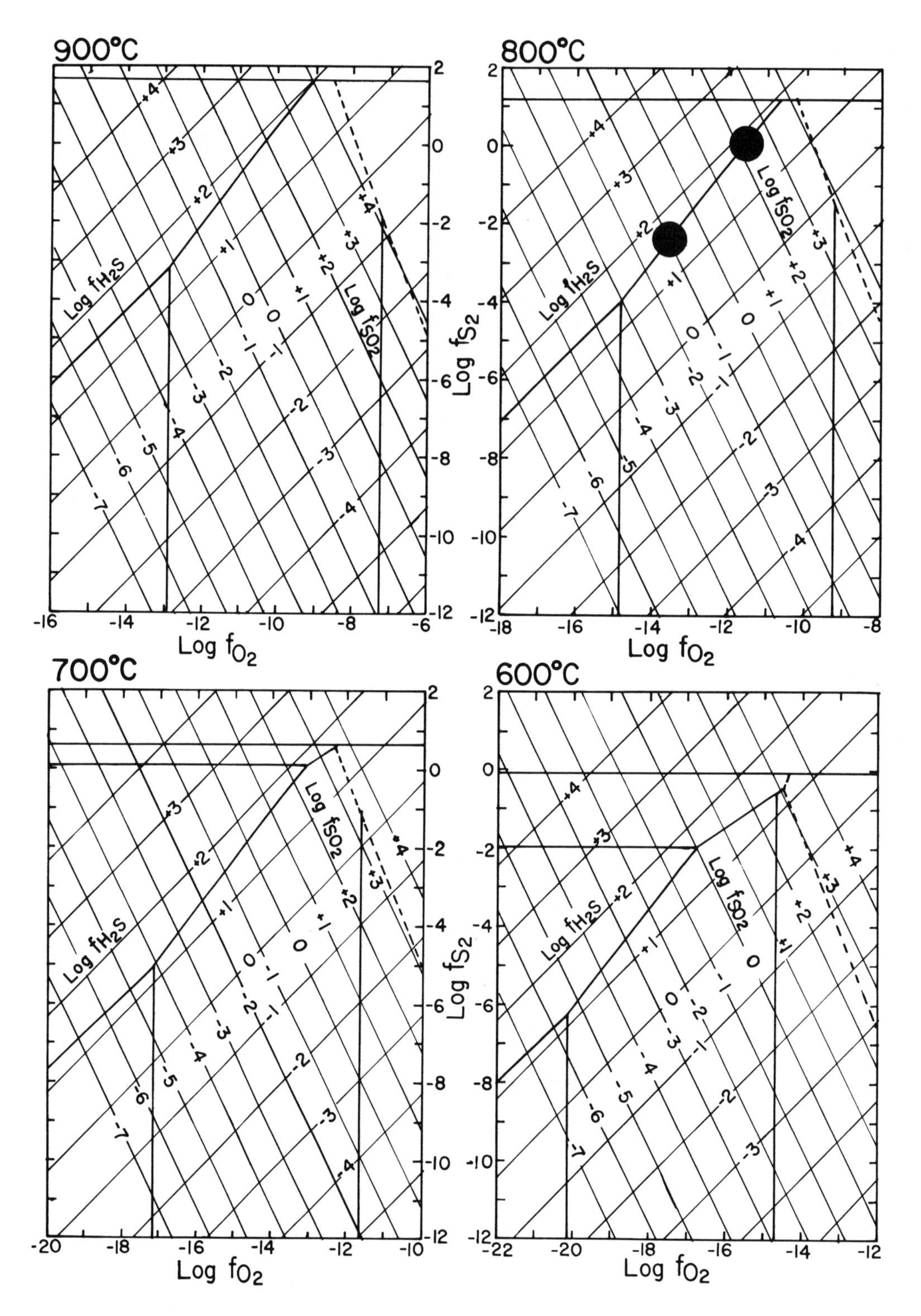

Figure 12.2. Isothermal log f_{O2} versus log f_{S2} diagrams contoured for log f_{SO2} and log f_{H2S}. Stability fields are the same as in Figure 12.1. Fugacity of H_2O is 2000 bars.

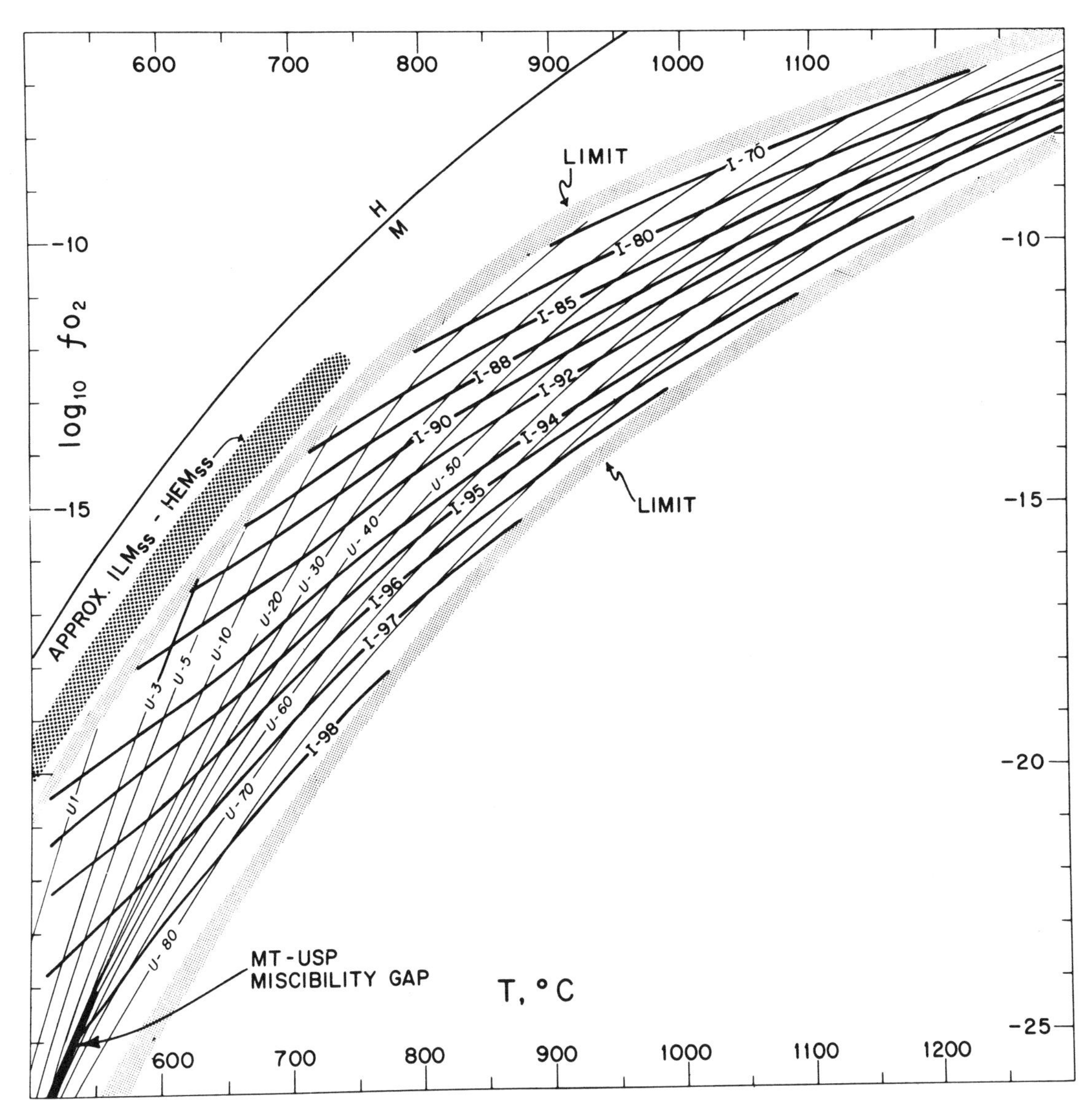

Figure 12.3. Temperature versus log f_{O2} grid for coexisting magnetite and ilmenite solid solution pairs. Reproduced from Spencer and Lindsley, 1981.

Note that of the geologically controlled variables, f_{H2O} has a relatively small range of probable values compared to f_{O2} or f_{S2}, even for a constant temperature.

Calculate and plot the 1 bar and 100 bar isopleths for hydrogen on all isotherms on Figure 12.2.

A large number of additional reactions may be plotted on these coordinates, but let us consider only a few:

(a) Calculate the one bar isopleth for f_{SO3} and plot it on Figure 12.2. Need we be concerned with SO_3 as a species in magmatic gases?

(b) Mueller (1972) has clarified the mixing properties of biotite, and Wones (1972) has provided an equation interrelating the chemical parameters regarding the stability of biotite relative to Kspar + magnetite with emphasis on the Fe end-member annite, $KFe_3AlSi_3O_{10}(OH)_2$: (Program BI in Appendix.)

$$\log f_{H_2O} = 7409/T + 4.25 + .5 \log f_{O_2} + 3 \log X_{Fe} - \log a_{KAlSi_3O_8} - \log a_{Fe\ O} \quad (4)$$

Calculate and plot on Figure 12.1 the annite + magnetite + Kspar boundary for f_{H2O} = 2000 bars at 600°C. (T is in °K and X = 1 for Fe_3 in the biotite formula as written a few lines above.)

If half of the Fe is replaced by Mg (X = 0.5), how much does the curve shift? Note that this equation only accounts for substitution for Fe in the octahedral sites. If other elements are present, such as F- and Cl- substituting for OH-, additional terms are required.

(c) Next we have a buffer system you might mine! The reaction for the sulfidation of bornite (nominally, Cu_5FeS_4) + pyrrhotite to intermediate solid solution (disordered chalcopyrite, nominally $CuFeS_{1.8}$) gives f_{S2} = -6.2 and - 4.1 for 700 and 800°, respectively. Plot the reaction in the 700 and 800°C isotherms in Figure 12.1 Write and plot the reaction where bornite and intermediate solid solution coexist with magnetite instead of pyrrhotite.

COMPOSITION OF A GAS IN EQUILIBRIUM WITH A MAGMA

Let us consider the pyrrhotite + magnetite + ilmenite assemblage at 800°C that we discussed above. The fugacities of gases are: $\log f_{H2O} = 3.0$; $\log f_{O2} = -13.5$; $\log f_{S2} = -2.3$; $\log f_{H2} = + 0.56$; $\log f_{SO2} = -0.9$; $\log f_{H2S} = 1.6$. Figure 12.4 shows the variation of f_{O2} with temperature and also superposes various volcanic systems. Note that the example we are discussing is similar to conditions derived from the St. Helens pumice and well within the $H_2S > SO_2$ field. But also note that there are systems for which $SO_2 > H_2S$, such as the Julcani (Peru) vitrophyres (Figure 12.4).

In order to quickly evaluate the fugacities of sulfurous gases, it is convenient to contour S_2, SO_2, and H_2S on $\log f_{O2}$ - t(°C) diagrams similar to those of Buddington and Lindsley (1964). Figures 12.5, 12.6, and 12.7 are examples of such diagrams for sulfide saturated assemblages containing pyrrhotite or pyrite. Program "SG" (Appendix I) provides a quick means to compute the fugacities of sulphurous gases.

CO_2 may be an important species, but we have no way to evaluate it except to note that in volcanic gases, unoxidized geothermal systems, and ore deposits the CO_2 seems to exceed H_2S by a factor of from 10 to 100. In mafic rocks containing graphite one may calculate f_{CO2} readily, but such magmatic environments are much more reducing than those of interest here. Someday we may have enough data on things like the stability of a carbonate component in apatite or scapolite, or the compositions of fluid inclusions to evaluate the fugacity of CO_2 accurately, but not today. Whitney (in press) discusses the covariation of the fugacities of S_2, H_2S, and SO_2 with O_2 and temperature and the probable trends of magmatic evolution. This area of investigation is important for the development of skarns, but it cannot be dealt with here.

Volcanic gases interact with the surrounding rock, and they also contain components not well modeled on the basis of current data. Furthermore, the data base for gases from silicic magmas is poor because the explosive nature of the eruptions makes sampling difficult. Table 12.1 gives some typical analyses, mostly from more mafic systems.

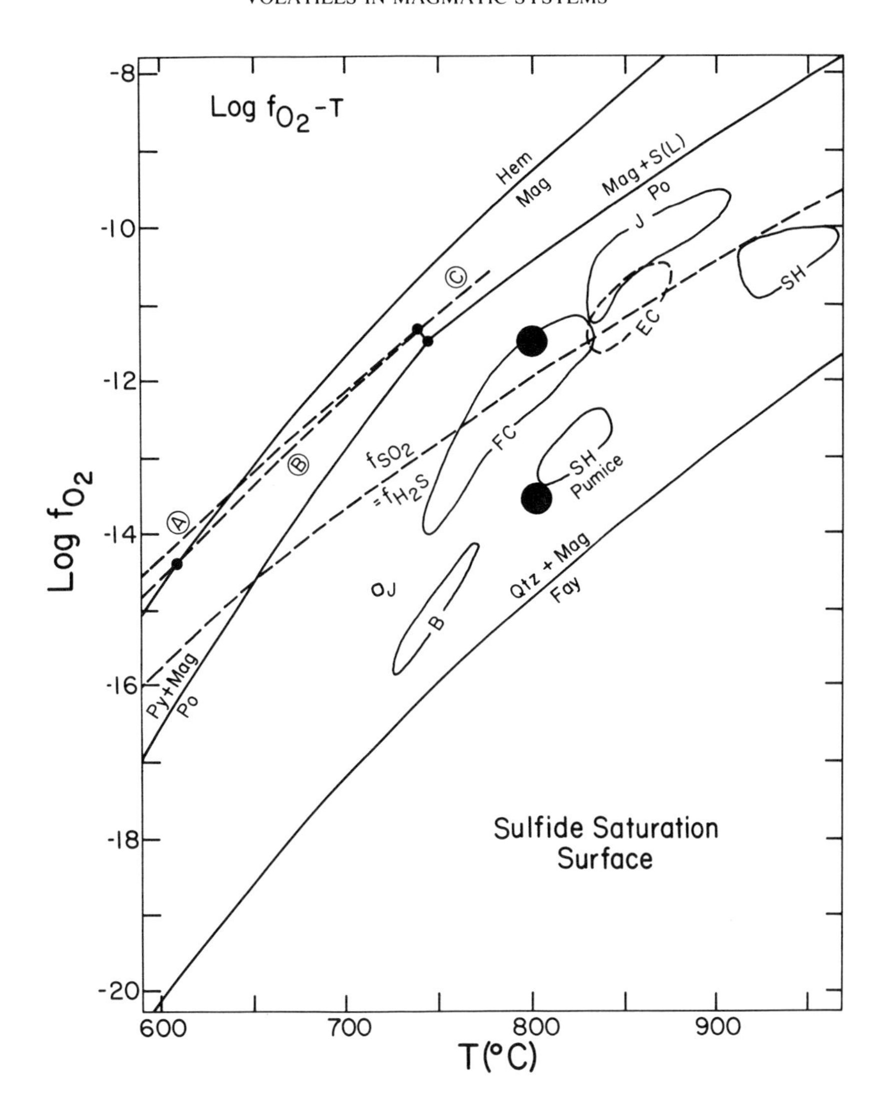

Figure 12.4. Log f_{O_2} - t diagram for sulfide saturated equilibria. Reactions involving $FeSO_4$ are only approximately known and are labeled as follows: A. $FeSO_4$ + S(L) = pyrite; B. $FeSO_4$ = magnetite + pyrite; C. $FeSO_4$ = magnetite + S(L). Conditions for various pyrrhotite-bearing volcanic rocks shown. B = Bishop Tuff, FC = Fish Canyon Tuff, J = Julcani vitrophyre; SH = St. Helens ash; EC = El Chichon ash. See Whitney, 1984 for complete description of techniques and sources of data. The upper and lower solid circles respectively indicate the oxidized and reduced conditions discussed in the text.

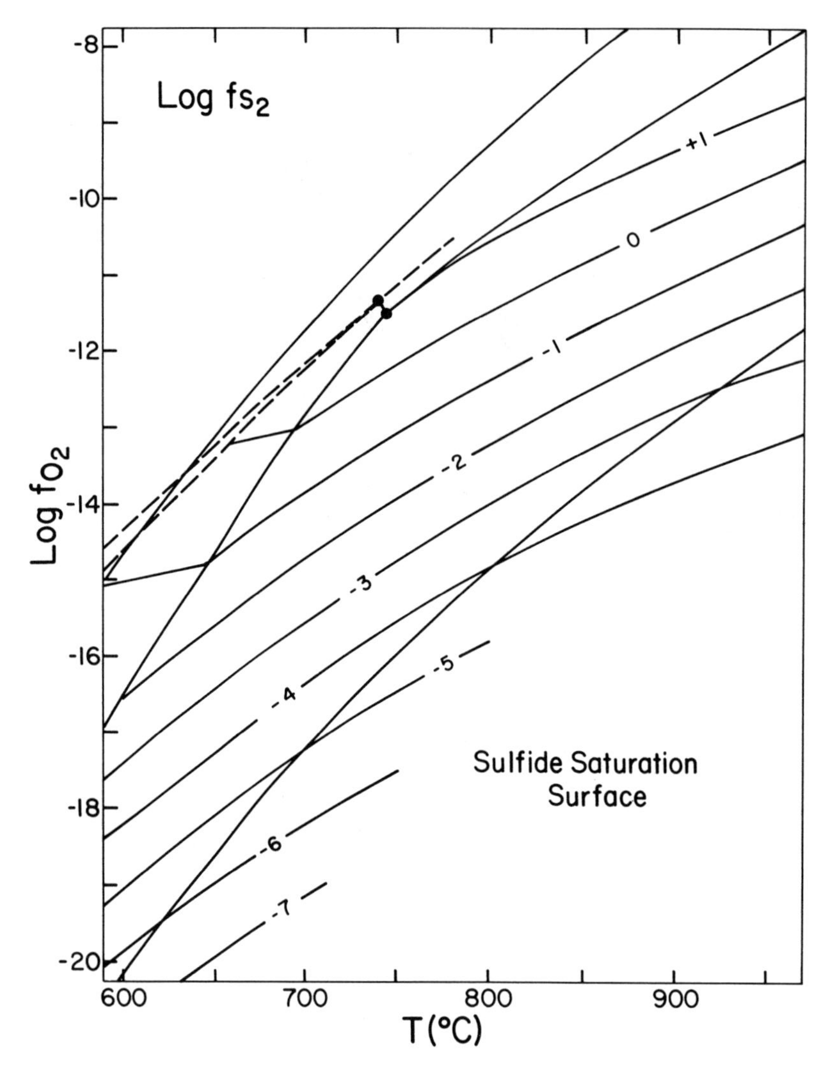

Figure 12.5. Log f_{O2} - t diagram portraying contours of constant log f_{S2} (isopleths) on the sulfide saturation surface. Reaction lines are the same as in Figure 12.4.

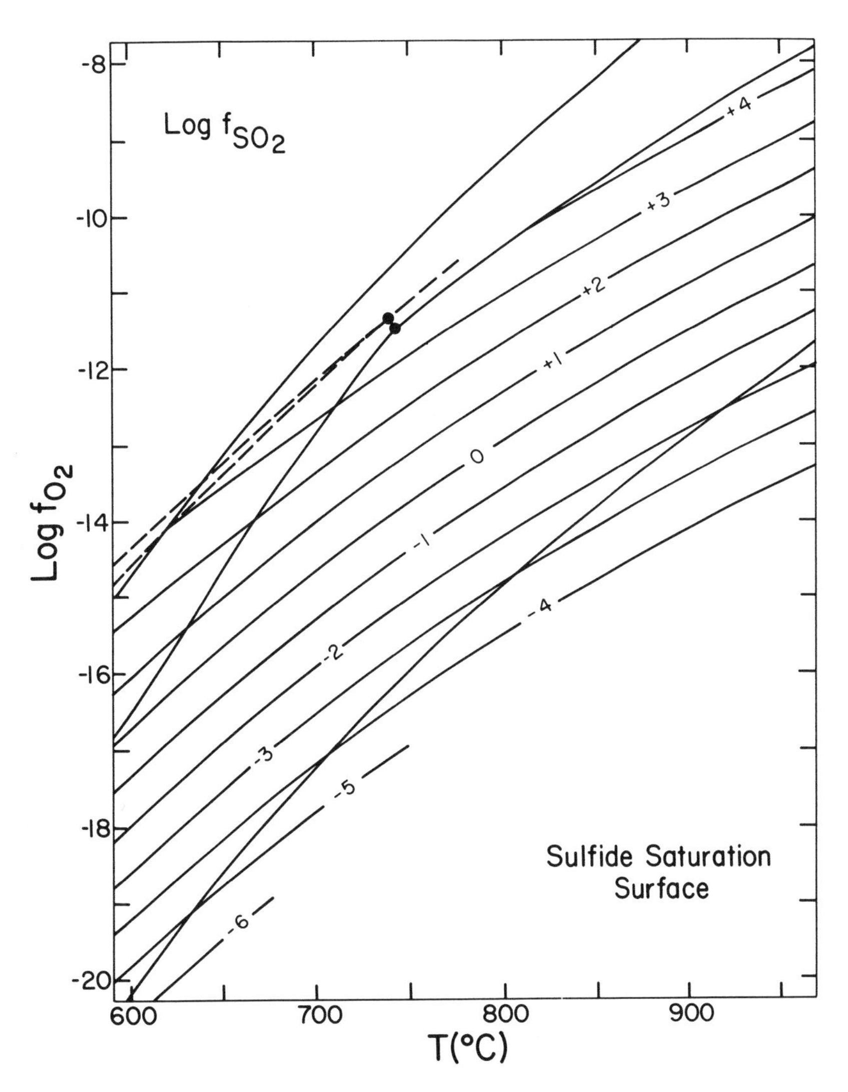

Figure 12.6. Log f_{O2} - t diagram portraying contours of constant log f_{SO2} on the sulfide saturation surface. Reaction lines are the same as in Figure 12.4.

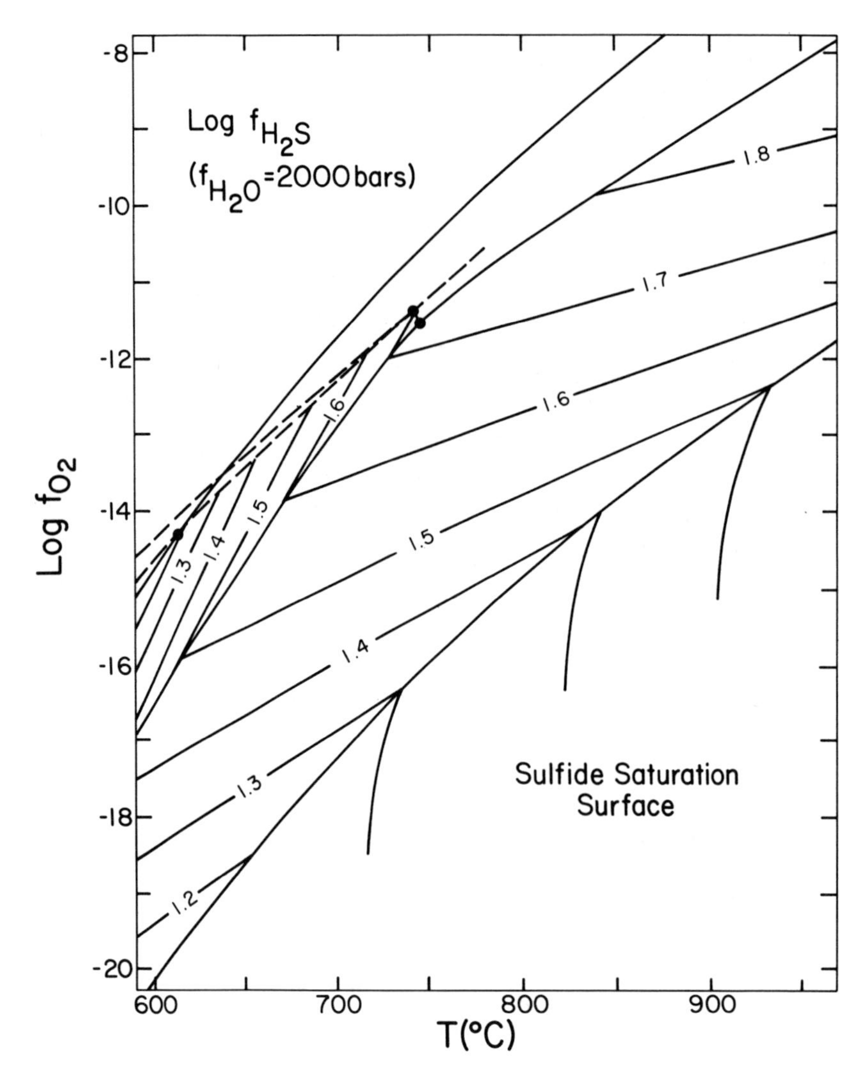

Figure 12.7. Log f_{O2} - t diagram portraying contours of constant log f_{H2S} on the sulfide saturation surface for a water fugacity of 2000 bars. To convert to other water fugacities add the term log (f_{H2O}/2000). Reaction lines are the same as in Figure 12.4.

Table 12.1 — Analyses of typical volcanic gases

		millimoles/mole dry gas												
	t(°C)	x_g	CO_2	S_t	$\bar{n}_S$*	HCl	HF	NH_3	H_2	Ar	O_2	N_2	CH_4	CO
White Island,	620	165	617	297	+3.2	59.0	0.66	0.13	24.1	0.004	<0.01	3.3	0.04	0.59
New Zealand	115	25	820	156	-1.8	3.2	0.21	0.06	0.6	0.07	<0.1	10.2	8.90	0.01
Mt. Ngauruhoe, New Zealand	520	101	696	179	+1.6	31.0	2.6	0.31	28.4	0.15	<0.01	42.4	0.01	0.45
Mt. St. Helens Washington, USA	686	27	589	106	+4.8	21.9	8.6	0.60	197	0.98	<0.006	76.1	<0.08	nd
Mt. Poas, Costa Rica	870	205	510	344	+4.2	32.1	4.9	0.095	90.4	0.26	0.021	18.2	<0.006	<0.009

References: Giggenbach, 1973; D. Sheppard, personal communication, 1983.

* n - is the average oxidation state of the sulfur.

Let us take two magmatic gas compositions, compare their bulk compositions, and ask what their chemistries will be like if they are cooled isochemically to 250°C. We shall choose a magmatic temperature of 800°C and a partial pressure of water of 500 bars; this is not very different from the liquidus and vaporus of Burnham's (1979) hornblende-biotite granodiorite. We shall designate two initial redox states involving the pyrrhotite + magnetite assemblage: one designated "O" (sulfur oxidized) and the other "R" (sulfur reduced). Both are indicated in Figures 12.1, 12.2, and 12.4. Complete Table 12.2 using data from this chapter and from Table 8.1 as necessary. This calculation is the gaseous and/or aqueous-species analog to the mineralogical species calculated as a petrologic norm.

In order to complete the 250°C portion of the chart, several assumptions are required involving reactions taking place during the cooling process. Assume: 1). H_2S remains as H_2S. 2). SO_2 disappears (i.e., the reaction effectively goes to completion) by the following reaction

$$4SO_2 + 4H_2O = nHSO_4^- + (3-n)SO_4^= + (2(3-n)+n)H^+ + H_2S \quad (5)$$

The proportions of $SO_4^=$ and HSO_4^- are controlled by the pH. 3). The H_2 reacts with the sulfate to yield H_2S, but a small amount of H_2 remains. 4). Although the mineralogy of deposits thought to originate through the processes described here is typically that of advanced argillic alteration and indicates pH values in the 3 to 4 range, we shall simplify the discussion for illustrative purposes by assuming a pH of 5.4 at 250°C so as to make comparison with previous discussions more straightforward. The log K_f^o for steam and liquid water at 250°C are respectively 21.75 and 20.23; that for gaseous H_2S is 4.056, (Robie et al., 1979).

The f_{H2O} for 500 bars pressure at 800° equals 458 bars (log f_{H2O} = 2.66). Had we chosen 2 kbar f_{H2O}, log f_{H2O} would only have changed to 3.20. Our calculation procedure will be to establish the bulk composition of the 800° steam and then try to estimate what that bulk composition will be like at 250°C. Use Figures 12.5, 12.6, and 12.7. Fugacity coefficients at 800°C are obtainable for major species by taking the ratios of the fugacities to the pressures in column one. These are derived from the modified Redlich-Kwong Model (Holloway, 1981, 1977; Flowers, 1979).

Remember that the 800° calculation uses mole fractions whereas the 250° calculation uses moles/kilogram H_2O.

The conversion from fugacities to partial pressures to moles per kilogram of water is not at all self evident, and we shall need some approximations here. Moreover, certain important constituents of the vapor, e.g., CO_2, HCl, HF, metals, alkalis, and so on, are neglected, as are probable fluid - mineral reactions during cooling. We shall probably have a reasonable conclusion concerning the redox pattern that emerges, but our simplified chemistry (neglecting CO_2 and rock - fluid interactions during cooling) obviates making a meaningful calculation of pH, thus the hydrolysis trajectory through pH-t space is very difficult to calculate (some of the factors are shown in Figures 12.2 and 12.4. Evidence from ore deposits suggests that the pH for such a magmatic fluid would be fixed at about 4 by wallrock reactions, but in order to illustrate the redox state in terms of an already-established chemistry, we shall use the pH of 5.4 that was employed for Figure 8.3, assuming that the K-feldspar + Kmica + quartz + K^+ ion buffer is appliciable. Regardless of the approximate nature of the calculation, the principles remain valid.

Table 12.2 -- Comparison of the chemistries of quenched gases from oxidized and reduced I-type magmas.

	800° oxidized	800° reduced	250° oxidized	250° reduced
$\log f_{H_2O}$	2.66	2.66	0*	0*
$\log P_{H_2O}$	2.70	2.70	1.53	1.53
X_{H_2O}	.74			
$\log f_{O_2}$	-11.5	-13.5		
X_{O_2}	nil	nil		
$\log f_{H_2}$	-1.00	+0.22		
X_{H_2}	.0001			
$\log f_{S_2}$	0.00		--	--
X_{S_2}	0.001			
$\log f_{H_2S}$	1.10		--	--
$\log P_{H_2S}$	1.14			
X_{H_2S}	0.02			
m_{H_2S}	--	--		
$\log f_{SO_2}$	2.3		--	--
$\log P_{SO_2}$	2.2		--	--
X_{SO_2}	0.24		--	--
$\log f_{SO_3^=}$	-3.54		--	--
$X_{SO_3^=}$	nil		--	--
$m_{Total\ SO_4}$	--	--		
P_{Total}	673 bars			

* Note that we have changed standard states for H_2O, from 1 bar vapor at 800°C to liquid water at 250°C. Remember also that m is molal concentration, or moles per kilogram of solvent, i.e., H_2O. The dashes, --, indicate that no value need be calculated.

Plot the position of the 250°C solution on Figure 8.3. Note that that the compositions in Figure 8.3 are in the vicinity of the maximum f_{S2} along any given contour of total sulfur regardless of whether the fluid began as the reduced or oxidized alternative. Why is this so?

Is native sulfur stable under these conditions?

The high f_{S2} leads to minerals (or mineral assemblages) having high sulfidation states. Give some examples (see Figure 8.6): Alunite also is expected in such environments, especially as pH becomes lower. All in all, this environment is similar to the "D veins" so characteristic of late-stage copper porphyry systems (Gustafson and Hunt, 1975).

Sketch on the 800° isotherm in Figure 12.2 the trend that one might expect the residual magmatic environment to follow as gas of the composition noted in Table 12.2 is removed from each of the two starting environments. The trends will be away from the component removed. The "R" system will move to the right and down; the "O" will move to the left and down. Eventually the paths will merge along a near-vertical trend along which the ratios of the sulfur-bearing volatiles in the gas will be the same as they are in the liquid. We cannot calculate this quantitatively without a great deal more experimental data, but the character of the pattern is suspected to show a final bornite + magnetite assemblage with neither fayalite nor hematite nor pyrrhotite. Interestingly, this bornite + magnetite assemblage is found in the deep zones of some porphyry copper deposits.

If a granitic magma (containing magnetite and pyrrhotite) degasses at 900°C with a log f_{O2} of -11, what would be the fugacities of the various species (assume f_{H2O} is 2000 bars)?

If the same magma cooled to 750°C and log f_{O2} of -15, what would be the fugacities of the various species again assuming f_{H2O} is 2000 bars?

Estimate the approximate amount of sulfur in the gas phase in each case, assuming that all significant gas species are represented by H_2O, S_2, SO_2, and H_2S. (Since this is an estimate, you may use the Lewis-Randall rule ($f_i = X_i f_i$; where f_i is the fugacity of the pure component i at the total pressure of the the mixture; and assuming that f_i is approximately equal to the total pressure, which may be approximated by Total f_i + ...).

Considering the approximate fugacity coefficients derivable at 800°C from Table 12.2, how far off are the approximations likely to be?

If the sulfur content of the melt is approximately 500 ppm (based on analysis of glass inclusions), what is the bulk distribution coefficient (in weight) for each case?

$ If sulfur is dominantly dissolved in the melt as HS^-, is the sulfur content of sulfide saturated melts likely to change drastically over the normal magmatic range? Why or why not?

$ In the above cases, what weight percent water would be lost from the melt before the melt ran out of sulfur?

SEPARATION OF A VOLATILE PHASE FROM A CRYSTALLIZING MAGMA

The separation of a volatile phase from a crystallizing magma has been quantitatively discussed by several authors (Whitney, 1975a, 1975b summarized by Burnham, 1979). The following synopsis is based on these sources.

At the time of intrusion, an ascending magma will reach some pressure at which a vapor phase will begin to exsolve. The approximate water content for many silicic melts is probably

2 to 4 weight percent based on crystallization temperatures and phase assemblages (Whitney, 1975a, Naney and Swanson, 1980, Burnham, 1979). In these cases, vapor saturation will occur at about 1.3 kb. If CO_2 is present is significant amounts, the depth will be increased. High sulfur fugacities may also cause the zone of initial vapor saturation to be deeper.

As the magmatic volatiles rise, they will interact with other waters forming a hydrothermal plume. The movement of fluids in such a system has been modeled by Henley and McNabb (1978) for porphyry copper deposits. Other authors (e.g., Cathles, 1981) have also reviewed the movement of meteoric fluids around such cooling stocks.

$ If a granitic magma of the composition represented in Figure 12.8 is intruded to low pressures, at what confining pressures will volatile separation begin assuming 4 weight percent H_2O in the original magma and the two temperatures shown?

Assuming that the vapor-saturated liquid phase contains about 6.5 weight percent water at 2 kb and 750°C, 6 weight percent water at 2 kb and 900°C, and 0 weight percent at 0 kb for both temperatures, sketch the water content of vapor saturated liquid as a function of P on Figure 12.8, and Figure 12.9. Note that the water content difference between the saturation surface and the solubility in the melt, in the absence of any other significant water-bearing phase, is a measure of the percent crystallization.

Assuming that the boundaries marking changes in the crystalline assemblages in Figure 12.9 are approximately representative of constant percent of crystallization, contour Figure 12.9 with lines of constant percent crystallization (or percent melt).

> How do the amounts of crystallization over the same temperature interval compare between the vapor-saturated and the vapor-absent case situations?
>
> If the gases evolved were actually those discussed in the previous box, and the initial water content of the magma was 4 percent, at what temperature would the melt become effectively depleted in sulfur (about 99% removed) assuming a single stage of volatile separation?

The process of vapor separation can be considered to be composed of two parts. The first is spontaneous vesiculaton due to oversaturation of the melt phase in response to declining confining pressure. Since vesiculation is fairly fast, such a process is effectively instantaneous, although some amount of oversaturation must occur before nucleation begins. The second portion of volatile separation occurs due to increasing water content in the melt due to crystallization of anhydrous phases. This process is somewhat slower, and depends on crystal nucleation and growth.

On Figure 12.8 suppose that a magma composed of aphyric melt at 4 kb and 900°C, with 8 weight percent water were suddenly intruded to a confining pressure of 1 kb. If the vapor phase were free to escape at a confining pressure of 1 kb, how much water (in terms of weight $ percent of the original melt) would be evolved immediately? How much more would evolve upon attainment of an equilibrium crystal assemblage?

$ Considering the previous boxes, at which stage would sulfur be effectively lost from the melt?

As an igneous body cools, vapor generation will continue in the upper portion of the body as crystallization occurs. The equilibrium conditions are shown for a 2 km square stock in Figures 12.10 and 12.11 with an initial temperature of approximately 800°C. If latent heat and convection are ignored, t_1 would be about 1,250 years and t_2 would be 12,500 years after intrusion. Latent heat could approximately double the time required to reach these states, while convective upwelling could extend the time much longer if it were continuously active. (See Whitney, 1975b, for detailed development of this model)

> Considering the previous discussion of percent crystallization, sketch the approximate percent crystallization on Figures 12.10 and 12.11.

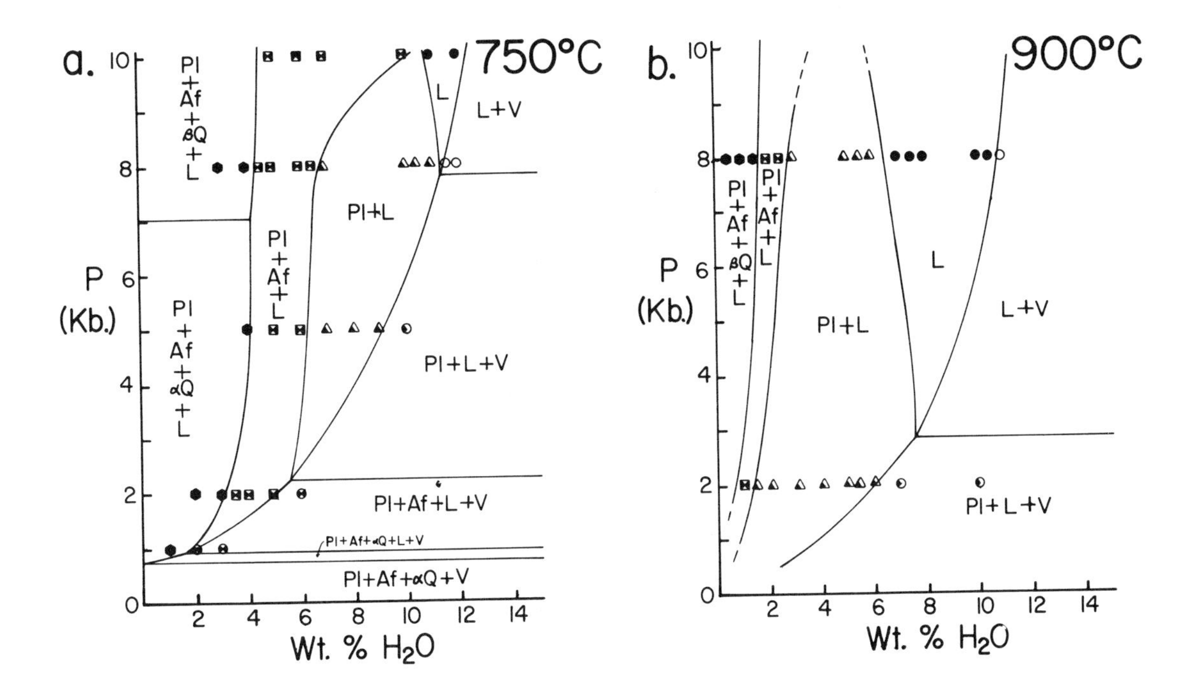

Figure 12.8. Isothermal pressure versus X_{H2O} phase assemblage diagrams for the synthetic two-feldspar granite (i.e. quartz monzonite), from Whitney, 1975b. Pl = plagioclase, Af = alkali feldspar, Q = quartz, L = silicate melt, V = hydrous fluid. Symbols represent the results of individual experiments.

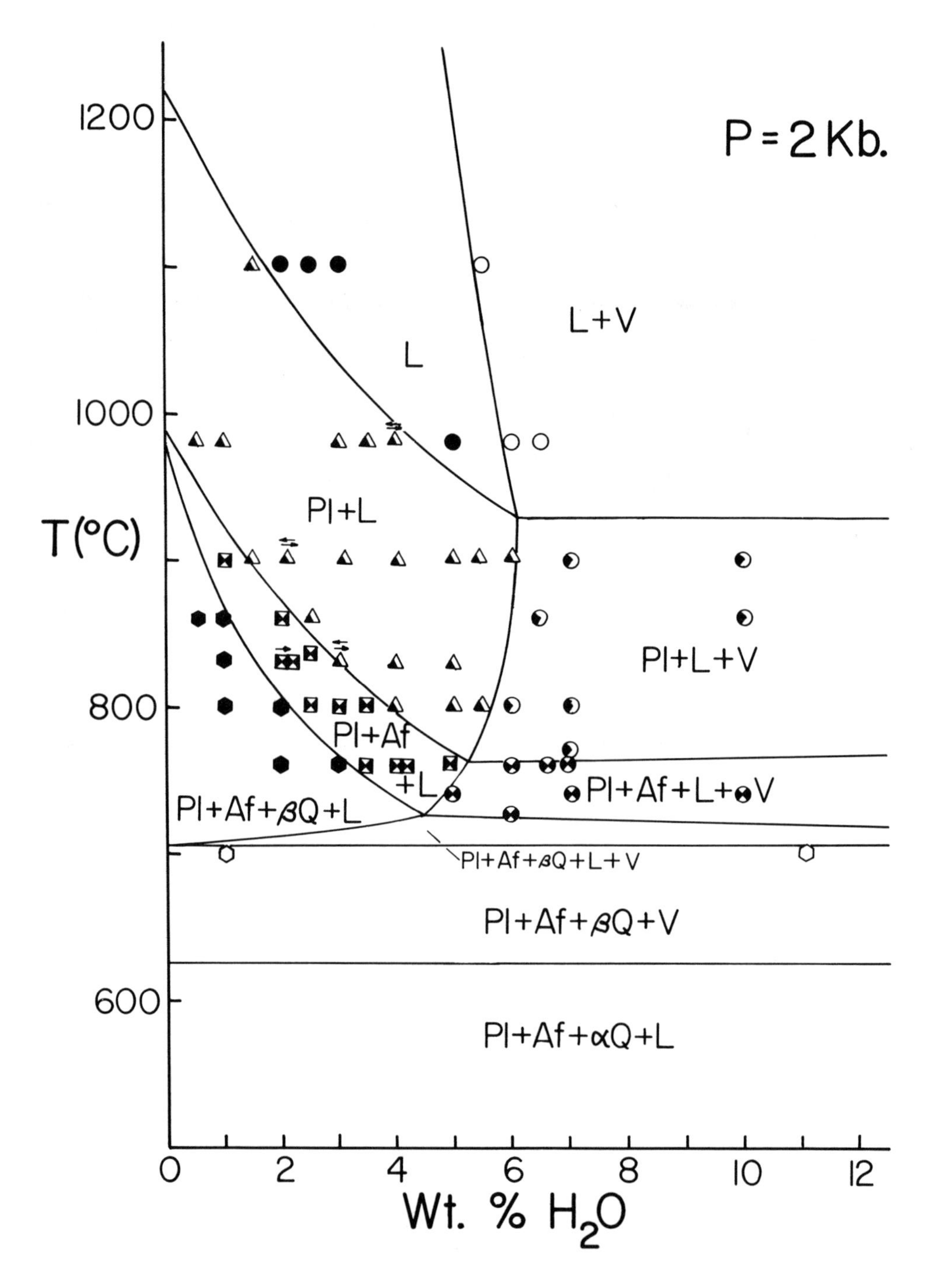

Figure 12.9. Isobaric temperature versus X_{H2O} diagram for the same compositiion as in Figure 12.8. Taken from Whitney, 1975a. Symbols represent the results of individual experiments.

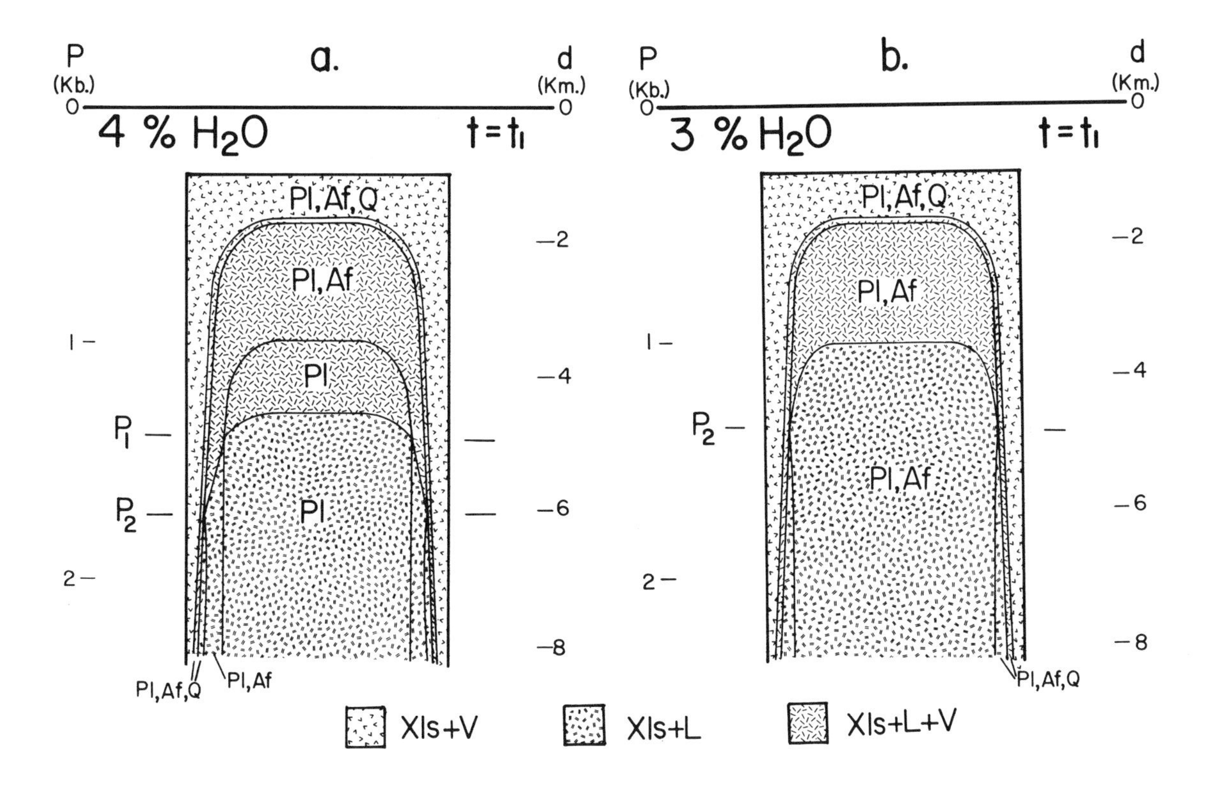

Figure 12.10. Equilibrium crystallization model for a 2 km square stock. Base on the model of Whitney, 1975b. Composition is the same as in Figures 12.8 and 12.9. If latent heat and convection are ignored, the time would be about 1,250 years after intrusion.

Upwelling of new magma into a stock attached to a larger batholith, either through periodic re-injection or convective overturn, may greatly increase the amount of volatile material available and also renew the concentration of gases strongly partitioned into the gas phase. Figure 12.12 represents a model for a convection system. Through time, the overall volatile content of the entire magma may decrease as part is vented through stocks extending into the vapor-saturated region. Thus, case 12.12a may evolve into 12.12b.

Chlorine has a bulk distribution coefficient between melt and vapor phase that is of the same order magnitude as sulfur (Kilinc and Burnham, 1972).
Fluorine, on the other hand, is not as strongly fractionated into the vapor; Although its exact values are unknown, its concentration in the volatile phase is probably not much different than that in the melt.

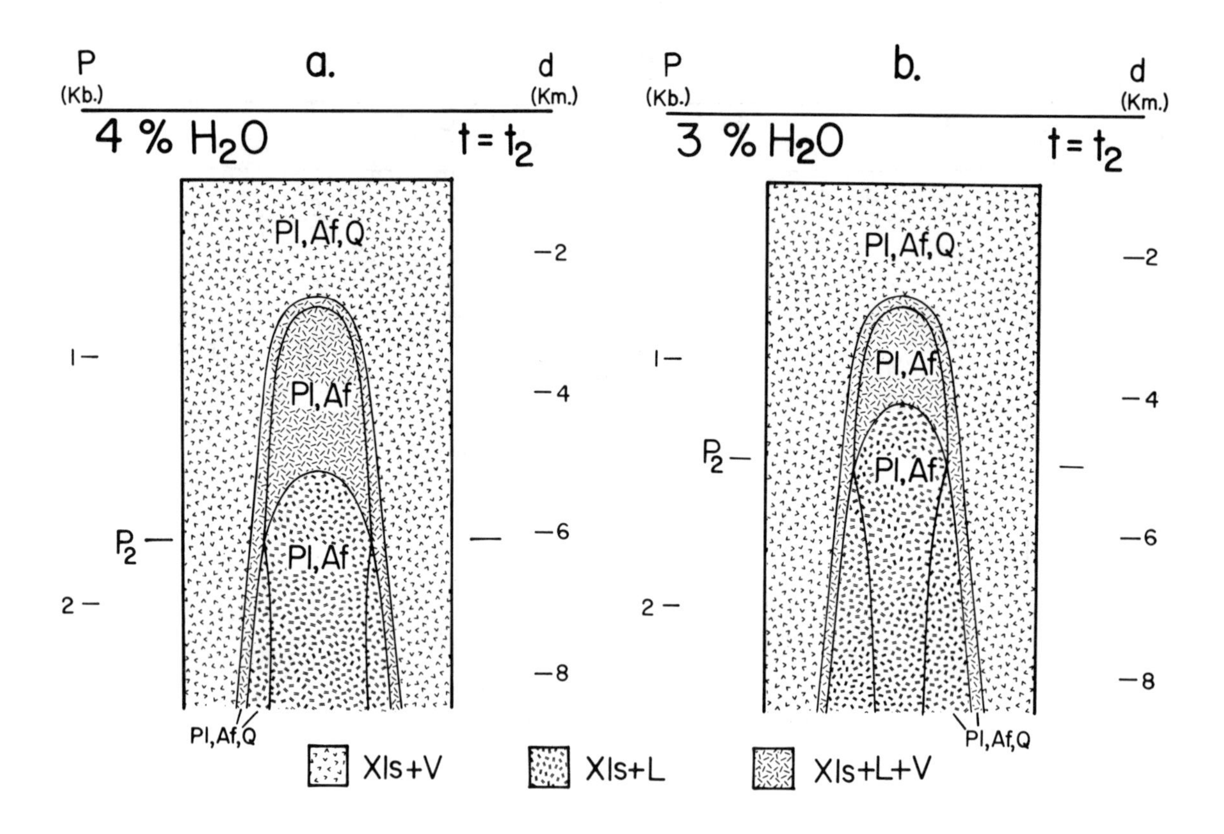

Figure 12.11. Equilibrium crystallization model for a 2 km square stock, based on the model of Whitney, 1975b. The model is the same as Figure 12.10 except that the times is ten times that in 12.10. If latent heat and convection is ignored, the time would be about 12,500 years.

> Discuss how chlorine, sulfur, and fluorine would vary in concentration and distribution in the vapor phases, vapor saturated melt near the top of the stock, and underlying parent magma, in the cases shown in Figures 12.10, 12.11 and 12.12.

ADDITIONAL RESEARCH REQUIRED

To better describe the contributions of magmatic volatiles several studies are needed. First we need more data on the volatile content of natural melts, especially glass inclusions which may not have degassed under near-surface conditions. Second, we need to know distribution coefficients for chlorine and fluorine between the melt, vapor phase, and hydrous miner-

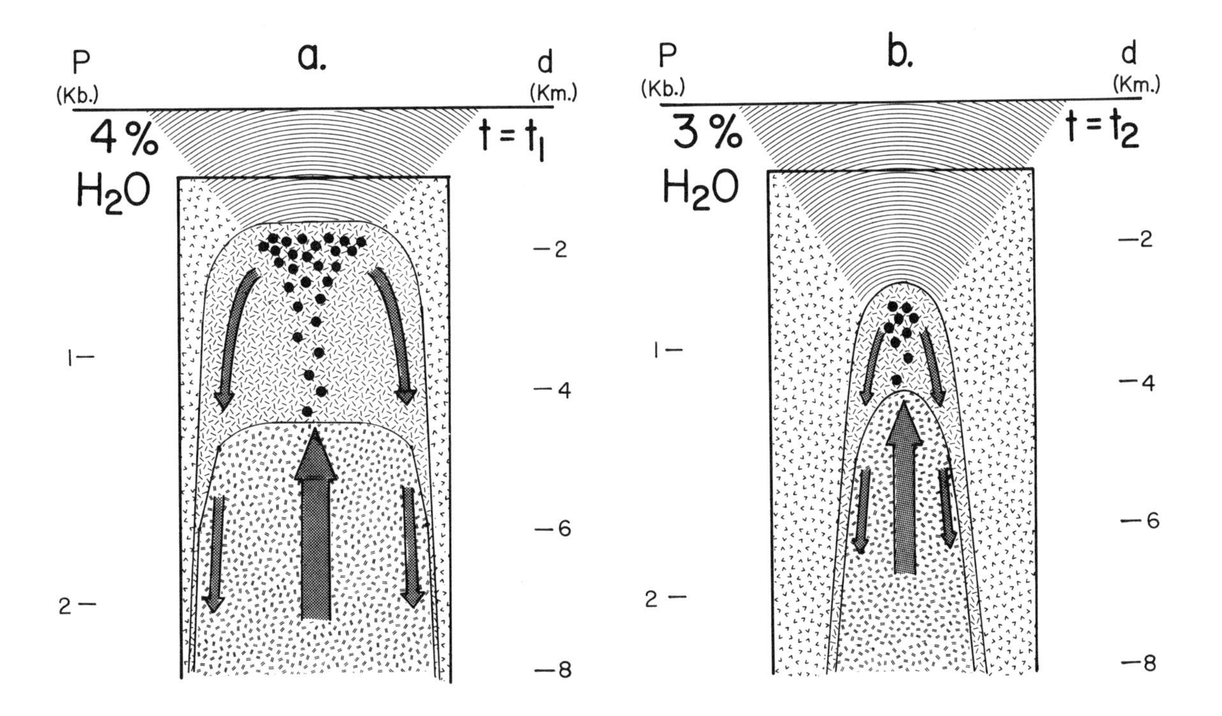

Figure 12.12. Crystallization model showing the possible effects of convection on crystallization model of 12.10 and 12.11. With time, the stock may lose water from the entire body and conditions move from case a to case b.

als. Values involving apatite are especially needed due to the widespread occurence of apatite and its resistance to re-equilibration. We also need to know more about the composition of chloride solutions equilibrated with granitic compositions at high temperatures to determine the possible degree to which heavy metals can be transported in such solutions. We will then be better able to evaluate the over-all composition of magmatic volatile phases and their role in the protore stage of porphyry and skarn systems.

REFERENCES

Barton, P.B., Jr., and Skinner, B.J., 1979, Sulfide mineral stabilities; in Barnes, H.L., Geochemistry of Hydrothermal Ore Deposits, 2nd Edition: p. 278-403.

Buddington, A.F., and Lindsley, D.H., 1964, Iron titanium oxide minerals and synthetic equivalents: Journal of Petrology, v. 5, p. 310-357.

Burnham, C.W., 1979, Magmas and Hydrothermal Fluids; in Barnes, H.L., ed., Geochemistry of Hydrothermal Ore Deposits, 2nd Edition: p. 71-136.

Cathles, L.M., 1981, Fluid Flow and genesis of hydrothermal ore deposits: Economic Geology, 75th Anniversary Volume, p. 424-457.

Flowers, G.C., 1979, Correction of Holloway's (1977) adaption of the modified Redlich-Kwong equation of state for calculation of the fugacities of molecular species in supercritical fluids of geologic interest: Contributions to Mineralogy and Petrology, v. 69, p. 315-318.

Franck, E.U., 1956, Hochverdichteter wasserdampf III, ionendissoziation von HCl, KOH, und H_2O in uberkiritschem wasser: Zeitsch. physikal Chemie, v. 8, p. 107-206.

Giggenbach, W.F., 1983, The chemical and isotopic composition of gas discharges from New Zealand andesitic volcanoes: Bulletin Volcanologique, in press.

Gustafson, L. B., and Hunt, J. P., 1974, The porphyry copper deposit at El Salvador, Chile: Economic Geology, v. 70, p. 857-912.

Henley, I.A., and McNabb, A., 1978, Magmatic vapor plumes and ground-water interaction in porphyry copper emplacement: Economic Geology, v. 73, p. 1-20.

Holloway, J.R., 1977, Fugacity and activity of molecular species in supercritical fluids; in Thermodynamics in Geology: D.G. Fraser, editor, D. Rediel, Dordrecht, Holland, p. 161-181.

Holloway, J.R., 1981, Compositions and volumes of supercritical fluids in the earth's crust; in Fluid Inclusions: Applications to Petrology: L.S. Hollister and M.L. Crawford, editors, Mineralogical Assocation of Canada, Calgary, p. 13-38.

Kilinc, I.A., and Burnham, C.W., 1972, Partitioning of chloride between a silicate melt and coexisting aqueous phase from 2 to 8 kilobars: Economic Geology, v. 67, p. 231-235.

Mueller, R.F., 1972, The stability of biotite: A discussion: American Mineralogist, v. 57, p. 300-315.

Naney, M.T., and Swanson, S.E., 1980, The effects of Fe and Mg on crystallization in granitic systems: American Mineralogist, v. 65, p. 639-653.

Quist, A.S., and Marshall, W.L., 1968, Electrical conductances of equeous sodium chloride from 0° to 800° and at pressures to 4000 bars: Journal of Physical Chemistry, v. 72, p. 684-703.

Rau, H., 1976, Energetics of defect formations and interactions in pyrrhotite, $Fe_{1-x}S$ and its homogenity range: Journal of Physical Chemistry Solids, v. 37, p. 425-429.

Robie, R.A., Hemingway, B.S., and Fisher, J.R., 1979, Thermodynamic properties of minerals and related substances at 298.15 K (25° C) and one bar (105 pascals) pressure and at higher temperature (revised, 1979): U.S. Geological Survey Bulletin 1452, 456 p. (originally printed in 1978).

Spencer, K.J., and Lindsley, D.H., 1981, A solution model for coexisting iron-titanium oxides: American Mineralogist, v. 66, p. 1189-1201.

Stormer, J.C., Jr., 1983, The recalculation of multi-component Fe-Ti oxide analyses for determination of temperature and oxygen fugacity: American Mineralogist, v. 78, p. 586-594.

Toulmin, P., III, and Barton, P.B., Jr., 1964, A thermodynamic study of pyrite and pyrrhotite: Geochimica et Cosmochimica Acta, v. 28, p. 641-671.

Whitney, J.A., 1975a, The effects of pressure, temperature, and X_{H2O} on phase assemblage in four synthetic rock compostions: Journal of Geology, v. 83, p. 346-358.

Whitney, J.A., 1975b, Vapor generation in a quartz monzonite magma: A synthetic model with application to porphyry copper deposits: Economic Geology, v. 70, p. 346-358.

Whitney, J.A., 1977, A synthetic model for vapor generation in tonalite magmas and its economic ramifications: Economic Geology, v. 72, p. 686-690.

Whitney, J.A., 1984 (in press), Fugacities of sulfurous gases in pyrrhotite-bearing silicic magmas: American Mineralogist, v. 78, p. 69-78.

Wones, D.B., 1972, Stability of biotite: A reply: American Mineralogist, v. 57, p. 316-319.

Chapter 13
HIGH TEMPERATURE CALCULATIONS IN GEOTHERMAL DEVELOPMENT

The large scale development of geothermal resources usually involves low to medium pressure steam separation from production wells where flashing to steam and water also occurs within the wells themselves. The temperature changes due to such flashing processes and the pH changes attendant on gas removal in the steam phase are the principal processes leading to mineral deposition in wells, separator plants, and waste water disposal pipelines or channels. In this chapter we will examine the chemistry of waters resulting from steam (+ gas) separation with special reference to the common problems of silica and calcite scaling. In earlier chapters the effects of boiling on the deposition of metals and metal sulfides were examined. We shall also briefly touch on the calculation of the pH and composition of steam condensate; these data are frequently required in geothermal development studies where problems of pipeline corrosion, disposal of condensate or control of H_2S emmission are under consideration.

pH CALCULATIONS AND STEAM SEPARATION

In Chapter 6 we showed that removal of 90% of the dissolved CO_2 from an initially single phase fluid resulted in a pH increase of about 1 unit. While this may be useful as a rule of thumb, a more refined procedure is required for geothermal development calculations where multistage boiling leads to relatively greater gas removal and the pH of the resultant liquid begins to be internally buffered by weak acid - base equilibria. The most important buffering reaction is still that of carbonic acid-bicarbonate ion, but significant contributions are also made by the boric acid-borate, ammonium-ammonia and silicic acid-silicate reactions.

As before, we recognize Δ_t (or alkalinity or HTOT, Chapter 7) as a controlling parameter for our fluid. Initially in the total discharge

$$\Delta_t = m_{HCO_3^-} + m_{H_2BO_3^-} + m_{HS^-} + 2m_{CO_3^=} + m_{OH^-} - m_{H^+} - m_{NH_4^+} \quad (1)$$

and after loss of a steam fraction y (with ' indicating flashed liquid and vapor compositions)

$$\Delta'_t \quad = \quad \Delta_t/(1 - y) \quad = \quad m'_{HCO_3^-} + m'_{H_2BO_3^-} + \ldots \quad (2)$$

Write equivalent expressions for alkalinity or HTOT. (See Chapter 7)

Each of the base terms that make up Δ'_t are dependent on pH'_t and the total concentration of each component remaining.

For example, for a non-volatile component such as silica, for which $m'_{\Sigma SiO2} = m_{\Sigma SiO2}/(1-y)$,

$$m'_{H_3SiO_4^-} \;=\; m'_{\Sigma SiO2} \,/\{\, (a_{H^+}K_{H_4SiO_4}/\gamma_{H_3SiO_4^-}) - 1\} \quad (3)$$

while for a volatile component like CO_2 and its conjugate bases

$$m_{\Sigma C} = m_{CO_2} + m_{HCO_3^-} + m_{CO_3^=} + m_{CaHCO_3} +$$

$$= (m'_{CO_2,l} + m'_{HCO_3} + m'_{CO_3^=} + \ldots)(1 - y) \;+\; y(m'_{CO_2,v}) \quad (4)$$

Dividing through by $m'_{CO_2,l}$ gives

$$\frac{m'_{\Sigma C}}{m'_{CO_2,l}} = \{1 + (m'_{HCO_3}/m'_{CO_2}) + (m'_{CO_3^=}/m'_{CO_2}) + \ldots\}(1 - y) \;+\; y\,B_t \quad (5)$$

where B_t is our old friend, the gas distribution coefficient at the steam separation temperature. Each of the acid-base terms remaining is pH_t' dependent while the expression itself is dependent on the steam fraction separated, initial total dissolved carbonate in the discharge, and the separation temperature.

Similar expressions can easily be written for the other volatile components like H_2S, NH_3 and H_3BO_3 which contribute to Δ. Combined with expressions like (3) for the other non-volatile components in the modified charge balance equation (2) we have a complex pH dependent expression which may be solved iteratively to obtain pH_t' values for the residual liquid phase.

> For a given total discharge composition (ΣC, ΣH_2S, ΣSiO_2, and ΣB) draw a flow chart for the calculation of pH_t' at any specified steam separation temperature. Assume that Δ_t is known from previous calculations and that the steam fraction is calculated separately.

These expressions are easily combined into a hand calculator pH program such as that outlined in Appendix I, or into the larger programs (e.g. ENTHALP) described in Chapter 7.

Henley and Singers (1982) have shown that conventional cyclone separators of the type installed at Wairakei (Bangma, 1961) achieve a quasi-single stage equilibrium distribution of gases between the separated water and steam phases. As a result, the chemistry of flash separated waters at any chosen conditions may be calculated by iterative solution of a set of pH dependent mass balance equations for the major dissolved constituents.

The ENTHALP computer program (Truesdell and Singers, 1971) has been adapted for the routine calculation of these data (Singers, et al., in prep). The output includes the pH_t of the water resulting from flash separation at any pressure of interest, the concentrations of the principal dissolved species (e.g. bicarbonate, silicate ions), and the gas concentrations in the separated steam. In addition, the program may be used to calculate these data for successive stages of steam separation from a given well and further, up to five well discharges may be processed together to obtain the phase chemistry of steam and water derived from multiple flash plants.

\$ As an exercise, outline the additional input data and mass balance equations you would need to program the flashed water composition resulting from steam separation from the combined discharge from three wells.

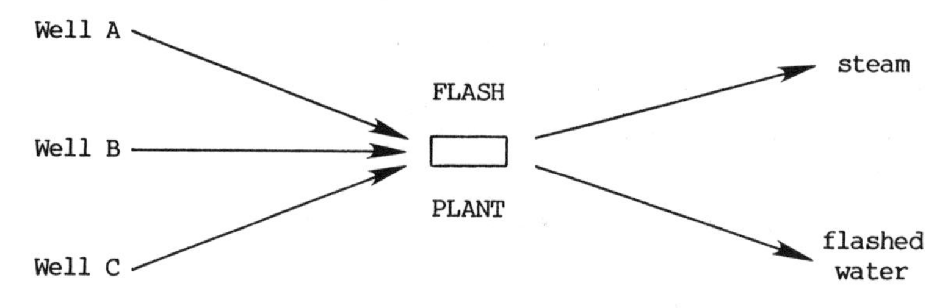

SILICA DEPOSITION

Removal of steam from a geothermal discharge that is initially saturated with quartz at the aquifer temperature leads to an increase in silica concentration in the residual water. If the new silica concentration exceeds that of amorphous silica saturation, a silica scale may subsequently deposit during flow to disposal sites. Excessive scale formation may reduce flow rates and lead to costly removal operations -- as at Wairakei.

Using the appropriate geothermometer data in Tables 3.1 and 3.2, calculate and draw saturation curves for quartz and amorphous silica in water as a function of temperature.

If a reservoir fluid initially at 265°C is separated into liquid and vapor at the wellhead, find the separation temperature below which the residual water would be supersaturated

$ with amorphous silica. If the BR22 discharge, whose analysis was given in Table 2.3, was separated at 10 b.g. and 4.5 b.g. successively, would the residual water be super or under-saturated with respect to amorphous silica?

Waters may often be transmitted well below this base temperature without significant scale formation. At BR22, for example, field trials using double flashed waters have shown only minor scale formation at 130°C. How do you account for this apparent contradiction of the thermodynamics?

Silica saturation may be expressed by the formation of silicic acid

$$SiO_2 + 2H_2O = H_4SiO_4$$

and the amorphous silica geothermometer is based directly on this relation.*

---

* Note that the activity of water should be included in the equilibrium constant. For most waters $\gamma_{H2O} \simeq 1$ but in highly saline waters such as in the Imperial Valley, where Cl = 155000 mg/kg, $\gamma_{H2O} \simeq 0.7$ and the solubility of amorphous silica at 150°C is reduced by 50% relative to that in pure water.

If silicate ion forms by dissociation of H_4SiO_4, more silica may enter solution. The solubility of amorphous silica is therefore pH dependent with the total silica in solution equal to silicic acid plus silicate ions.

$$m_{\Sigma SiO_2} = m_{H_4SiO_4} + m_{H_3SiO_4^-} \quad (6)$$

Unless pH > 10 (approx.), other silicate species are negligible. Recalling your earlier calculation of the concentration of silicate ion as a function of pH, you may (using equation (13) in Chapter 7) derive a new pH dependent equation for amorphous silica solubility in dilute geothermal waters. Setting $\gamma_{H4SiO4} = 1$ for dilute solutions,

$$m_{\Sigma SiO_2} = m_{H_4SiO_4}\{1 + (K_{H_4SiO_4} \times \gamma_{H_2SiO_4^-} / a_{H^+})\} \quad (7)$$

m_{H4SiO4} may be substituted by the rearranged amorphous silica geothermometer equation (Table 3.2) to obtain the pH dependent silica solubility directly in mg/kg.

$ Using this equation, compare the 155°C solubility of amorphous silica ($\Sigma SiO2$) at pH = 6.5 with that at pH = 3, 6.5, and 9.

pH = 3, ΣSiO_2 = mg/kg; pH = 6.5, ΣSiO_2 = mg/kg; pH = 9, ΣSiO_2 = mg/kg

Using the procedures discussed in the previous section the pH_t of BR22 water, after two stages of steam separation, is found to be 8.1. What is the solubility of amorphous silica in this water?

The solubility is in fact higher than we first thought, so that the apparent danger of silica scaling has been decreased considerably. Calculate the solubility of amorphous silica in this water if it were subsequently flashed to 1 b.a in a weirbox. Are you surprised to find that under this condition voluminous white silica scale occurs in the weirbox?

Recognition of this effect has been beneficial in the design of the Broadlands power station in New Zealand where considerable concern existed concerning deposition in reinjection pipelines. Use of the modified ENTHALP program for available exploration well data showed that most of the wells when separated at a design pressure of 4.5 b.g. would be undersaturated with amorphous silica (Henley 1983).

---

* The Broadlands field to which we have referred frequently in the text has been recently renamed Ohaaki. We have retained the earlier geographic name in previous chapters to allow more ready cross reference to the existing literature on the geochemistry of this field.

The reservoir fluid at Ohaaki (Broadlands) is similar to that at Wairakei but characterised by much higher gas content. Reservoir temperatures in the proposed production field range from about 240 to 280°C, and the discharges of a significant number of wells contain excess reservoir steam. Each of these three factors contribute to the higher pH, broader pH range, and higher silica content of Broadlands flashed waters compared to Wairakei. The range of silica contents and pH for the Broadlands wells are shown in Figure 13.1 calculated for single and double flash conditions. The separation pressures chosen independently for power production are compatible with the optimum separation pressure difference required to maximize water pH and amorphous silica solubility. From the figure it is clear that the majority of wells are undersaturated with respect to silica following double flash separation at 155°C (4.5 b.g.) but cooling during flow to reinjection wells may lead to the waters from some flash plants becoming just supersaturated. A few wells, such as BR21, are characterised by high source temperatures and gas contents and, as a result, have relatively high saturation ratios at the final flash temperature of 155°C. The possible effect of reduction in field enthalpies due to exploitation are also shown by the silica - pH vectors in Figure 13.1

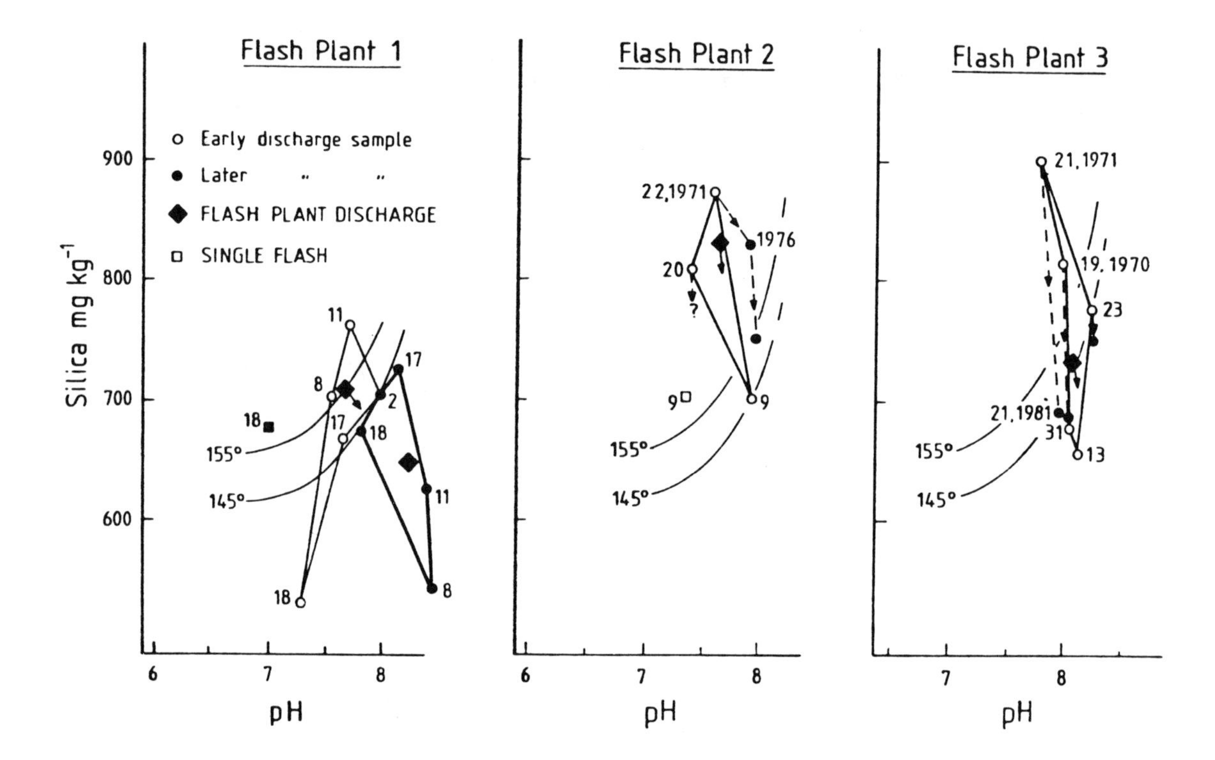

Figure 13.1. Calculated residual water pH and silica concentrations at 155°C (solid diamonds) for proposed two stage flash plants in the Ohaaki geothermal field. Steam separation occurs at 13.2 and 5.4 b.g. pH-silica data for individual wells are shown by the circle symbols; open circles are based on the discharge chemistry observed during exploitation and solid circles on discharge chemistry observed following sustained field tests. The trends with time between these data sets are illustrated by the arrows and indicate how residual flash plant waters will change during field production. For flash plant 1 the expected pH and silica concentration changes are expected to be of the order +0.8 and 80 mg/kg respectively. The square symbols show the residual water compositions resulting $ from single stage flash to 155°C for wells 9 and 18. Why should two stages of flash result in a higher pH?

Solubility curves for silica are shown at 155°C and at 145°C allowing the estimation of the increase in supersaturation as a result of cooling during pipeline transport to disposal wells. (Reproduced from Henley, 1983)

In fields such as this where steam is to be separated from up to five wells simultaneously at common flash plants, proximity is a principal factor in selecting the feed wells and the site of plant. In some cases adoption of only this criterion may lead to the choice of well combinations which would produce flashed water of unnecessarily high saturation ratio, while other combinations would optimise the pH and amorphous silica solubility of the flashed waters. Figure 13.1 shows such a test case for well combinations at Ohaaki.

Although the computational methods developed above are generally useful, the specific results may not apply to other systems. Where higher reservoir temperatures occur, silica contents are higher than at Ohaaki so that supersaturations must be higher after flashing at the same pressures. Gas contents also vary between fields effecting both the pH of the separated fluid and the solubility of amorphous silica.

Rates of Silica Scaling

The gas content and salinity of the field, its reservoir temperature, and the choice of flash pressures determine whether, after multiple flash separation of steam, residual waters are supersaturated or undersaturated with silica when discharged to inland waterways, evaporation ponds or reinjection wells. For waters supersaturated with silica, the rate at which silica deposits form becomes an important factor in the design and economics of geothermal development.

The deposition rate is a complex function of the following:

1. Precipitation mechanism and rate constant
2. Temperature, pH, salinity, and degree of supersaturation
3. Reynold's number and flow regime of the discharging fluid

Quantities like 'adhesion' may be considered consequences of these three.

Two mechanisms appear to dominate silica deposition:

1. Molecular deposition. Dissolved silica (H_4SiO_4, $H_3SiO_4^-$) bonds directly onto a growth surface. The rate of deposition is a function of temperature, supersaturation and the surface density of ionised silinol (-Si-OH) groups, which itself is a function of salinity and pH. It is found that the deposition rate increases with pH until silicate ion has increased the solublity of silica sufficiently to render the solution undersaturated (Weres et al., 1980; Weres and Yee,1982). Fluid flow rates may contribute to the deposition rate through the rate of supply of supersaturated solution to the growth site both through the bulk flow of fluid and the hydrodynamic control on the boundary layer thickness in the flow regime under consideration.

2. Homogeneous nucleation. Dissolved silica may proceed through 'condensation' reactions of the form

$$2\ H_4SiO_4 = H_6Si_2O_7 + H_2O$$

to produce high molecular weight polymers. This polymerisation process proceeds spontaneously in supersaturated solutions to produce some 10^{16} to 10^{20} particles/litre. The most important variables in controlling this process are again supersaturation and temperature and the surface tension of the silica-water interface, itself a function of the pH, temperature and salinity of the system. Normally polymerisation is preceeded by an apparent 'induction period' during which no net loss of 'monomeric' silica is observable; the length of this induction period is related to temperature and supersaturation but in some cases, (e.g. at Ohaaki) it may be shortened through the presence of a catalyst of some kind.

The induction period depends on the degree of supersaturation and is the time needed for homogeneous nucleation of particles. The induction period ends when the particles have reached a critical size where the suface area is sufficiently large that the collision rate with nutrient molecules is no longer rate determining. Where the supersaturation is greater than about 2, no induction period is generally observed. Usually molecular deposition produces a hard, vitreous, high density (ρ = 2.1 + 2.3 gm/cc), often dark coloured precipitate quite distinct from the low density white silica 'floc' (ρ_{dry} $\sim$0.95 gm/cc) which accumulates in weirboxes and drains at Wairakei Broadlands, Cerro Prieto and elsewhere.

Changes in fluid flow rates or supersaturation of the bulk flow in a pipe or channel may be marked by alternating laminae of vitreous and white silica and these same patterns are observable in sinters around active hot springs as well as those associated with epithermal gold-silver deposits.

The mechanism and rates of silica deposition are discussed by Weres, et al. (1980), Weres and Yee (1982), Mackrides, et al. (1980), Bohlman et al., (1980), and Rimstidt and Barnes (1981). The reader is referred to these articles for further information and examples of the theoretical calculation of silica deposition rate in supersaturated solutions. The prediction of silica scaling rates is obviously complex but determination of supersaturation is an important guide. Figure 13.2 shows observed deposition rates at Cerro Prieto and Ohaaki as a function of supersaturation.

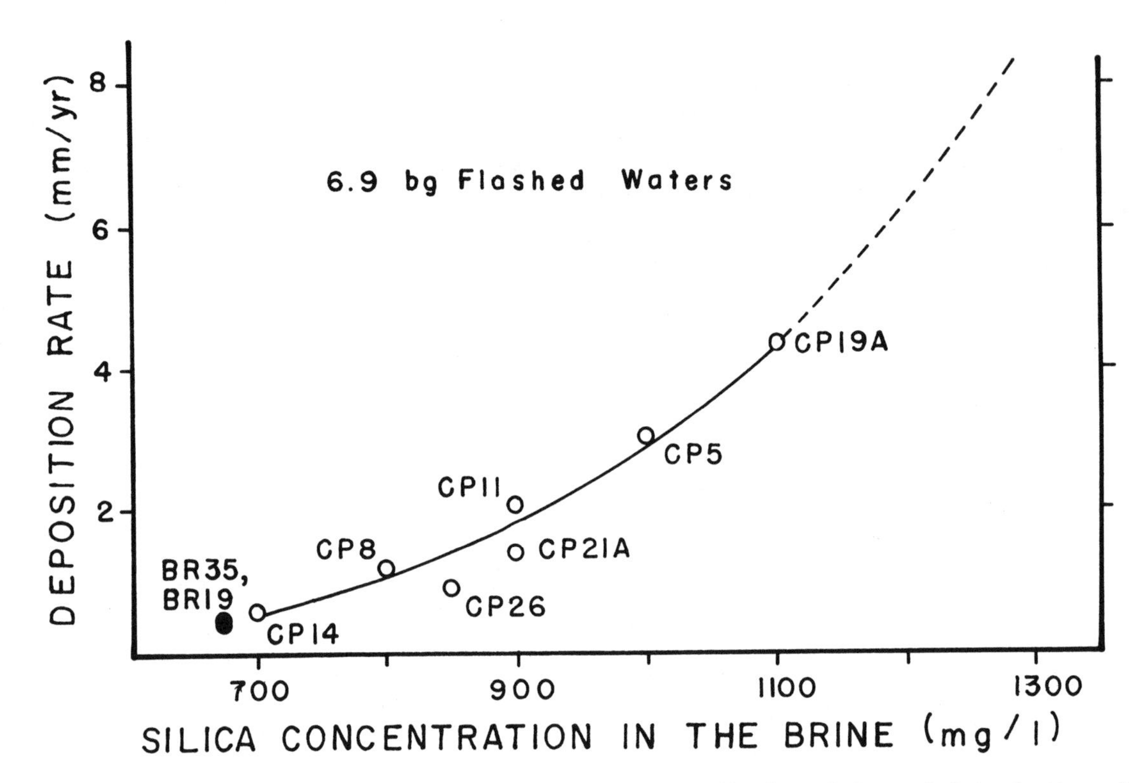

Figure 13.2. Observed silica deposition rates in the Broadlands and Cerro Prieto Geothermal Fields. (Modified from Hurtado, CFE (Mexico) Internal Report)

Various methods have been suggested for the control of silica deposition in geothermal waters. These include the controlled precipitation of a calcium silicate (possibly useful as a filler for wallboards, etc.) (Rothbaum and Anderton, 1975) and acidification to a pH ∿5.

How would acidification effect the rate of deposition of silica from solution?

CALCITE DEPOSITION

Deposition of calcite has been troublesome in a number of discharging wells at Wairakei, Ohaaki (Broadlands), Kawerau, Imperial Valley, and particularly at Kizeldere, Turkey.

Ellis (1963) has measured the solubility of calcite in solutions containing CO_2 and as shown in Figure 13.3 the mineral solubility decreases with temperature. The experimental data

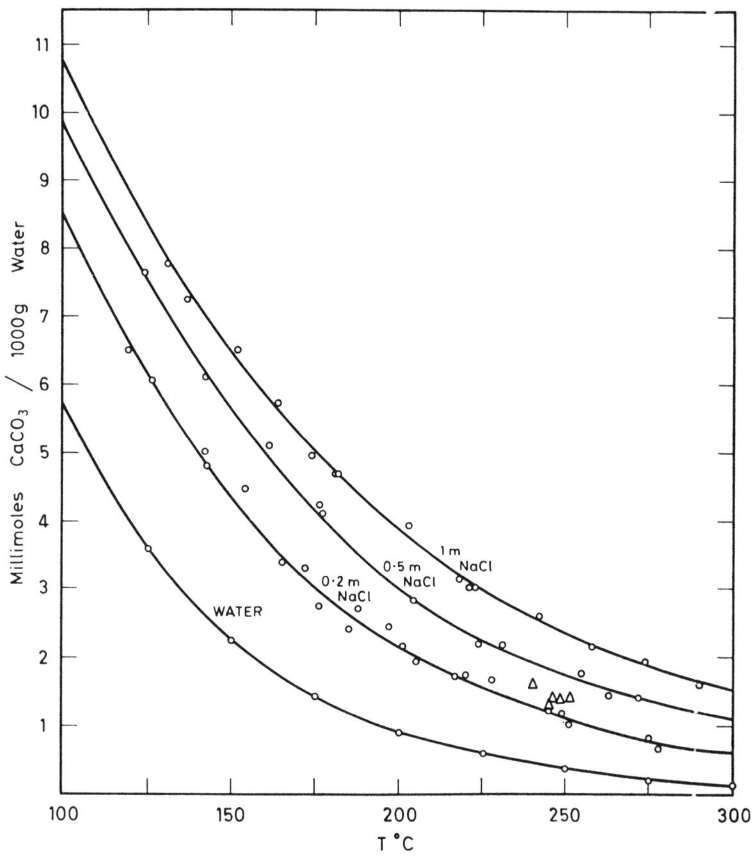

Figure 13.3. Solubility of calcite in water and in sodium chloride solutions at a carbon dioxide pressure of 12 atmospheres (12.2 bars). Triangles mark solubilities in Wairakei thermal waters and circles the experimental data. (From Ellis, 1963 with permission)

for calcite solubility in Figure 13.3 may be used to derive equilibrium constants ($K_{CHC'}$) for the solubility reaction over the temperature range shown.

$$CaCO_3 + CO_{2,g} + H_2O = Ca^{++} + 2HCO_3^-$$

By inserting appropriate values for the activity coefficients of the ionic species (γ_i) and fugacity coefficient, α, for CO_2, and since each mole of calcite that dissolves gives 1 mole of Ca^{++} but 2 moles of HCO_3^-, the expression for the equilibrium constant is as follows

$$\log K_{CHC'} = m_{Ca^{++}} m^2_{HCO_3^-} \gamma_{Ca^{++}} \gamma^2_{HCO_3} / \alpha_{CO_2} P_{CO_2}$$

$$= 4m^3_{CaCO_3} \gamma_{Ca^{++}} \gamma^2_{HCO_3} / \alpha_{CO_2} P_{CO_2} \qquad (8)$$

For most geothermal conditions the fugacity coefficient of $CO_2 \simeq 1$ and, in the absence of other data, we may ignore the possibility (for the time being) of additional ion pairing reactions like

$$Ca^{++} + HCO_3^- = CaHCO_3^+$$

$$Ca^{++} + CO_3^= = CaCO_3$$

$$Ca^{++} + OH^- = CaOH^+$$

Data obtained in this manner are shown in Table 13.1 and from them we may obtain equilibrium constants for other forms of the calcite solubility reaction by adding appropriate constants for the gas soluibility or for weak acid dissociation.

Table 13.1

			100	150	200	250	300	Ref
$CaCO_3 + H_2O + CO_{2(g)}$	$= Ca^{++} + 2HCO_3^-$	$(K_{CHC'})$	-7.56	-8.53	-9.55	-10.71	-12.19*	(1)
H_2CO_3	$= CO_{2(g)} + H_2O_{(1)}$	(K_H)	1.97	2.08	2.08	1.98	1.8	
$HCO_3^- + H^+$	$= H_2CO_3$		6.42	6.77	7.23	7.75	8.29	(2)
HCO_3^-	$= H^+ + CO_3^=$		-10.16	-10.39	-10.78	-11.29	-11.89	(3)
$CaCO_3$	$= Ca^{++} + CO_3^=$			-10.07			-14.0*	
$CaCO_3 + H_2CO_3$	$= Ca^{++} + 2HCO_3^-$	(K_{CHC})						

References (1) recalculated from Ellis, (1963) (0.2 m NaCl solutions)
(2) Read, (1975)
(3) Plummer and Busenberg, (1982)

* Extrapolated

The solubility data of Ellis (1963), Holland and Borcsik (1965), Segnit, et al. (1962) and Malinin (1971) show that increasing salinity does increase the solubility of calcite (see review by Holland (1978)). This is probably due to additional complexes forming in solution, but increasing ion association at high temperatures introduces an error into the calculation of ionic strength and therefore of the individual ion activity coefficients used in reducing the experimental data to equilibrium constants.

Complete Table 13.1 and use it to obtain an equilibrium solubility product of calcium carbonate expressed by the following reaction equation,

$$CaCO_3 + H^+ = Ca^{++} + HCO_3^-$$

We have used data derived from this table in previous chapters; firstly, for the rapid estimation of aquifer fluid pH by assuming calcite saturation and later to obtain values of the relative activity ratios ($a_{Ca^{++}}/a_{H^+}$) of fluids at Broadlands and Wairakei.

From the reaction equation it is quite clear that removal of CO_2 to a steam phase during flashing leads to supersaturation with respect to calcite even though the accompanying temperature drop itself leads to a calcite saturation increase. Since most reservoir fluids are close to saturation with calcite, carbonate scaling inside production casing is a possibility for all geothermal wells and calcite may deposit in the reservoir itself during exploitation, leading to a slow decline in output.

The fluid flow regime developed within a discharging well during flashing may be important in determining whether or not calcite deposits may form. Where flashing leads to an annular flow regime, more effective transfer of CO_2 to the steam phase may occur than where a disperse (bubbly) two phase flow regime is formed. In the first case higher saturation states are reached and m_{H2CO3} is buffered by the steam composition.

In well KJ-9 at Krafla, Iceland, Stefansson (1980) shows that a 3.9 mg/kg decrease in calcium content occurred in the well fluid between the reservoir and wellhead corresponding to deposition of .16 gm calcite/sec, or 1 m^3 through the 180 day flow test. Subsequently the volume of scale actually measured in the well liner was 1.1 m^3!

The state of calcite saturation with respect to steam loss may also be shown with respect to the solubility product of calcite. Figure 13.4 (from Arnorrson, 1978) shows such data for a number of Icelandic wells calculated using a modified version of the ENTHALP program discussed in Chapters 7 and 13. The most serious calcification occurred in wells at Leira and

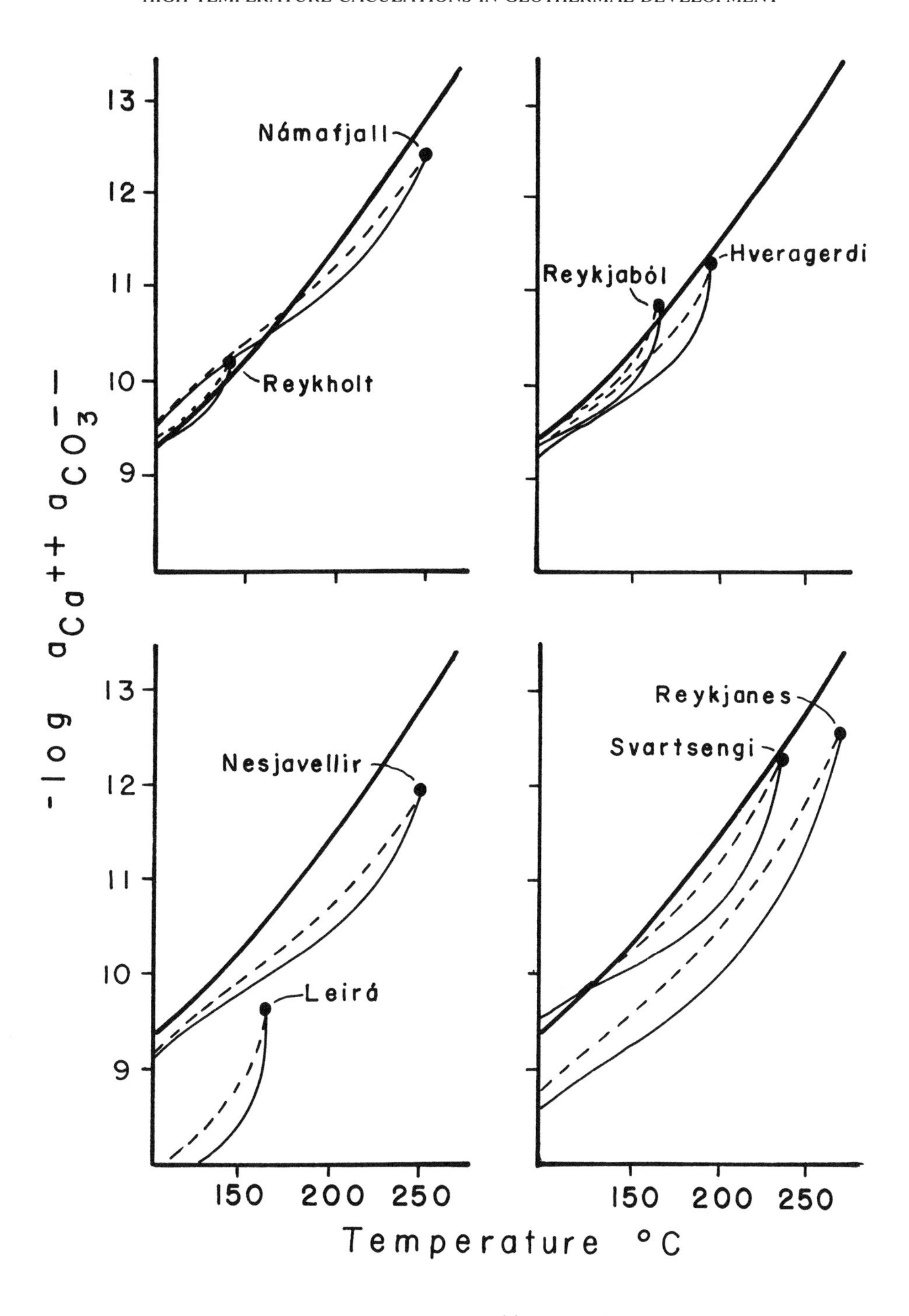

Figure 13.4. The computed activity product of Ca^{++} and $CO_3^{=}$ in geothermal water during one step adiabatic flashing in relation to the calcite solubility curve (thick solid curve). The solid lines assume maximum degassing, but the dashed lines 1/5 of maximum degassing. Calcite solubility data from Helgeson (1969). Solutions supersaturated with respect to calcite lie below the calcite solubility curve. (Reproduced from Arnorrson, 1978).

Hvergerdi and the data shown suggest that it may also be a problem at Svartsengi where high supersaturation occurs. High supersaturations at Reykjanes have as yet not resulted in deposition of calcite in the wells but production decline may in this case relate to calcite deposition in the reservoir.

Aragonite has been found in some well precipitates and was particularly abundant at Kizeldere. Its formation may relate to high supersaturations. Tulloch (1982) has shown that calcite in a number of New Zealand wells is unusually high in magnesium, so that it is also possible that Mg exercises some kinetic control.

Various means have been suggested to combat or remove calcite scaling in wells to avoid the need for an expensive reaming-out operation using a drillrig. One interesting possibility is the EFP (Equilibrium Flash Production) system (Kuwada, 1982) in which CO_2 separated at the wellhead is returned under pressure down the production well acting both as a pumping system and as a chemical means of preventing scale formation.

Explain the chemical basis for the operation of the EFP system.

As an additional factor it is interesting to note that many New Zealand wells which have shown calcite scaling have evidence of inflow of near surface waters during their prolonged exploitation. Such near surface waters, at lower temperatures, may act as an additional source of calcium and magnesium in the well and therefore lead to a higher deposition potential than would be expected from unmodified aquifer fluid.

As a postscript - if rapid calcite deposition occurs within a discharging well, how might it affect the estimation of reservoir temperatures or alteration mineral stabilities using chemical data?

STEAM CONDENSATES

The chemistry of environments containing steam condensates is an important consideration in geothermal development, particularly in terms of corrosion. Much of the basis for selection of steam pipe material is empirical, but some progress is currently being made in establishing a thermochemical basis for such choices (Giggenbach, 1979).

Consider steam flowing through a pipeline. Heat loss through the pipe wall leads to the formation of condensate which accumulates and flows along the pipe wall. We need to know the pH of this liquid phase in order to consider what corrosion products may form. Again we take an example from Ohaaki. The BR22 steam separated at 10 b.g. contains

1165	mmoles/100	moles	steam		CO_2
22.1	"	"	"	"	H_2S
and 4.1	"	"	"	"	NH_3

(Note that because NH_3 is a very soluble gas relative to CO_2 and H_2S, it will play a more important role in these calculations, than in our previous calculations of aquifer chemistry.)

Now calculate the composition and pH of condensate formed at 175°C (7.92 b.g):

Gas concentrations expressed in the above units may be used directly with Henry's Law coefficients, K_H (see Table 4.4) to obtain the concentrations of dissolved gas in the condensate as follows. First calculate the partial pressure of each gas.

P_{gas}	=	P_{total} x mole fraction of gas		
P_{H_2O}	=			8.92 bars
P_{CO_2}	=	8.92 x (1165 x 10^{-5})	=	0.104 bars
P_{H_2S}	=		=	bars
P_{NH_3}	=		=	bars

and if X = mole fraction of gas dissolved in the liquid

X_{CO_2}	=	$P_{CO_2}/K_{H_2CO_3}$	=	1.51 x 10^{-5}
X_{H_2S}	=		=	
X_{NH_3}	=		=	

Charge balance in this environment requires that

$$m_{NH_4^+} + m_{H^+} = m_{HCO_3^-} + m_{HS^-} + m_{OH^-} \quad (10)$$

Since m_{H2CO3}, etc., are fixed by the gas partial pressures, substitution of appropriate equations gives :

$$a_{H^+}\{m_{NH_3}/K_N\,\gamma_{NH_4^+} + 1/\gamma_{H^+}\} - 1/a_{H^+}\{(K_C m_{CO_2}/\ HCO_3^-) + (K_S m_{H_2S}/\gamma_{HS^-}) + (K_W/\gamma_{OH^-})\} = 0 \quad (11)$$

(Subscripts for each of the weak acid dissociation constants have here been abbreviated: C = H_2CO_3, S = H_2S, N = NH_4, W = H_2O)

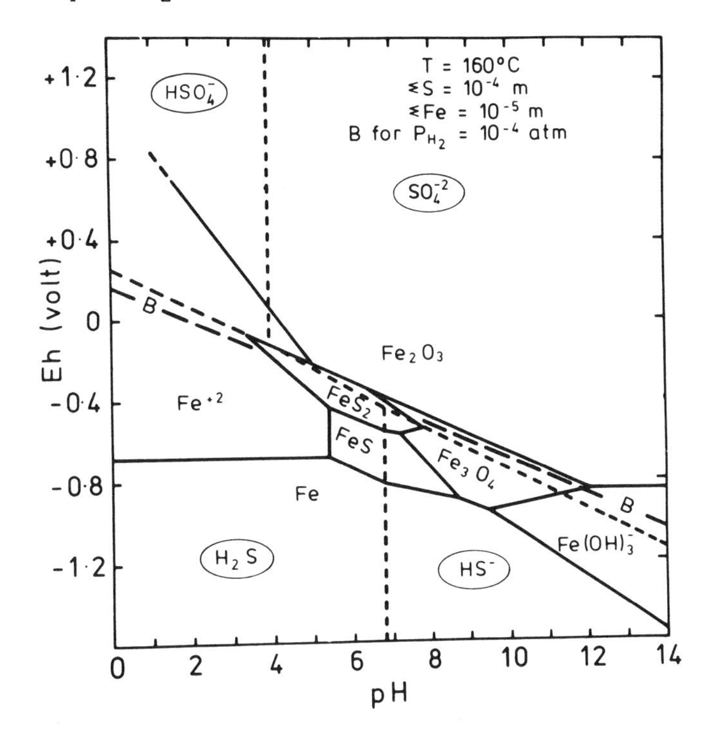

Figure 13.5. EH-pH diagram for the system Fe-S-O-H at 160°C. The partial pressure of H_2 for Ohaaki condition is indicated by line B. (Reproduced from Braithwaite and Lichti, 1979)

This equation may be solved numerically or graphically to obtain the pH of the condensate. (For the small gas concentrations involved, ionic strengths are low and all the actvity coefficients may be initially taken as 0.98). Now complete the calculation of pH condensate for the BR22 steam.

Apart from pipeline and other condensate corrosion problems, calculations of condensate chemistry are becoming particularly important where control of gas emission from cooling towers is required (e.g. Geysers California) using direct contact with cooling waters or shell and tube condensers (Weres, 1983). Glover (1982) has written an HP41C program to handle problems such as these. In these applications, as with many of the scaling problems discussed above, kinetic factors are important. Weres (op cit) has emphasized that reaction rates for dissolution of CO_2 and conversion to HCO_3^- are pH dependent and of the same order as residence times in the condenser so that H_2S and NH_3 exert a greater effect on condensate chemistry than would occur for complete CO_2 equilibration.

One of the principal reasons for calculating pH and species concentrations in steam condensate is to provide a framework for the interpretation of corrosion products on steam lines -- a prerequisite for the determination of corrosion rates. Figure 13.5 gives an example from the Ohaaki field where the relatively high total sulfur content results in the formation of Mackinawite (Fe, Ni)S, pyrrhotite ($Fe_{1-x}S$) and troilite (FeS) at the condensate pH of 5.7 (Braithwaite and Lichti, 1979). By contrast at Wairakei, the separated steam contains much less sulfur so that iron oxides are the stable corrosion products. Using related diagrams it was possible to trace the path of the corrosion process in the Wairakei steam transmission lines (Giggenbach,1979).

REFERENCES

Arnorsson, S., 1978, Precipation of calcite from flashed geothermal waters in Iceland: Contributions to Mineralogy and Petrology, v. 66, p. 21-28.

Bangma, P., 1961, The development and performance of a steam-water separator for use in geothermal bores: U.N. Conference on New Sources of Energy, Rome, p. 60-78.

Bohlman, E. G., Mesmer, R. E., and Berlinski, P., 1980, Kinetics of silica deposition in simulated geothermal brines: Society of Petroleum Engineers Journal (August, 1980), p.239-248.

Braithwaite, W. R., and Lichti, K. A., 1980, Surface corrosion of metals in Geothermal Fluids at Broadlands; in Casper, B. A., and Pinchback, T. R. (eds), Geothermal Scaling and Corrosion: American Society of Testing and Materials, p. 81-112.

Ellis, A. J., 1959, The Solubility of Calcite in Carbon Dioxide Solutions: American Journal of Science, v. 257, p. 354-365.

Ellis, A. J., 1963, Solubility of calcite in sodium chloride solutions at high temperatures: American Journal of Science, v. 261, p. 259-267.

Giggenbach W. R., 1979, Application of mineral phase diagrams to geothermal corrosion: New Zealand Geothermal Workshop, University of Auckland.

Glover, R. B., 1982, Determination of the Chemistry of some geothermal environments: N.Z. D.S.I.R. Chem. Div. Report, C.D. 2323.

Henley, R. W., 1983, Chemistry and Silica Scaling Potential of Multiple Flash Geothermal Waters: Geothermics, v. 12, p. 307-321.

Henley, R. W., and Singers, W. A., 1982, Geothermal gas separation in conventional cyclone separators: New Zealand Journal of Science, v. 25, p. 37-45.

Holland, H.D., 1979, The solubility and occurrence of non-ore minerals; in Barnes, H.L.(ed), Geochemistry of Hydrothermal Ore Deposits: Wiley, New York, p. 461-508.

Holland, H.D., and Borcsik, M., 1965, On the solution and deposition of calcite in hydrothermal systems; in Symposium Problems of post-magmatic ore deposition: Prague, v. 2, p. 418-421.

Kuwada, J. T., 1982, Field Demonstration of the EFP System for Carbonate Scale Control: Geothermal Resources Council Bulletin, p. 3-9.

Mackrides, A. C., Turner, M., and Slaughter, J., 1980, Condensation of silica from supersaturated silica solutions: Journal of Colloid and Interface Science, v. 73, p. 345-367.

Plummer, L.N., and Busenberg, E., 1982, The solubilities of calcite,aragonite and vaterite in CO_2 - H_2O solutions between 0 and 90°C, and an evaluation of the aqueous model for the system $CaCO_3$ - CO_2 - H_2O: Geochimica et Cosmochimica Acta, v. 46, p. 1011 - 1042.

Read, A.J., 1975, The first ionization constant of carbonic acid from 25 to 250°C and to 2000 bars: Journal of Solution Chemistry, v. 4, p. 53 - 70.

Rimstidt, J. B., and Barnes, H. L., 1981, The kinetics of silica-water reactions: Geochimica et Cosmochimica Acta, v. 44, p. 1683-99.

Rothbaum, H. P., Anderton, B. H., Harrison, R. E., Rhode, A. G., and Slatter, A., 1979, Effect of silica polymerisation and pH on silica scaling: Geothermics, v. 8, p. 1-20.

Rothbaum, H. P., and Anderton, B. H., 1975, Removal of silica and arsenic from geothermal discharge waters by precipitation of useful calcium silicates: U.N. Symposium on Development and Use of Geothermal Resources, San Francisco, p. 1417-1426.

Stefansson, V., 1980, The Krafla Geothermal Field, N.E. Iceland; in Rybach, L. and Muffler, L. J. P., (eds.), Geothermal Systems: Principles and Case Histories: p. 273-294.

Segnit, E. R., Holland, H. D., and Biscardi, C. J., 1962, Solubility of calcite in aqueous solutions, I: Geochimica et Cosmochimica Acta, v. 26, p. 1301-1331.

Singers, W.A., Henley, R.W., and Giggenbach, W.F., GEODATA-ENTHALP: Revisions and Users Guide; N.Z. D.S.I.R., Chemistry Division Report, in prep

Truesdell, A.H., and Singers, W.A., 1971, Computer calculation of downhole chemistry in geothermal areas: N.Z. D.S.I.R., Chemistry Division Report CD2136.

Tulloch, A. J., 1982, Mineralogical observations on carbonate scaling in geothermal wells at Kawerau and Broadlands: N.Z. Geothermal Institute Workshop, p. 131-135.

Weres, O., Yee, A., and Tsao, L., 1980, Kinetics of Silica Polymerisation: Lawrence Berkely Laboratory Report, LBL-7033, 119 p.

Weres, Oleh, 1983, Partioning of hydrogen sulfide in the condenser of Geysers Unit 15: Geothermics, v. 12, p. 1-16.

Weres, O., and Yee, A., 1982, Equations and type curves for predicting the polymerisation of amorphous silica in geothermal brines: Society of Petroleum Engineers Journal (Feb., 1982), p. 9-16.

Plate 5: Sketch of the Commodore Mine, Creede, Colorado by Pan Eimon. The portals tap the southern end of the Amethyst vein system which between 1890 and 1983 had produced more than 70 million ounces of silver plus substantial amounts of lead, zinc, copper and gold. The view represents a zone that was probably 400 to 600 meters below a late Oligocene thermal spring field.

Chapter 14
HIGH-TEMPERATURE CALCULATIONS APPLIED TO ORE DEPOSITS

Throughout these short-course notes we have presented situations that have possible application to mineral deposits. The principles and procedures involved in interconverting between compositional and other thermodynamic parameters and in dealing with mixing, boiling, and cooling are the same whether one is dealing with a geothermal system or the formative stages of an ore deposit. The problems of silica and calcite deposition in geothermal wells and pipes are almost perfect analogs to some ore veins. In this chapter we shall apply some of these calculations to a few types of deposits as examples, not as a definitive chemical characterization for any specific deposit.

THE MINERAL WORLD: WHY THE RECORD IS SO HARD TO INTERPRET

Most geologists are familiar with the idea of reconstructing ancient environments from the information recorded in rocks. In ore deposits this requires the application of thermodynamic tools to try to reconstruct an ancient geochemical environment. By contrast, in the active geothermal systems we can deal with features that can be measured directly. In fact, the real strength of the comparison of geothermal with hydrothermal processes is that the geothermal scene provides opportunities to check one's computations by actually measuring temperatures, solution concentrations and so on; and the high level of agreement between observation and prediction signals hope for solution of many of the existing enigmatic problems.

A number of factors, which we can touch upon only briefly here, create situations which cause problems in the application of thermodynamics to real rocks and fluids. Among these are: metastable phases and disequilibria in general, incompletely characterized phases, surface energies inherent in fine-grained materials, overlooked chemical species, and inadequate or inaccurate thermodynamic data.

Metastability

The world should be thankful for metastability! Otherwise, each of us would vanish into puffs of CO_2, HN_3 and steam, plus damp piles of graphite, apatite, and so on. In fact, if equilibrium were the rule, all records of previous events would be erased in order for the equilibrium of the moment to prevail. The basic tools of the geothermal/ hydrothermal geoscientist - the geothermometers, the mineral parageneses, the fluid inclusions, the isotopic distributions, and on and on - would be erased if equilibrium truly prevailed. On the other hand, our ability to use "simple" thermodynamics depends on the interpretation of what we observe in terms of some previous equilibrium system that has been "frozen" for us to examine (see also discussion in Barton et al., 1963). Reactions of importance to geologic processes range widely with regard to the rates at which they approach equilibrium. Figure 14.1 compares rates and shows that most homogeneous reactions in solution (e.g., $H_2S = H^+ + HS^-$ or $Cu^+ + H^+ = Cu^{++} + 1/2\ H_2$) are geologically "instantaneous", although others (e.g., $SO_4^= + 4H_2 + 2H^+ = H_2S + 4H_2O$) are quite sluggish. The precipitation or dissolution of minerals procedes at different rates, the less soluble minerals tending to take longer to equilibrate, and as we have emphasized elsewhere, the activity of silica is especially important in its influence over other reactions. Solid-state reactions (e.g., the homogenization of compositionally banded sphalerite, or the reaction of covellite with chalcopyrite to yield bornite + pyrite) have wide ranges of rates with the hard, high-melting-point minerals tending to be the most sluggish.

The activity - activity diagram provides a convenient way to assess the consequences of a given disequilibrium. Among the sulfides, pyrite justifiably has the reputation of being highly recalcitrant, thus we shall use pyrite to exemplify a procedure for dealing with metastability. Figure 14.2 contains two a_{S2} - a_{FeS2} diagrams that illustrate situations in which the supersaturation with respect to pyrite permits the metastable formation of idaite (Cu_5FeS_6) or the metastable assemblage covellite + chalcopyrite (Fig. 14.2a), and the precipitation of the metastable assemblage of pyrrhotite + low-iron sphalerite (Fig. 14.2b). Metastable equilibria follow the same mass action rules as other equilibria and are potentially quite useful. Metastability can also have a strong influence on mineral solubilities, as various examples throughout this book illustrate.

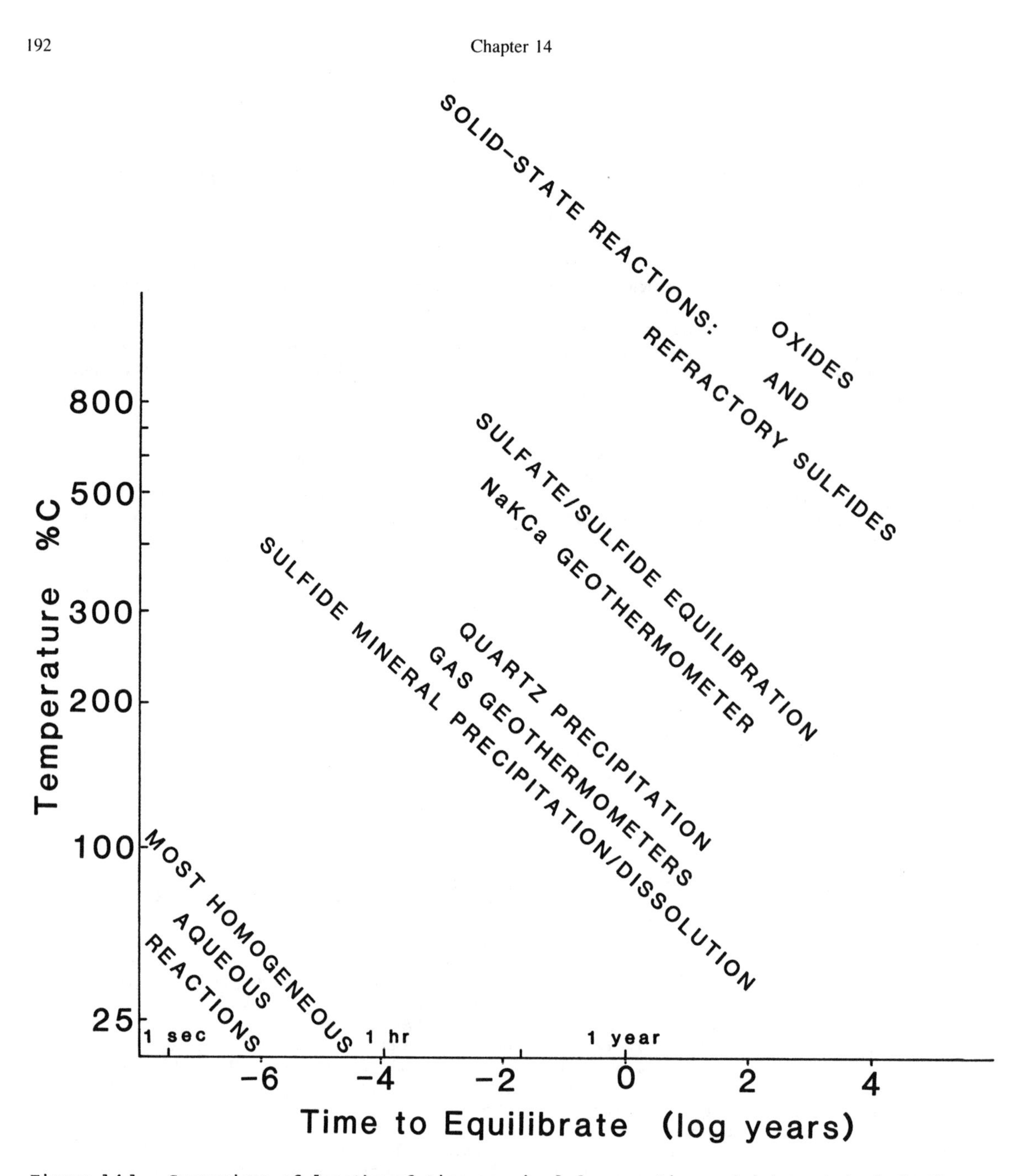

Figure 14.1. Comparison of lengths of time required for reactions of interest to hydrothermal/geothermal processes to reach equilibrium. This treatment has no intention of providing more than a very qualitative comparison of times to equilibrate (and realizing that most reaction rates are concentration-dependent, and thus "never" reach true equilibrium).

Inadequately Characterized Phases

Geothermal systems and low-temperature mineral deposits share the problem of "messy" minerals. Data are abundant, and probably even reasonably accurate for the "tidy" minerals such as quartz, pyrite, galena, anhydrite, and so on; but the mixed-layer clays, the zeolites, tetrahedrite, iron sulfide "gels", the chlorites, the amorphous clays, vermiculite, and so on present serious problems with regard to thermodynamic characterization. These problems also spill over into laboratory studies where they confuse, blind, or mislead the experimentalist.

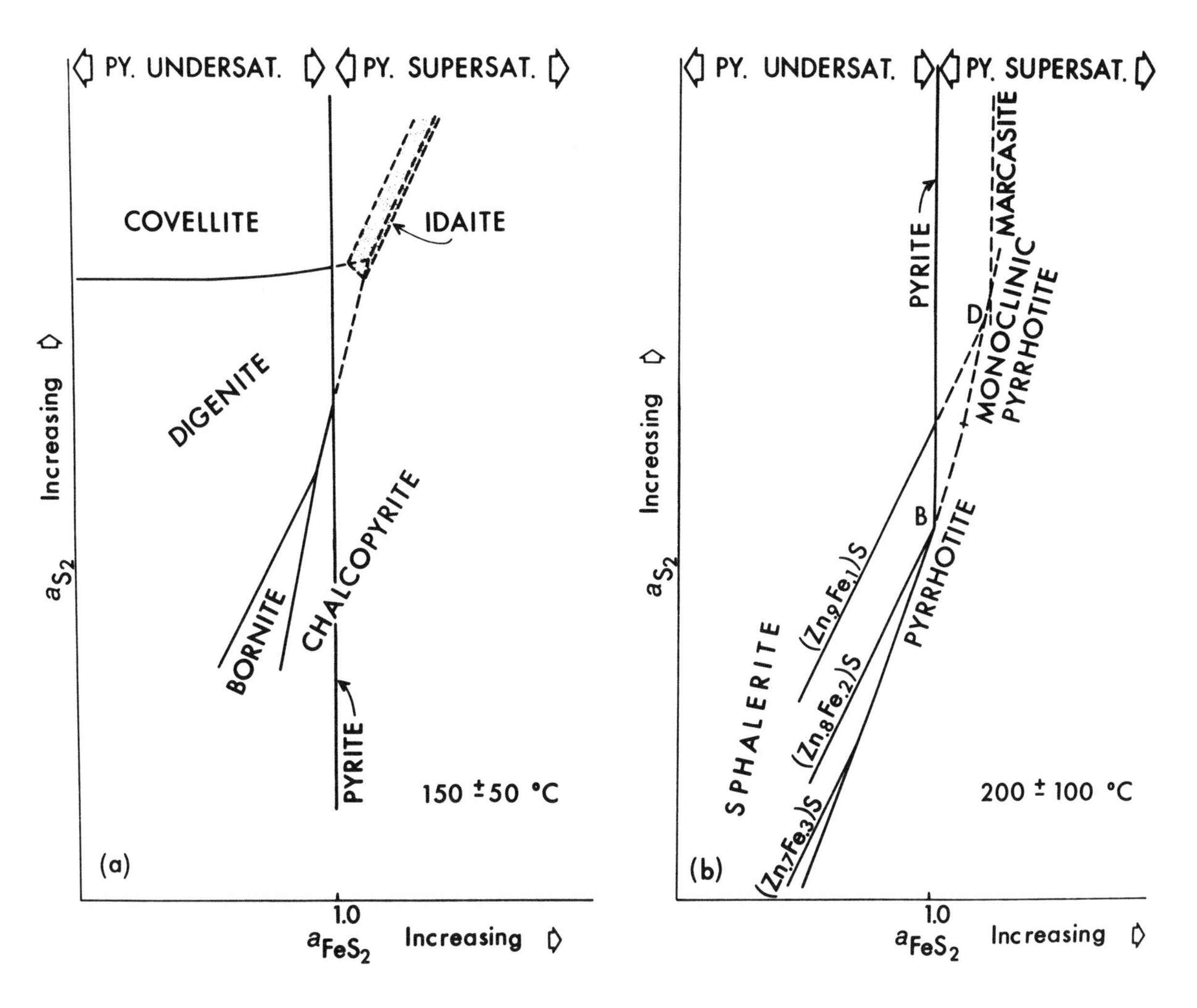

Figure 14.2. Schematic a_{S2} - a_{FeS2} diagrams illustrating the effect of pyrite saturation on phase assemblages.
The sketches are made using a log - log base. From Barton and Skinner (1979).
Figure 14.2a. The Cu-Fe-S system showing the stabilization of idaite, and of chalcopyrite + covellite in the pyrite-supersaturated region.
Figure 14.2b. The Fe-Zn-S system showing contours of composition of sphalerite in equilibrium with pyrrhotite, and their metastable extentions into the pyrite-supersaturated region. It shows how low-iron sphalerite can form in metastable equilibrium with pyrrhotite.

In Chapter 8 we noted how Barton et al. (1977) and Giggenbach (1980, 1981) identified chlorite as a key phase in buffering chemical environments. But they did not obtain a real understanding of chlorite; they merely dealt with the chemical potential for a hypothetical, hydrous iron-silicate. Such a procedure is valid for their applications, but their numerical solutions have little direct transfer value to the problems of chlorite elsewhere.

Surface Energy

Mineral surfaces possess an energy that contributes to modifying the position of equilibrium and provides a driving force for recrystallization into coarser-grained, less energetic, and more stable products; the surface energy may be considered an aspect of Metastability. The surface energy is a function of both sides of the crystal surface; the energy terms decrease in the following order: a crystal surface in a vacuum, the same crystal surface immersed in water, the crystal surface in contact with another crystal, a twin surface within a single crystal. The crystallographic orientation of the surface is also important, but numerical data for calculation of crystal morphologies in heterogeneous media are absent (although we do know, for example, that the surface energy drive for pyrite to form striated cubes and pyritohedra is much greater than that for galena to form any sort of crystal form.) In a summary on recrystallization, Ewers (1967) notes that surface energies in vacuum are from a few hundred to a few thousand ergs cm^{-2}. (1 erg = 10^{-7} J)

Let us calculate the surface energy for 1 mole of galena as cubic crystals having particle sizes (cube edge dimensions) ranging down to 5.9 Å, the å of a single unit cell (1 Å = 10^{-10} meter). The molar volume of galena is 31.49 cm^3, thus a cube containing 1 mole would be 3.158 cm on an edge; the surface area of such a cube would be 6 x $(3.158)^2$ = 59.832 cm^2. Multiplying the area by Bondi's (1953) estimate of 520 ergs cm^{-2} for galena we get 3.111 x 10^4 ergs or 0.003111 joules mol^{-1}, a rather inconsequential amount.

[The area of cubic particles = 6 x $edge^2$; the number of such particles per mole = (molar vol.)/$(edge)^3$: thus Area = 6 x (molar vol.) x $(edge)^{-1}$.]

For free-standing galena:

particles/mol	cell edge (cm)	surface energy (joules/mol)
1	3.158	0.003111
3.149 x 10	1	0.009731
3.149 x 10^4	0.1	0.09731
3.149 x 10^7	0.01	0.9731
3.149 x 10^{10}	0.001	9.731
3.149 x 10^{13}	0.0001 (= 1 micron)	97.31
3.149 x 10^{16}	0.00001	973.1
3.149 x 10^{19}	0.000001	9,731
1.882 x 10^{22}	0.000000118716 (= 2 x å)	82,680
3.149 x 10^{22}	0.0000001 (= 10 Å)	97,310
1.505 x 10^{23}	0.000000059358 (= å)(Z=4)	165,360

At 25°C a surface energy increase of 5702 joules increases the solubility product by 1 log unit; thus we note that at about 1 micron the surface energy term becomes appreciable, amounting to 0.02 log units in K_{sp}. If estimates for the energies of corners and edges could be added, the surface energies would be substantially larger for incipient crystals. It makes one wonder how a crystal ever starts in the first place, but fortunately for the writers, that issue is not our question here. $ What is the surface energy of one mole of equant gold grains each of which is a cube 0.1 micron on edge? Assume that the surface energy of gold is 1000 ergs/cm^2 (this energy figure is probably too large, but the overestimate of surface energy should help compensate for the neglect of edges and corners). The cell edge of gold is 4.07 x 10^{-8} cm; and the unit cell contains 4 Au.

$ If gold dust in a placer has grains 0.1 micron in diameter, how much more soluble are such fine grains than is a nugget a centimeter in diameter? Express the result as a percent supersaturation.

$ What other factors might influence the relative solubility of gold in an active stream placer, or for that matter, in a sheared gold-bearing vein?

Consider the effect of grain shape. If one has tetragonal symmetry for the grains with a = c/100 (an acicular crystal), or a = 100c (a plate-like crystal), the surface areas per mole are respectively 3.1 and 7.3 times larger than for a similar mass having for isometric symmetry. These factors carry over to the surface energies only to the extent that the freely formed, "equilibrium" crystals would have had approximately equant morphologies, i.e., an acicular crystal of an isometric mineral such as cuprite; the shape argument would not apply to a strongly non-isometric phase such as chrysotile for which the surface energies of the edges may be very different from other surfaces.

The bottom line on grain size for gold is that: (1) surface energy alone is insufficient to drive solution/reprecipitation processes to form large nuggets, and (2) the standard thermodynamic data for gold are probably appliciable to the fine-grained gold deposits of the Carlin type.

Overlooked Chemical Species

Failure to include all of the pertinent phases or solution species clearly can be serious. For example, in none of the preceeding pages have we mentioned siderite, $FeCO_3$; yet siderite is a common gangue mineral in ore deposits and $FeCO_3$ is a major component in ankeritic carbonate which is a common mineral in sediments. Regarding metal solubilization and transport, none of the current computations considers any organic complexes, or the possibility of such ligands as NH_3, CN^-, or SCN^-.

Inaccurate or Incomplete Data

This is obviously a source of possible error, as the discrepancies noted in Chapter 6 for the kaolinite + Kmica reaction demonstrate. Sometimes surprisingly large errors can be found. For example, Barton (1969) found that the basic thermodynamic data (the standard free energies and enthalpies of formation) for orpiment (As_2S_3) and realgar (As_2S_2) were off by about 80,000 joules per mole; that corresponds to an error of 14 log units in solubility product constant! Such large errors are not anticipated, but even the most extensively studied systems often have discordant values by different researchers. J. L. Haas and coworkers have devised a computer-based appraisal system to derive the "best fit" data from the whole array of information available on either simple or complex systems (Haas and Fisher, 1976).

APPLICATIONS TO MINERAL DEPOSITION

Several of the examples chosen in earlier chapters apply to epithermal deposits; similar geochemical relationships also apply to the submarine environments in which massive sulfide deposits form, and to a great extent to the formative environments for the Mississippi Valley-type deposits. However, the interpretations of geologic processes are highly divergent for various deposit types. Let us consider here the factors which determine just which metals are precipitated in which proportions to yield the precious- and base-metal ores, and perhaps which might be responsible for differences between barren and productive hydrothermal systems.

Chapter 9 has dealt with factors influencing metal solubilities; reviewing these we note that temperature, pH, f_{H2} (which also represents f_{O2}), a_{Cl^-} and f_{H2S}, are the most important and that pressure and ionic strength are comparatively negligible except as we consider the region near the critical curve. These factors are complexly interrelated, and definitive answers are difficult to make quantitative (as well as difficult to describe). Our goal in this exercise is to show how to sort out and appraise the role of several geochemical variables that may influence mineral solubilities. Complete Table 14.1, supplying the differentials of solubility, C, with respect to some important geochemical parameters.

Table 14.1 -- Factors influencing mineral solubilities.

Conditions: 250°C, pH = 5.0, log f_{H2} = -3.0, log a_{Cl^-} = 0.0, log f_{H2S} = -2.0.
Assume that the activity coefficients do not vary.

	MINERAL				
	Galena	Pyrite	Gold	Silver	Argentite
species considered					$AgCl_2^-$
dlog C/dpH					-1
dlog C/dlog a_{H2}					0
dlog C/dlog a_{Cl^-}					2
dlog C/dlog f_{H2S}					-0.5
dlog C/dt	+0.03	-0.02	-0.01	+0.05	+.06

1. The first step is to identify and list the dominant metal-bearing species in solution under the conditions stated for Table 14.1; this information is available in Chapter 9. The conditions listed for Table 14.1 provide enough information to decide among alternative species.

2. Next, write the balanced reaction expressing solubility in terms of as many variables as necessary. Be sure to write the reactions in terms of the chemical species identified in the differentials above; that is, use H_2S, H^+, H_2, and Cl^-, not $S^=$, O_2, OH^- or others not already listed in the table headings or "species considered" row. An example for argentite is given below.

3. For some minerals, some of the variables will not appear in the solubility reaction; therefore, differentials involving these variables will be assigned 0.0 values in Table 14.1 as they do not influence solubility.

4. From the stoichiometric coefficients in the solubility reaction determine all of the differentials (except those involving temperature). For example, the equation for the dissolution of argentite is:

$$Ag_2S + 4\ Cl^- + 2\ H^+ = 2\ AgCl_2^- + H_2S \qquad (1)$$

for which the log of the equilibrium constant has the form:

$$\log K = 2 \log a_{AgCl^-} + \log f_{H\ S} - 4 \log a_{Cl^-} + 2\ pH \qquad (2)$$

Thus because H_2 does not appear in the reaction, dlog C/dlog a_{H2} must be 0.0. The a_{AgCl2^-} term represents the major silver-bearing species in solution; therefore, the a_{AgCl2^-} term (divided by the activity coefficient) also constitutes the "C" in the differentials. But we have noted above that each activity coefficient is assumed to remain constant, thus the log change in activity is exactly the same as the log change in concentration and thus the dlog C term can represent changes in either activity or concentration.

Consider the differential with respect to hydrogen ion, dlog C/dpH. From equation (2) it is obvious that dlog C/dpH = -2.0/2.0 = -1.0; for each log unit that the pH increases (assuming that everything else is held constant) the solubility of argentite decreases by one log unit. Similarly, dlog C/da_{Cl^-} = 4.0/2.0 = 2.0; for each log unit that a_{Cl^-} increases, the solubility of argentite increases by 2.0 log units.

What about conditions in which the species used for the equation may not be the major species? For example, if the pH is greater than about 7 the dominant sulfide species is HS^- instead of H_2S. Substituting HS^- for H_2S in equations (1) and (2), and assuming that the total sulfide (i.e., HS^- + H_2S) concentration remains constant, the derivative of C with respect to pH becomes -0.5; a pH increase of 2 units would decrease C by 1.0 log unit.

5. The only derivative not capable of evaluation by inspection of the balanced equations is dlog C/dt. We could make a computation from the enthalpy or entropy changes for the reaction; or more simply, we can calculate the equilibrium constant at two temperatures and thus evaluate dlog K/dt. Note that the changes in log K's with t are not very linear (log K vs 1/t would be a better approximation) and thus care should be taken not to use too wide a temperature range. These reactions have been evaluated using information given in Chapter 9 plus the free energy for the reaction 4 Ag + S_2 = 2 Ag_2S. HOWEVER, there is a serious flaw in the application of dlog C/dt, the bottom row of this tabulation: by relating the oxidation state solely to f_{H2} we have implicitly assumed that the that the H_2 content of the fluid is the dominant redox buffer. But you can calculate that the H_2 concentration is only a few parts per million, which may prove to be ineffective as a buffer, as we have discussed in Chapter 8. Obviously we need to identify the redox buffer functioning for the deposit being considered, but that is a site-specific geologic/mineralogic issue. One is hard pressed to identify such buffers, but perhaps the CO_2/CH_4 ratio (or organic carbon as at Cerro Prieto), the N_2/NH_4 ratio, or ferrous/ferric minerals (such as the chlorite mentioned preiously) might be called upon. Finally, although the first four differentials might be considered individually, all are locked irrevocably to temperature-dependent equilibrium constants; thus consideration of temperature alone is impractical.

Table 14.1 provides a quantitative (or perhaps "pseudo-quantitative" would be a better adjective) basis to let the geologist get at substantial questions concerning the influence of geologic factors (lithologic changes, mixing of solutions, cooling, etc.) on mineral solubility, and thus on the probability and character of mineralization as a function of geologic environment. Although the chemistry is complicated and tedious, and the data base incomplete, the difficult questions are mostly geological, not chemical.

Next let us consider some oversimplified geologic scenarios for the solution residing in Table 14.1:

a. What is expected if the solution encounters a limestone (pH rises, the other variables remain unchanged)?

b. What happens if a kaolin-bearing shale is encountered (pH decreases)?

c. What happens if the solution enters an evaporite (Cl^- increases)?

d. What happens if the solution is diluted by heated meteoric water (t constant)?

e. What happens if the solution enters redbeds (H_2 decreases)?

f. How might pyrite, galena, gold, and argentite be deposited simultaneously?

g. Under what process will gold replace pyrite?

The simple solubility equations used in constructing Table 14.1 can easily be modified to tie them more closely to an observed geologic environment. Let us use as an example the deposition of gold in an environment (perhaps represented by a deposit such as Goldfield, Nevada) where the pH of the depositing solution is buffered by the formation of alunite from kaolinite:

$$3\ Al_2Si_2O_5(OH)_4 + 2\ K^+ + 2\ H^+ + 4\ HSO_4^- = 2\ KAl_3(SO_4)_2(OH)_6 + 6\ SiO_2 + 3\ H_2O \qquad (3)$$

$ 1. Write the equation for the equilibrium constant in terms of a single H^+ (i.e, divide the equation by 2) and add it to the equation you wrote for the dissolution of gold for step 2 following Table 14.1. (If done properly the H^+'s will cancel out.)

$ 2. What is the derivative of gold solubility with respect to log a_{SiO2}? Would you expect higher gold solubility in an alunite + kaolinite-bearing system in which chalcedonic silica textures suggest that a_{SiO2} had been higher than that for quartz?

$ 3. If the K-feldspar + Kmica + quartz buffer were substituted for the alunite buffer what would be the d log C/d log a_{SiO2}?

It is important to recognize that the role of silica here is not as a complexing agent for gold, but as a participant in pH-buffering reactions.

Sulfate has two possible roles in the above reaction: as a component in alunite, and as a possible redox participant in a reaction involving H^+, H_2S and O_2, but it is quite expectable for the sulfate to function near equilibrium with respect to the alunite reaction and at the same time be far removed from redox equilibrium.

Remember that most geologic processes will occur so as to modify more than a single parameter at a time. For example, the mixing of oxygenated (and nitrogenated) ground waters will lower temperature, dilute chloride and sulfide concentration, partly oxidize the system, change pH, and so on. These processes and their influences have been dealt with in earlier chapters as general exercises; they are particularly complicated for such situations as boiling or the venting of hot brines into cold sea water. Their application to specific ores or geothermal systems requires more time and information that is practical here; but the principles and procedures involved are nevertheless valid, and the success of application of the calculations to geothermal environments bodes well for the geochemical modeling of ore deposits.

$ Figure 9.2 shows the solubility window for gold. The conditions we have been discussing with regard to Table 14.1 apply to only part of this window; which part? Note that in other parts of the diagram the same changes in pH and f_{H2} (remember that f_{H2} and f_{O2} are opposite

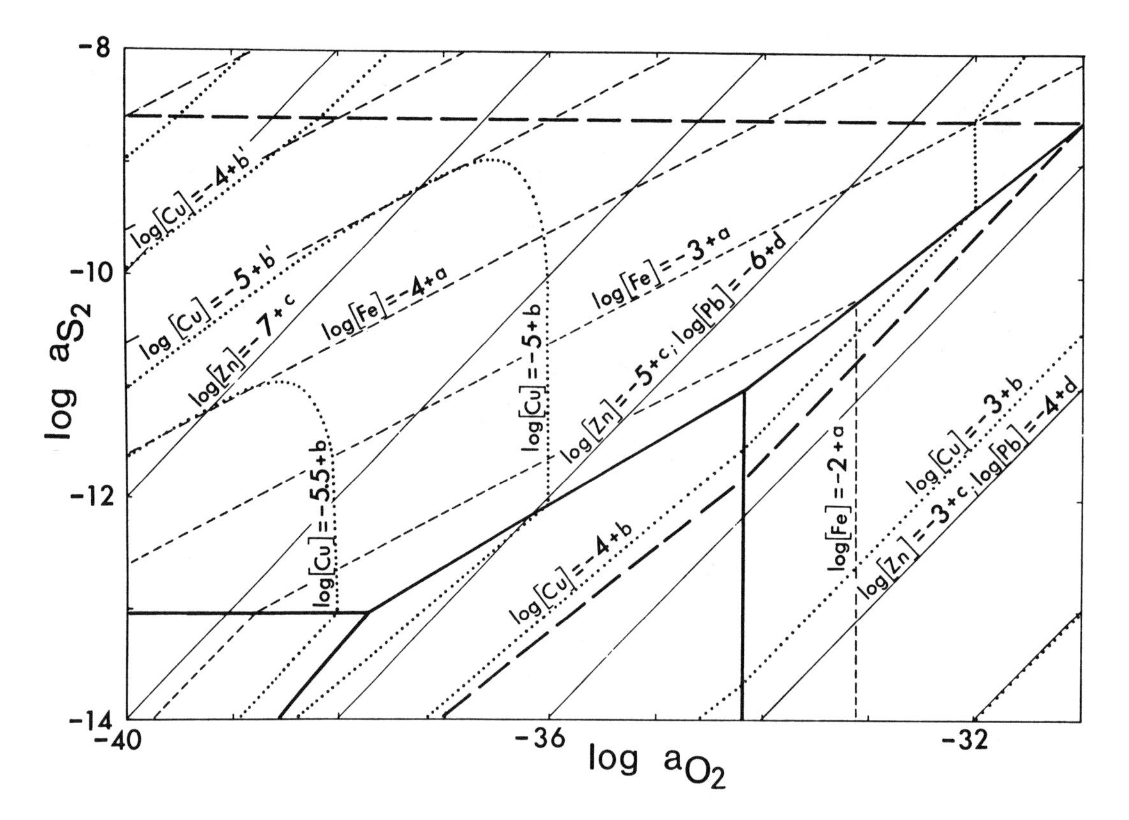

Figure 14.3. Log a_{S2} - log a_{O2} diagram showing contours of metal concentrations. The logs of the metal concentrations for solutions in equilibrium at 250°C with galena, sphalerite, chalcopyrite or bornite, and pyrite, hematite, magnetite, or chlorite are given (the letters added are explained in the text). This figure is on the identical base as Figure 8.3 and is very similar to that given in Figure 8.2a. From Barton et al. (1977).

sides of the same coin) that in the discussion above had promoted gold solubility may now diminish solubility.

One additional caveat deserves note; this discussion has used lead as a representative for base metals; it probably is also a reasonable model for zinc inasmuch as zinc also is probably transported as chloride complexes and it possesses the same 1:1 metal:sulfur ratio as galena (thus the differentials in Table 14.1 will be identical for zinc and lead, which explains why zinc and lead are closely associated in ores despite their contrasting crystal chemistry and petrologic affiliation. Copper, however, is a more complex problem in that (for solutions reacting with chalcopyrite or bornite) one must deal either with both iron and copper in solution or include an iron mineral to fix the level of iron. The latter alternative has been followed in Figure 14.3 by calculating copper concentrations in the presence of pyrite, pyrrhotite, Fe-chlorite, or hematite. A high solubility for iron depresses copper, and vice versa. Copper forms a bisulfide complex in addition to the chloride complex; each may have a region of dominance (separated by a region of lower solubility), as described by Crerar and Barnes (1976) and as plotted on the log a_{S2} - log a_{O2} diagram in Figure 14.3. In constructing Figure 14.3 Barton et al. (1977) were not certain of the absolute values to be assigned to each solubility, yet they were confident of the relative solubilities in S_2 - O_2 space; therefore they used the undefined multiplier (the letters added to the logs of the metal concentrations) in their representation of solubilities.

$ Figure 14.3 shows contours of solubility for iron, lead, and zinc as well as copper. If galena is observed to have been replaced by pyrite, what sort of process might have been responsible (in the context of Figure 14.3)?

$ If chalcopyrite is seen to have replaced a pyrite + pyrrhotite assemblage, what inferences can be made regarding the changes in f_{O2} and f_{S2}?

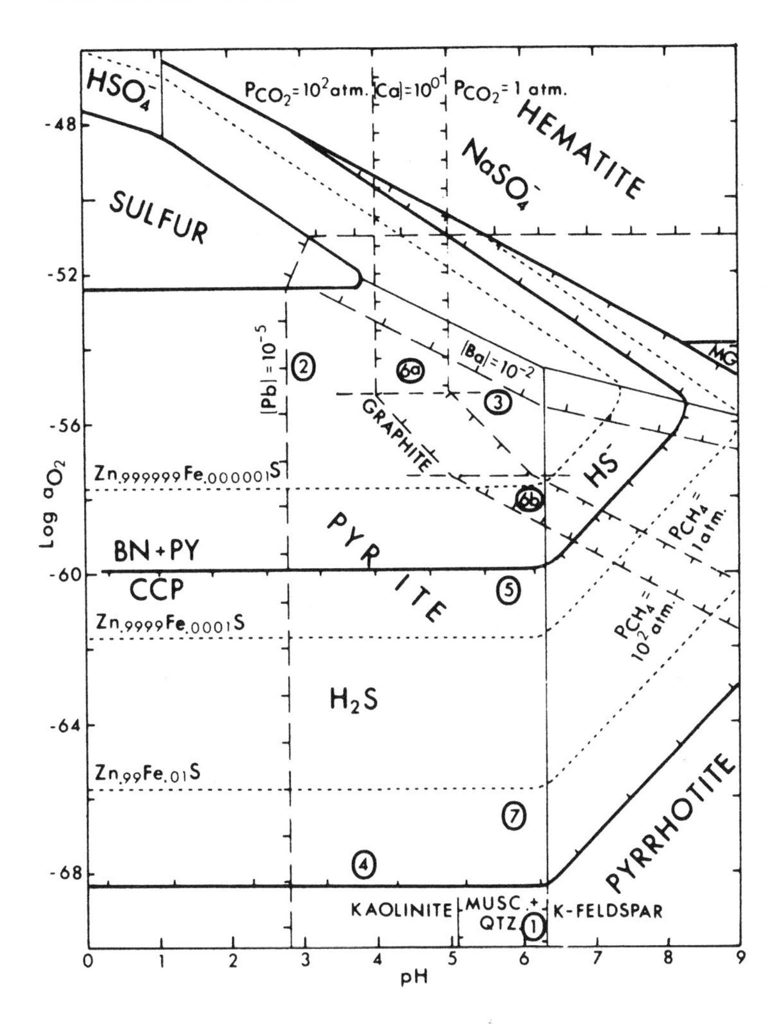

Figure 14.4. Log a_{O2} – pH diagram for the chemical environment of the Mississippi Valley-type ores.
Calculated for 100°C, 3 mol/kg NaCl, log a_{K^+} = -1.5. The concentration units for Ca and Ba are moles/kg H_2O. Abbreviations: MG = magnetite; BN = bornite; ccp = chalcopyrite; py = pyrite. The numbers are discussed in the text. From Barton (1981).

It is textures (on scales ranging from individual polished sections to ore districts) that reveal sequences in time and space; interpretations of those sequences are the principal keys we have to decipher geologic processes and to predict when and where additional processes might have occurred.

THE PUZZLE OF THE MISSISSIPPI VALLEY-TYPE ORES

Thus far we have discussed situations for which we believe that we have the correct answer within our grasp, if only we could find the handle. Next we will discuss a situation in which no answer is yet in clear sight (at least not to the authors): the Mississippi Valley-type ores. These deposits constitute a major class of ores known from many parts of the world and form in a physico-chemical environment quite distinct from that of the volcanic-hosted hydrothermal ore deposits and geothermal systems which have been the focus of most of this volume. They are characterized by: formation at low temperatures (for the most part in the 90-150°C range) from high-salinity fluids; simple mineralogy (sphalerite, galena and dolomite; minor pyrite/marcasite and chalcopyrite; and locally quartz, barite, fluorite, calcite, K-feldspar and Kmica); often associated with organic matter, especially petroleum;

low silver and very low gold content; and epigenetic deposition in orogenically inactive settings in platform carbonate (or rarely sandstone) hosts. The puzzle is presented in Figure 14.4 (from Barton, 1981) which shows 7 numbered criteria, each of which should help limit the environment.

1. The pH should be near the K-spar + Kmica (musc.) boundary.

2. The solubility of galena is at least 2 ppm Pb (= 10^{-5} mol/kg) to avoid the necessity of impossibly large volumes of solution. Even if we had chosen .2 ppm Pb, the conclusion would not have changed.

3. The solubility of barite is less than 1300 ppm Ba (10^{-2} mol/kg).

4. Pyrite, rather than pyrrhotite, magnetite, or hematite is the stable iron mineral; the label (4) applies to the entire perimeter of the pyrite field.

5. Chalcopyrite, rather than bornite + pyrite, is the stable copper-iron mineral.

6. The presence of calcite (or dolomite inasmuch as the activity of $CaCO_3$ cannot go very low in the presence of dolomite) in an environment where the concentration of Ca^{++} is 100 mol/kg fixes the P_{CO2} or P_{CH4} as a function of pH. Neither geologic reconstruction of the probably sedimentary cover, nor the observations of fluid inclusions suggest that the gas pressures could have exceeded 100 atm. But fluid inclusions do show the presence of a hydrocarbon-like gas (presumably methane) at pressures greater than 1 atm. (E. Roedder, personal communication). Thus we show bracketing gas pressures of CO_2 between 1 and 100 bars (the 6a curves); <u>or</u> the pressure of CH_4 between 1 and 100 bars (the 6b curves). Curve 6b choice is almost required by the associated petroleum.

7. The iron content of sphalerite ranges widely, but most of it is in the 1 to 3 mole percent FeS range.

The hachure marks on curves 2, 3, 6a and 6b point in the direction that the chemical environment is supposed to lie. Where is the chemical environment for the Mississippi Valley ores?

As one goes through the above list, more and more of the diagram is identified as "off limits". Finally we note that <u>no</u> environment is permitted! It is a good thing that someone had already found these deposits because geochemists might have proven that they couldn't exist!

There is one more pertinent point that compounds the problem; hydrothermal leaching of sulfides is common. Leaching interspersed with deposition of galena, without destruction of associated pyrite, chalcopyrite, or sphalerite, means that <u>both</u> lead and sulfur had to be removed in the same solution without calling on a chemistry that, for example, would have oxidized pyrite to hematite or chalcopyrite to bornite. This observation provides a substantial problem for any interpretation that would seek to circumvent the incompatible chemical environments by calling on disequilibrium: as by mixing solutions (Anderson, 1975), or extracting sulfur from organic matter (Skinner, 1967), or reacting metastable sulfate + organic matter mixtures (Barton, 1967). If a solution capable of leaching galena can exist, the same solution could also function as an ore-bringer. Barnes (1979), among many others, has discussed the problems of ore transport at length, considering bisulfide and organic complexes in addition to the chloride complexes examined by Anderson (1975). Obviously the definitive word on these enigmatic deposits has yet to be written.

REFERENCES

Anderson, G.M., 1975, Precipitation of Mississippi Valley-type ores: Economic Geology, v. 70, p. 937-942.

Barnes, H.L., 1979, Solubilities of ore minerals: <u>in</u> Geochemistry of Hydrothermal Ore Deposits (2nd ed.): edited by H. L. Barnes, Wiley and Sons, New York, p. 404-460.

Barton, P.B., Jr., 1967, Possible role of organic matter in the precipitation of the Mississippi Valley ores; in Genesis of Stratiform Lead-Zinc-Barite-Fluorite Deposits: Economic Geology, Monograph 3, edited by J. S. Brown, p. 371-378.

Barton, P. B., Jr., 1969, Thermochemical study of the system Fe-As-S: Geochimica et Cosmochimica Acta, v. 33, p. 841-857.

Barton, P. B., Jr., 1981, Physical-chemical conditions of ore deposition; in Chemistry and Geochemistry of Solutions at High Temperatures and Pressures: edited by D. T. Rickard and F. E. Wickman, Pergamon, p. 509-526.

Barton, P. B., Jr., Bethke, P. M., and Roedder, E., 1977, Environment of ore deposition in the Creede Mining District, San Juan Mountains, Colorado: Part III. Progress toward interpretation of the chemistry of the ore-forming fluid in the OH vein: Economic Geology, v. 72, p. 1-24.

Barton, P. B., Jr., Bethke, P. M., and Toulmin, P., III, 1963, Equilibrium in ore deposits: Mineralogical Society Special Paper No. 1, p. 171-185.

Bondi, A., 1953, The spreading of liquid metals on solid surfaces: Chemistry Reviews, v. 52, p. 417-458.

Crerar, D., and Barnes, H. L., 1976, Ore solution chemistry V. Solubilities of chalcopyrite and chalcocite assemblages in hydrothermal solutions at 200° to 350°C: Economic Geology, v. 71, p. 772-794.

Ewers, W. E., 1967, Physico-chemical aspects of recrystallization: Mineralium Deposita, v. 2, p. 221-227.

Giggenbach, W. F., 1980, Geothermal gas equilibria: Geochimica et Cosmochimica Acta, v. 44, p. 2021-2032.

Giggenbach, W. F., 1981, Geothermal mineral equilibria: Geochimica et Cosmochimica Acta, v. 45, p. 393-410.

Haas, J. L., Jr., and Fisher, J. R., 1976, Simultaneous evaluation and correlation of thermodynamic data: American Journal of Science, v. 276, p. 525-545.

Skinner, B.J., 1967, Precipitation of Mississippi Valley ores: a possible mechanism; in Genesis of Stratiform Lead-Zinc-Barite-Fluorite Deposits: Economic Geology Monograph 3, edited by J. S. Brown, p. 363-370.

(courtesy Australian Geologist)

APPENDIX I: SAMPLE PROGRAMS FOR HP41 CALCULATORS

Sample programs and abbreviated program descriptions are given for:

ION	Program description and listing	
LIN	" " " "	
BAL	" " " "	
PKN	" " " "	
PH	" " " "	and sample problems
PBS	" " " "	" " "
AU	" " " "	
STEAM	" " " "	
FS, BI, SL	" " " "	
SG	" " " "	

Copies of bar codes for these programs may be obtained from R. W. Henley (enclose $5.00 U.S. for handling and postage).

ION -- A SIMPLE ACTIVITY COEFFICIENT PROGRAM

(programmed by RWH)

Program Description

This program uses the extended Debye-Hückel function to calculate individual ion activity coefficients, γ, from input ionic strength (I = millimoles/kg), Debye-Hückel A and B values for specified temperature and $\mathring{a}$ and z values for the ionic species required. For a discussion of the equation and values of the input parameters see Chapter 1.

$$\log \gamma_i = \frac{-A z^2 \sqrt{I}}{1 + \mathring{a} B \sqrt{I}} + b I$$

The program uses a value of b = 0.04, but this is readily changed by substitution at step 051.

Output

Activity coefficient for selected I, A, B, z and $\mathring{a}$. To subsequently obtain γ for another ionic species at the same temperature, simply input a new value at a = ?; for the same ionic species at new temperatures insert 0 at step a = ?

PROGRAM LISTING

ION

```
01♦LBL "ION"
02 FIX 3
03 "I=ME/KG?"
04 XEQ 01
05 1000
06 /
07 STO 01
08 SF 02
09♦LBL "P"
10 "a=?"
11 XEQ 01
12 X=0?
13 GTO "ABC"
14 SF 01
15 STO 02
16 "Z="
17 XEQ 01
18 STO 03
19 FS? 02
20 GTO "ABC"
21 FS? 01
22 GTO C
23♦LBL "ABC"
24 CF 01
25 "A="
26 XEQ 01
27 STO 04
28 "B="
29 XEQ 01
30 STO 06
31♦LBL C
32 RCL 06
33 RCL 02
34 *
35 RCL 01
36 SQRT
37 STO 05
38 *
39 1
40 +
41 1/X
42 RCL 03
43 X↑2
44 *
45 RCL 05
46 *
47 RCL 04
48 *
49 CHS
50 RCL 01
51 .04
52 *
53 +
54 10↑X
55 CF 02
56 "GAM="
57 ARCL X
58 AVIEW
59 PSE
60 GTO "P"
61♦LBL 01
62 PROMPT
63 RTN
64 .END.
```

LIN — CURVE FITTING

Program Description

For a set of data points (x_i, y_i), i - 1,2, ..., n, this program can be used to fit the data to any of the following curves:

1. Straight line (linear regression): $y = a + bx$.
2. Exponential curve: $y = ae^{bx}$ $(a > 0)$,
3. Logarithmic curve: $y = a + b \ln x$,
4. Power curve: $y = ax^b$ $(a > 0)$.

Remarks:

1. The program applies the least square method, either to the original equations (straight line and logarithmic curve) or to the transformed equations (exponential curve and power curve).
2. Negative and zero values of x_i will cause a calculator error for logarithmic curve fits. Negative and zero values of y_i will cause a machine error for exponential curve fits. For power curve fits both x_i and y_i must be positive, non-zero values.
3. As the differences between x and/or y values become small, the accuracy of the regression coefficients will decrease.

				SIZE: 016
STEP	INSTRUCTIONS	INPUT	FUNCTION	DISPLAY
1)	Set status and key in the program			
2)	Initialize the program			
	for STRAIGHT LINE		XEQ LIN	LIN
	or for EXPONENTIAL CURVE		XEQ EXP	EXP
	or for LOGARITHMIC CURVE		XEQ LOG	LOG
	or for POWER CURVE		XEQ POW	POW
3)	Repeat step 3) and 4) for i = 1,2... n input: x_i	x_i	ENTER	
	y_i	y_i	A	(i)
4)	If you made a mistake in inputting x_k and y_k, then correct by	x_k	ENTER	
		y_k	C	(k-1)
5)	Calculate R^2 and regression coefficients a and b		E	R2=(R^2)
			R/S	a=(a)
			R/S	b=(b)
6)	Calculate estimated y from regression, input x	X	R/S	Y = (y)
7)	Repeat step 6) for different x's			
8)	Repeat step 5) if you want the results again			

9)	To use the same program for another set of data, initialize the program by Y	A	LIN or EXP or LOG or POW
	then go to step 3)		
10)	To use another program, go to step 2)		

PROGRAM LISTING

LIN

```
 01♦LBL "LIN"
5 "LIN" GTO 13

 05♦LBL "EXP"
6 "EXP" GTO 13

 09♦LBL "LOG"
7 "LOG" GTO 13

 13♦LBL "POW"
8 "POW"

 16♦LBL 13
XEQ "INIT" STO 00
ASTO 08 ΣREG 10 CLΣ
AVIEW STOP

 24♦LBL C
X<>Y XEQ IND 00 Σ-
STOP

 29♦LBL A
X<>Y XEQ IND 00 Σ+
STOP

 34♦LBL 07
LN RTN

 37♦LBL 08
LN

 39♦LBL 06
X<>Y LN X<>Y RTN

 44♦LBL E
RCL 15 RCL 11 RCL 10
RCL 10 XEQ 09 STO 03
RCL 12 RCL 11 RCL 10
RCL 14 XEQ 09 RCL 03
/ STO 04 XEQ IND 00
STO 06 RCL 15 RCL 14
RCL 10 RCL 12 XEQ 09
RCL 03 / STO 05

 69♦LBL 03
RCL 04 RCL 12 *
RCL 05 RCL 14 * +
RCL 12 X↑2 RCL 15 /
STO 09 - RCL 13
RCL 09 - / "R2"
XEQ 88 RCL 06 "Y=a+bX"
AVIEW PSE "a" XEQ 88
RCL 05 "b" GTO 01

 98♦LBL 06
 99♦LBL 08
E↑X

101♦LBL 05
102♦LBL 07
RTN

104♦LBL 09
* STO 07 RDN *
RCL 07 - RTN

112♦LBL 00
"Y"

114♦LBL 01
"⊢=" ARCL X AVIEW
STOP FS? 55 STOP

121♦LBL 04
GTO IND 00

123♦LBL 08
RCL 05 Y↑X GTO 09

127♦LBL 06
RCL 05 * E↑X

131♦LBL 09
RCL 06 * GTO 00

135♦LBL 07
LN

137♦LBL 05
RCL 05 * RCL 06 +
GTO 00

143♦LBL 88
"⊢=" ARCL X AVIEW
STOP RTN

149♦LBL a
GTO IND 08

151♦LBL "INIT"
CLRG CF 00 CF 01
CF 02 SF 27 CF 29 RTN
.END.
```

BAL

(programmed by AHT)

Program Description

This program has a dual function; (a) provision of a simple ion balance to check analytical quality, and (b) input data store for geothermometer subprogram NKC. (a) Input data are prompted as follows and each item is followed by R/S; Na, K, Ca, Mg [OUTPUT R value for the estimation of the Mg-correction to the NaKCa geothermometer temperature] , Li [OUTPUT ΣCAT], HCO_3, Cl, SO_4, F [OUTPUT Σ AN]. All input data are analytical totals in mg/kg. The ion balance is calculated through BAL = 2(ΣCAT - ΣAN)/(ΣCAT + ΣAN)

Geothermometer temperatures are calculated through the expressions listed in Table 3.2.

Subprogram NKC; calculates TNAK with β =1/3, β = 4/3 and Log ($\sqrt{Ca}$/Na).

Subprogram NAK; follows NKC and calculates TNAK using in sequence the White/Ellis, Fournier and Arnorsson calibrations.

Subprogram SI; calculates silica temperatures for the following polymorphs - quartz conductive, quartz adiabatic, chalcedony, amorphous silica. Input data are SiO_2 (mg/kg) and this program may be used independent of BAL. It is useful to assign BAL, NKC and SI to user keys on the calculator. Coefficients included in subprogram SI are from Truesdell, 2nd UN Symposium on the Development and Use of Geothermal Resources, San Francisco, 1975, Vol. 1, Summary of Section III, Table 2. They may be readily changed to those of Table 3.2.

PROGRAM LISTING

BAL

```
01♦LBL "BAL"
FIX 2  0  STO 07  STO 04
.0435  "NA=" PROMPT
STO 01  XEQ 02  .02557
"K=" PROMPT  STO 02
XEQ 02  ST+ 04  .0499
"CA=" PROMPT  STO 03
XEQ 02  ST+ 04  .082264
"MG=" PROMPT  STO 09
XEQ 02  ST+ 04  RCL 04
/  100  *  "R="  XEQ 03
.144  "LI="  XEQ 01
"ΣCAT="  RCL 07  STO 08
XEQ 03  0  STO 07
.016393  "HCO3="  XEQ 01
.028206  "CL="  XEQ 01
.02082  "SO4="  XEQ 01
.05264  "F="  XEQ 01
"ΣAN="  RCL 07  XEQ 03
"BAL="  RCL 08  -  CHS
RCL 07  RCL 08  +  2  /
/  XEQ 03

 70♦LBL 01
PROMPT
```

```
 83♦LBL "NKC"
RCL 01  RCL 02  /  LOG
STO 05  RCL 03  SQRT
RCL 01  /  LOG  STO 06
3  /  .6867  XEQ 01
"T13="  XEQ 03  1.33333
*  2.7467  XEQ 01
STO 07  "T43="  XEQ 03
2.0602  +  "L Ca/Na="
ARCL X  AVIEW  STOP

114♦LBL "NAK"
RCL 05  .8573  +  855.6
XEQ 02  "TNK-WE="
XEQ 03  RCL 05  1.483  +
1217  XEQ 02  "TNK-F="
XEQ 03  RCL 05  .993  +
933  XEQ 02  "TNK-A="
XEQ 03

136♦LBL 01
+  RCL 05  +  2.47  +
1647  XEQ 02  RTN

145♦LBL "SI"
"SIO2?"  PROMPT  LOG
CHS  STO 06  5.768  +
1533.5  XEQ 02  "TQA="
XEQ 03  5.248  +  1326
XEQ 02  "TQC="  XEQ 03
```

```
 72♦LBL 02
* ST+ 07 ARCL X AVIEW
RTN

 78♦LBL 03
ARCL X AVIEW PSE RTN

4.655 + 1015.1 XEQ 02
"TCH=" XEQ 03 4.52 +
731 XEQ 02 "TAM="
ARCL X AVIEW STOP

177♦LBL 02
X<>Y / 273.15 - RTN

183♦LBL 03
FIX 1 ARCL X AVIEW
STOP RCL 06 RTN END
```

PKN

(programmed by RWH)

Program Description

Program PKN calculates ionic species concentrations for the dissociation of weak mono- and diprotic acids.

Input data are prompted as follows, each item followed by R/S

(a) proton number e.g. for H_2CO_3 n = 2,
(b) pK_n - the negative logarithm of the dissociation constants at specified temperature
(c) activity coefficient (γ) of the univalent base ion*
(d) the activity coefficient of the acid (usually 1 in dilute solutions)
(e) the total concentration ($m_{\Sigma A}$) of the specified component (10^3 moles/kg) e.g.

$$\Sigma C = H_2CO_3(app) + HCO_3^- + CO_3^=$$

N.B. For the dissociation of the ammonium ion, input γ_{NH4^+} for GAMMA ACID and γ_{NH3} for GAMMA 1-.

Output data are species concentrations in log molal units in the order (a) log m_{HA^-}, log m_{HA}, log $m_{A^=}$. For ammonium the output order is log m_{NH3}, log m_{NH4}.

Storage registers

01	02	03	04	05	06
K_1	K_2	γ^-	$m_{\Sigma A}$	a_{H^+}	$\gamma^=$

---

* For most geothermal purposes the approximation $\gamma^= = (\gamma^-)^4$ is satisfactory and introduces a useful program simplification.

PROGRAM LISTING

PKN

```
01♦LBL "PKN"
02 FIX 2
03 CF 00
04 "PROTONS=?"
05 PROMPT
06 1
07 -
08 X>0?
09 SF 00
10 "PK1="
11 XEQ 01
12 STO 01
13 FS? 00
14 XEQ 03
15 GTO "CC"
16♦LBL 03
17 "PK2="
18 XEQ 01
19 STO 02
20 RTN
21♦LBL "CC"
22 "GAMMA 1- =?"
23 PROMPT
24 STO 03
25 "GAMMA ACID =?"
26 PROMPT
27 STO 07
28 "CONC=MM/KG"
29 PROMPT
30 STO 04
31♦LBL 05
32 "PH="
33 XEQ 01
34 STO 05
35 RCL 03
36 *
37 RCL 01
38 /
39 RCL 07
40 /
41 1
42 +
43 FS? 00
44 XEQ 02
45 1/X
46 RCL 04
47 *
48 LOG
49 3
50 -
51 "ANION-1="
52 XEQ 03
53 10↑X
54 RCL 03
55 *
56 STO 07
57 RCL 05
58 *
59 RCL 01
60 /
61 LOG
62 "ACID="
63 XEQ 03
64 FS? 00
65 XEQ 04
66 GTO 05
67♦LBL 01
68 PROMPT
69 CHS
70 10↑X
71 RTN
72♦LBL 02
73 RCL 03
74 3
75 Y↑X
76 1/X
77 RCL 02
78 *
79 RCL 05
80 /
81 +
82 RTN
83♦LBL 04
84 RCL 07
85 RCL 02
86 *
87 RCL 05
88 /
89 RCL 03
90 X↑2
91 X↑2
92 /
93 LOG
94 "ANION 2="
95 XEQ 03
96 GTO 05
97 RTN
98♦LBL 03
99 ARCL X
100 AVIEW
101 PSE
102 RTN
103 END
```

Note: To correct a programming error, change step 56 to STO 09 and step 84 to RCO 09.

PH

(programmed by RWH)

Program Description

PH is an iterative program written for the HP41C to illustrate the calculation procedure used in ENTHALP (Truesdell and Singers, 1974). The program calculates pH and species distributions for high temperature solutions using analytical data for water samples including pH, obtained at room temperature. PH is intended as a core program to allow a first order determination of pH_t values for water samples, but additional routines may be inserted, together with requisite data, to allow specific calculations e.g. calculation of the pH of a reservoir fluid, calculations giving water and steam compositions and species following single step steam separation.

The version of the program given below contains a routine for the calculation of pH under aquifer conditions. The additional data required are,

y, LIQ - the steam fraction removed from the total discharge to produce the water sample

y, VAP - the steam fraction removed for the steam sample analysis

CO_2 in steam (mm/kg)

H_2S in steam (mm/kg)

NH_3 in steam (mm/kg)

t = the temperature, in °C, at which the calculation is to be performed.

Generally y is calculated from the enthalpy measurements and if no excess enthalpy effects occur in the well, y is compatible with t. However, this is not a requirement of the program, so that for a given total discharge composition based on measured discharge enthalpy, pH etc. may be calculated for a number of different temperatures e.g. T_{quartz}, T_{NaKCa}, etc. This facility, as well as ready access to storage registers (allowing change of concentrations and HTOT) makes the HP41C program very useful for research calculations, whereas manipulation of these quantities and calculation temperature is more complex in the larger 'main frame' based programs such as ENTHALP.

Dissociation constants are automatically calculated using temperature dependent equations given by Arnorsson et al. (1982) and activity coefficients using an extended Debye-Hückel equation. A and B values for the latter are given in Chapter 1, Table 1.2; the b value used is .04 (see step 190). Estimation of ionic strength (I) is a manual operation, and for optimum results the program may be reiterated manually with successive introduction of I values (if this is to be the normal mode of operation additional loops may be incorporated in the program, but this is unnecessary for most purposes). Normally $I \simeq m_{Na^+} + m_{K^+} + 2m_{Ca^{++}} + 2m_{Mg^{++}}$.

Intially the program calculates an ionisable-proton total (HTOT) for the solution at the analysis temperature.

$$HTOT = 2m_{H_2CO_3} + m_{HCO_3^-} + m_{H_2S} + m_{HBO_2} + m_{SO_4^=} + m_{NH_4^+} + 2m_{H_4SiO_4} + m_{H_3SiO_4^-}$$

The HTOT value is used as the iteration base in the succeeding calculation using dissociation constant and activity coefficient data for the higher temperature (s) required via successive calculations of HTAQ - the intermediate proton total. The pH to commence each loop is estimated by comparing last HTAQ with HTOT using pH (last completed calculation) as the initial value. You are required to initiate the iterative calculation by a reasonable guess for the pH. In the version presented here a maximum of 30 loops is allowed (loop control no.

step 262; REG 35), failure to converge to HTOT - HTAQ < .01, results in the message "NOT CONVERGED".

Convergence is controlled by routine (LBL 06)

$$a_{H^+,2} = a_{H^+,1} \times 10^F \left(1 - \frac{HTAQ}{HTOT}\right)$$

where $a_{H^+,2}$ is the next value for iteration and $a_{H^+,1}$ the last value of a a_{H^+} in the LOOP routine. The rate of convergence, F, is inserted independently at the start of the program. F may also be changed at any time by intercepting program operation and returning to 161 BB.

pH < 6	F = 20	pH 6 - 7.5	F = 10
pH 7.5 - 8.5	F = 5	pH > 8.5	F = 2

Use of excessive values of F results in oscillation and non-convergence.

A flow chart for the program is provided below.

Necessary Accessories Optional printer

Operating Limits and Warnings: Since the use of a printer is optional, Flag 21 is used. If changing to another program using a printer, following use of pH without this accessory, check that Flag 21 is set.

This program uses alphanumeric labels for subroutines so care should be taken to avoid coincident labels in other programs in your calculator. For example, if you have installed the HP41C program BLACKJACK a labelling coincidence may lead to an unwelcome random error in your output.

References:

Arnorsson, S., Sigurdsson, S., and Svavarsson, H., 1982, The chemistry of geothermal waters in Iceland: Calculation of aqueous speciation from 0°C to 370°C: Geochimica et Cosmochimica Acta, v. 46, p. 1513-1532.

Truesdell, A.H., and Singers, W.A., 1974, Calculation of aquifer chemistry in hot-water geothermal systems: Journal of Research, U.S. Geological Survey, v. 2 (3), p. 271-278.

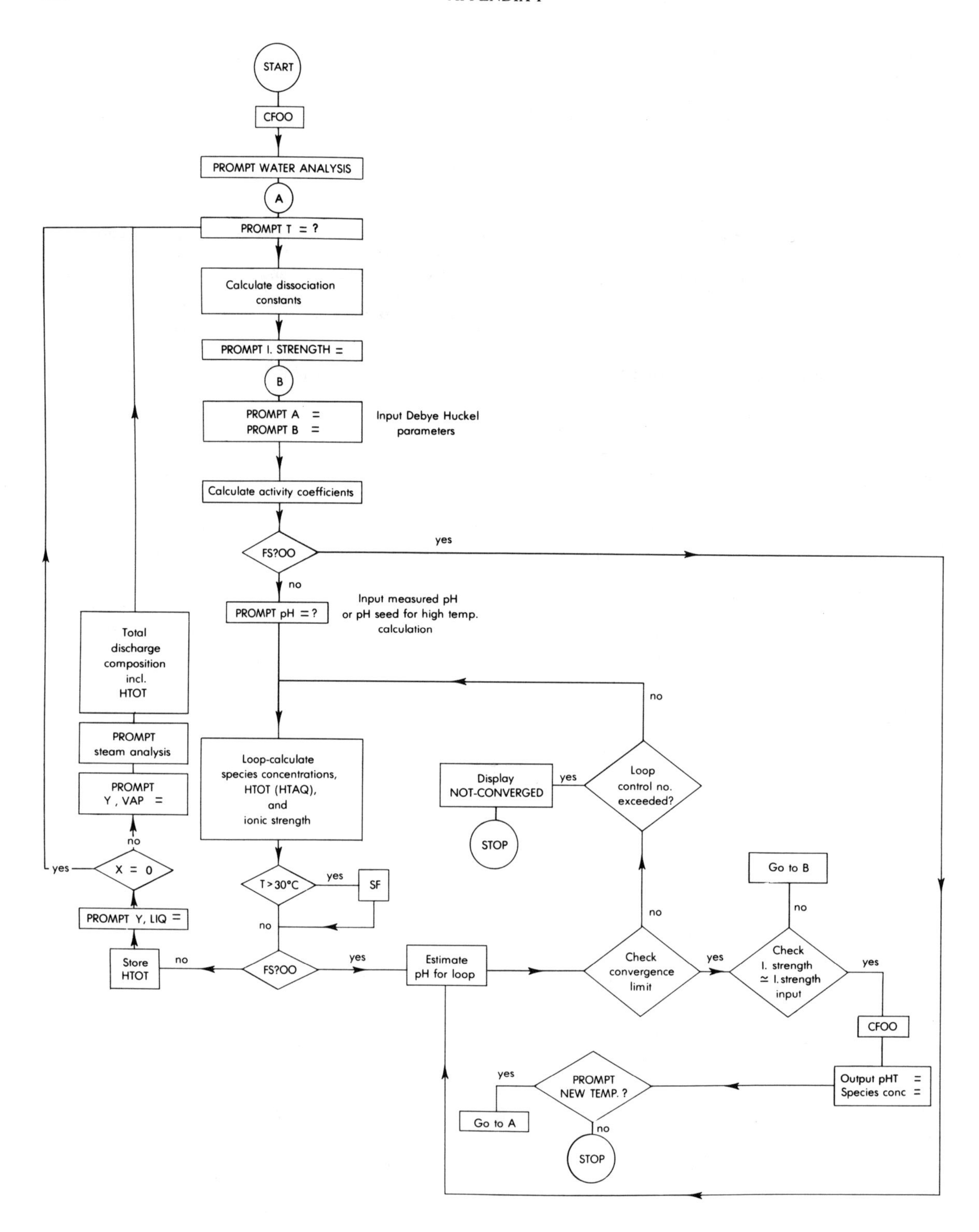
START
CFOO
PROMPT WATER ANALYSIS
A
PROMPT T = ?
Calculate dissociation constants
PROMPT I. STRENGTH =
B
PROMPT A =
PROMPT B =
Input Debye Huckel parameters
Calculate activity coefficients
FS?OO
yes
no
PROMPT pH = ?
Input measured pH or pH seed for high temp. calculation
no
Loop-calculate species concentrations, HTOT (HTAQ), and ionic strength
Display NOT-CONVERGED
yes
Loop control no. exceeded?
STOP
T > 30°C
yes
SF
no
FS?OO
yes
Estimate pH for loop
no
Check convergence limit
yes
Check I. strength ≃ I. strength input
no
Go to B
yes
CFOO
Output pHT =
Species conc =
PROMPT NEW TEMP. ?
yes
Go to A
no
STOP
Total discharge composition incl. HTOT
PROMPT steam analysis
PROMPT Y , VAP =
no
X = 0
yes
PROMPT Y, LIQ =
Store HTOT
no

Figure I:1

Table I PH.1 -- User Instructions for HP 41C.

STEP		INPUT	FUNCTION	DISPLAY
1.	Input sample number.	ALPHA Mode	R/S	SAMPLE NO.
2.	Input convergence rate factor.		R/S	RATE = 2-20?
3.	Input water sample analysis in the order requested.	mmoles/kg	R/S	INPUT ORDER C,S,B, SI, SO4, N
4.	Input analysis temperature.	t°C	R/S	T = ?
5.	Input ionic strength estimate.	I	R/S	ISTRENGTH = MM/KG
6.	Input Debye-Hückel A factor*	A	R/S	A = ?
7.	Input Debye-Hückel B factor*	B	R/S	B = ?
8.	Input sample pH.	pH	R/S	PH = ?
9.	Input Y, LIQ if YLIQ = 0 go to step 14.	Y,LIQ	R/S	Y,LIQ = ?
10.	Input Y,VAP	YVAP	R/S	Y, VAP = ?
11.	Input steam analysis as prompted	CO2	R/S	CO2 = MM/KG
12.		H2S	R/S	H2S = MM/KG
13.		NH3	R/S	NH3 = MM/KG
14.	Input calculation temperature required.	t°C	R/S	T = ?
15.	Follow steps 5 to 7 as prompted.			
16.	Input seed pH	pH	R/S	pH = ?
17.	Read calculated pH or non-convergence message. If non-convergence occurs return to step 1.			
18.	Input n > 0, go to step 19 n = 0, FINISH		R/S	NEW TEMP?
19.	Input new calculation temperature		R/S	T = ?
20.	Follow steps 4 through 20.			

* A & B values are listed in Table 1.3

Sample Problem 1.

Calculate the pH_t of the following:

A) the water sample at its separation temperature, 185°C, and at other selected temperatures

B) the reservoir fluid

BR22 (1.6.80) water line sample (10 b.g)

t°C	pH	Na	K	Ca	Mg	Cl	Total C as HCO3	H2S	H3BO3	SiO2	SO4	NH3	
20	7.39	870	188	1.2	.1	1432	216	9.2	171	705	2	6.1	ppm
						DATA INPUT	3.541	.271	3.904	11.75	0.021	.359	mmoles/kg

$Y_{water} = .19$ $Y_{steam} = .19$

CO_2 = 533 mmoles/kg H_2S = 10.3 mmoles/kg

NH_3 = A nominal value of 1.0 mmole/kg was used in the sample calculation prior to the availability of an analysis. The measured concentration is 2.35 millimoles NH_3/kg and this value changes pH_t by + 0.003.

Ionic strength ≈ 44 mmoles/kg at 20°C, 35.2 mmoles/kg at 265°C.

Solution:

A sample of the output from PH is shown on the following pages.

Sample Problem 2.

Calculate the aquifer pH for the following well discharge.

Cerro Prieto CPM5 (4/77)

pH	HCO_3	H_2S	B	SiO_2	SO_4	NH_3	
7.22	43	0	14.5	781	46	0	mg/kg
	.7049	0	1.3426	13.017	.4792	0	mm/kg

Enthalpy 1179.9 J/gm (taken at quartz temp - enthalpy).

Y_{water} = .2402 = Y_{steam}

CO_2 = 195.3 H_2S = 18.89 NH_3 = 1.0 mmoles/kg

Ionic strength ≈ 372 mmoles/kg at 20°C, 245 mmoles/kg at 269°C.

Table I PH.2

Storage Registers

0	Σ HCO_3 mmoles kg^{-1}	21	a_{H^+}
1	Σ H_2S "	22	pH
2	Σ B "	23	
3	Σ SiO_2 "	24	Accumulator
4	Σ SO_4 "	25	Reserved for ISG function
5	Σ NH_3 "	26	
6	y = steam fraction for steam sample*	27	m_{HS^-}
7	CO_2, steam intermediate storage in DIS CON Routine	28	$m_{BO_2^-}$
8	H_2S, steam	29	$m_{SiO(OH)^-}$
9	Dissociation Constant K_1 $CO_{2,aq}$	30	$m_{NH_4^+}$
10	" " K_2 $CO_{2,aq}$	31	$m_{HCO_3^-}$
11	" " K H_2S	32	$m_{CO_3^=}$
12	" " K HBO_2	33	H TOT - iteration base
13	" " K_1 H_4SiO_4	34	t°C - run temperature
14	" " K NH_4^+	35	Loop control number
15	" " K H_2O	36	Convergence Rate Factor
16	Activity Coefficients γ HS^-	37	
17	" " γ $SO_4^=$	38	
18	" " γ H^+	39	
19	" " γ HCO_3^-	40	
20	" " γ $CO_3^=$		

* The steam fraction for water sample is held temporarily in reg. 09.

PROGRAM LISTING

PH

```
 01♦LBL "PH"
FS? 55  GTO "START"
SF 21  CF 21

 06♦LBL "START"
"SAMPLE NO"  SF 12
AVIEW  STOP  FS? 55  PRA
CF 12  ADV  "RATE=2-20?"
AVIEW  PROMPT  STO 36
FS? 55  PRX
"INPUT ORDER"  AVIEW
"C,S,B,SI,SO4,N"  AVIEW
CF 00  .00501  STO 25
FIX 4

 29♦LBL 01
STOP  STO IND 25  ISG 25
GTO 01

 34♦LBL "DIS CON"
 35♦LBL "AA"
"T=?"  PROMPT  FS? 55
ADV  FS? 55  ADV  "T="
XEQ 14  STO 34
"DIS CONSTANTS"  FS? 55
PRA  273.15  +  STO 07
1/X  STO 08  2107
XEQ 02  6.38  +  RCL 07
19.13 E-3  XEQ 03
STO 09  RCL 08  2589
XEQ 02  4.4  +  RCL 07
20.36 E-3  XEQ 03
STO 10  RCL 08  1996
XEQ 02  1.2  +  RCL 07
X↑2  17.54 E-6  XEQ 03
STO 11  RCL 08  2623
XEQ 02  36.05  +  RCL 07
LN  6.41  XEQ 03  STO 12
RCL 08  2549  XEQ 02
RCL 07  X↑2  15.36 E-6
XEQ 03  STO 13  RCL 08
528  XEQ 02  5.12  +
RCL 07  17.93 E3  XEQ 03
STO 14  RCL 08  6553
XEQ 02  84.4  +  RCL 07
X↑2  8.32 E-6  *  +
RCL 07  LN  13.411  *  -
RCL 07  .0187  *  12.74
-  E↑X  RCL 07  219  /
+  1.3622  -  E↑X
RCL 07  /  STO 16  224.9
*  -  STO 14  10↑X
STO 15  RCL 08  82  *
32.19  -  RCL 07
16.35 E-3  *  -  RCL 07
LN  5.813  *  +  RCL 16
322  *  -  RCL 14  -
CHS  10↑X  STO 14
FS? 55  XEQ "PA"  TONE 4
ADV  ADV  GTO "ISTRENG"

173♦LBL "PA"
9.015  PRREGX  RTN

177♦LBL "ISTRENG"
"ISTRENGTH=MM/KG"
PROMPT  "ION STRENGTH="
XEQ 14
"ACTIVITY COEFFS"
FS? 55  PRA  1000  /
SQRT  STO 07  LASTX  .04
*  STO 08  "A=?"  PROMPT
RCL 07  *  CHS  STO 23
"B=?"  PROMPT  RCL 07  *
STO 24  4  *  1  +
RCL 23  /  1/X  STO 17
XEQ 04  STO 16  RCL 17
4  *  XEQ 04  STO 17
RCL 24  4.5  *  STO 07
2  *  1  +  RCL 23  /
1/X  XEQ 04  STO 18
RCL 07  1  +  RCL 23  /
1/X  RCL 08  +  10↑X
STO 19  RCL 07  1  +
1/X  4  *  RCL 23  *
XEQ 04  STO 20  FS? 55
XEQ "PB"  TONE 8  FS? 00
GTO 06  GTO 11

258♦LBL "PB"
16.020  PRREGX  RTN

262♦LBL 02
*  CHS  RTN

266♦LBL 03
*  -  10↑X  RTN

271♦LBL 04
RCL 08  +  10↑X  RTN

276♦LBL "LOOP"
277♦LBL 11
"PH=?"  PROMPT  ADV
"PH="  XEQ 14  STO 22
CHS  10↑X  STO 21
.03001  STO 35

289♦LBL "BB"
0  STO 24  RCL 01
RCL 00  RCL 03  +  2  *
+  RCL 02  +  STO 24
RCL 16  RCL 21  *
RCL 11  XEQ 05  RCL 01
*  STO 27  ST- 24
RCL 16  RCL 21  *
RCL 12  XEQ 05  RCL 02
*  STO 28  ST- 24
RCL 16  RCL 21  *
RCL 13  XEQ 05  RCL 03
*  STO 29  ST- 24
RCL 14  RCL 16  *
RCL 21  XEQ 05  RCL 05
*  STO 30  ST+ 24
RCL 09  RCL 10  *
RCL 20  /  RCL 21  /
RCL 21  +  RCL 19  *
RCL 09  XEQ 05  RCL 00
*  STO 31  ST- 24
RCL 21  RCL 19  *
RCL 09  XEQ 05  1/X
RCL 31  *  CHS  RCL 00
+  STO 32  2  *  ST- 24
XEQ 13  FS? 00  GTO 06
RCL 34  FS? 55  PRX
RCL 22  FS? 55  PRX
RCL 24  FS? 55  PRX
STO 33  "Y, LIQ= ?"
```

PROGRAM LISTING

PH (contd)

```
PROMPT  X=0?  GTO "AA"
FS? 55  PRX  1  -  CHS
STO 09  ST* 33
"Y, VAP= ?"  PROMPT
FS? 55  PRX  STO 06
GTO 15

400♦LBL 13
RCL 34  30  -  X>0?
SF 00  RTN

407♦LBL 05
/  1  +  1/X  RTN

413♦LBL 15
"CO2= MM/KG"  PROMPT
STO 07  2  *
"H2S= MM/KG"  PROMPT
STO 08  +  RCL 06  *
ST+ 33  RCL 06  ST* 07
RCL 06  ST* 08
"NH3=MM/KG"  PROMPT
RCL 06  *  STO 10
.00501  STO 25  SF 12
"TOTAL"  FS? 55  PRA
"DISCHARGE"  FS? 55  PRA
CF 12

445♦LBL 16
RCL IND 25  RCL 09  *
STO IND 25  ISG 25
GTO 16  RCL 07  ST+ 00
RCL 08  ST+ 01  RCL 10
ST+ 05  FS? 55  XEQ "PC"
GTO "AA"

461♦LBL "PC"
00.005  PRREGX  RTN

465♦LBL 06
RCL 24  FS? 55  PRX
RCL 33  /  1  -  CHS
RCL 36  *  10↑X  RCL 21
*  STO 21  RCL 24
RCL 33  -  X↑2  .0001  -
X<0?  GTO 12  ISG 35
GTO "BB"
"NOT CONVERGED"  PROMPT
STOP

493♦LBL 12
RCL 21  LOG  CHS  CF 00
SF 12  "PHT="  XEQ 14
CF 12  TONE 8  FS? 55
XEQ "PD"  "NEW TEMP?"
PROMPT  X>0?  GTO "AA"
STOP

510♦LBL "PD"
ADV  "SPECIES"  PRA
27.032  PRREGX  ADV  ADV
RTN

519♦LBL 14
ARCL X  AVIEW  RTN  STOP
.END.
```

SAMPLE PROBLEM

1 A

```
SAMPLE NO
BR22

RATE=2-20?
          10.0000    ***
INPUT ORDER
C,S,B,SI,SO4,N
T=20.0000

DIS CONSTANTS

R09=  3.8423-07
R10=  3.9793-11
R11=  7.6536-08
R12=  4.8885-10
R13=  9.6561-11
R14=  3.8263-10
R15=  6.7264-15


ION STRENGTH=44.0000
ACTIVITY COEFFS

R16=  0.8294
R17=  0.4676
R18=  0.8637
R19=  0.8336
R20=  0.4770

PH=7.3900
          20.0000    ***
           7.3900    ***
          31.5766    ***
```

```
T=185.0000
DIS CONSTANTS

R09=  1.0391-07
R10=  2.6368-11
R11=  1.4510-07
R12=  1.1207-09
R13=  1.6302-09
R14=  1.0053-06
R15=  4.3363-12


ION STRENGTH=44.0000
ACTIVITY COEFFS

R16=  0.7542
R17=  0.3197
R18=  0.8045
R19=  0.7604
R20=  0.3302

PH=7.3900
          31.0895    ***
          31.3315    ***
          31.4510    ***
          31.5118    ***
          31.5430    ***
          31.5592    ***
          31.5676    ***
PHT=7.2344
```

```
T=150.0000
DIS CONSTANTS

R09=  2.0222-07
R10=  4.6373-11
R11=  2.1997-07
R12=  1.2171-09
R13=  1.6820-09
R14=  2.9225-07
R15=  2.2391-12


ION STRENGTH=44.0000
ACTIVITY COEFFS

R16=  0.7761
R17=  0.3584
R18=  0.8218
R19=  0.7816
R20=  0.3688

PH=7.2000
          31.2766    ***
          31.4005    ***
          31.4732    ***
          31.5159    ***
          31.5410    ***
          31.5557    ***
          31.5644    ***
          31.5694    ***
PHT=7.0866
```

SAMPLE PROBLEM

1B

```
SAMPLE NO
BR22

RATE=2-20?
        10.0000    ***
INPUT ORDER
C,S,B,SI,SO4,N


T=20.0000
DIS CONSTANTS

R09=  3.8423-07
R10=  3.9793-11
R11=  7.6536-08
R12=  4.8885-10
R13=  9.6561-11
R14=  3.8263-10
R15=  6.7264-15


ION STRENGTH=44.0000
ACTIVITY COEFFS

R16=  0.8294
R17=  0.4676
R18=  0.8637
R19=  0.8336
R20=  0.4770

PH=7.3900
        20.0000    ***
         7.3900    ***
        31.5766    ***
         0.1900    ***
         0.1900    ***
TOTAL
DISCHARGE

R00=  104.1382
R01=  2.1765
R02=  3.1622
R03=  9.5175
R04=  0.0170
R05=  0.4808
```

```
T=265.0000
DIS CONSTANTS

R09=  1.4789-08
R10=  4.2888-12
R11=  2.5782-08
R12=  7.3949-10
R13=  6.5323-10
R14=  1.4577-05
R15=  9.2061-12


ION STRENGTH=35.2000
ACTIVITY COEFFS

R16=  0.7022
R17=  0.2408
R18=  0.7593
R19=  0.7091
R20=  0.2504

PH=6.2000
          229.2070    ***
          229.4892    ***
          229.6672    ***
          229.7855    ***
          229.8667    ***
          229.9237    ***
          229.9644    ***
          229.9937    ***
          230.0150    ***
          230.0305    ***
          230.0419    ***
          230.0503    ***
          230.0565    ***
          230.0611    ***
          230.0644    ***
PHT=6.0742

SPECIES

R27=  0.0909
R28=  0.0039
R29=  0.0105
R30=  0.0365
R31=  2.5168
R32=  3.6300-05
```

```
T=290.0000
DIS CONSTANTS

R09=  7.3365-09
R10=  2.1723-12
R11=  1.2389-08
R12=  6.2265-10
R13=  4.0035-10
R14=  3.4907-05
R15=  9.3218-12


ION STRENGTH=35.2000
ACTIVITY COEFFS

R16=  0.6686
R17=  0.1979
R18=  0.7319
R19=  0.6762
R20=  0.2071

PH=6.2000
          230.8267    ***
          230.6848    ***
          230.5619    ***
          230.4585    ***

              ↓

          230.0924    ***
          230.0878    ***
          230.0843    ***
          230.0817    ***
PHT=6.3507

SPECIES

R27=  0.0868
R28=  0.0066
R29=  0.0128
R30=  0.0090
R31=  2.4713
R32=  3.9300-05
```

SAMPLE PROBLEMS (contd)

1B (contd)

```
T=240.0000
DIS CONSTANTS

R09= 2.8670-08
R10= 8.0716-12
R11= 4.9160-08
R12= 8.6376-10
R13= 9.7275-10
R14= 6.2967-06
R15= 8.0925-12

ION STRENGTH=35.2000
ACTIVITY COEFFS

R16= 0.7290
R17= 0.2798
R18= 0.7812
R19= 0.7353
R20= 0.2895

PH=6.0000
          228.6741    ***
          229.1978    ***
              |
              v
          230.0596    ***
          230.0636    ***
          230.0664    ***
PHT=5.8171

SPECIES

R27= 0.0923
R28= 0.0025
R29= 0.0083
R30= 0.1197
R31= 2.6003
R32= 3.5000-05
```

2

```
SAMPLE NO
CPM5

RATE=2-20?
          20.0000    ***
INPUT ORDER
C,S,B,SI,SO4,N

T=20.0000
DIS CONSTANTS

R09= 3.8423-07
R10= 3.9793-11
R11= 7.6536-08
R12= 4.8885-10
R13= 9.6561-11
R14= 3.8263-10
R15= 6.7264-15

ION STRENGTH=372.0000
ACTIVITY COEFFS

R16= 0.6980
R17= 0.2142
R18= 0.8033
R19= 0.7126
R20= 0.2327

PH=7.2200
          20.0000    ***
           7.2200    ***
          28.1057    ***
           0.2402    ***
           0.2402    ***
TOTAL
DISCHARGE

R00= 47.4466
R01= 4.5374
R02= 1.0201
R03= 9.8903
R04= 0.3641
R05= 0.2402
```

```
T=270.0000
DIS CONSTANTS

R09= 1.2892-08
R10= 3.7569-12
R11= 2.2420-08
R12= 7.1536-10
R13= 5.9652-10
R14= 1.7297-05
R15= 9.3249-12

ION STRENGTH=245.0000
ACTIVITY COEFFS

R16= 0.5064
R17= 0.0615
R18= 0.6485
R19= 0.5251
R20= 0.0710

PH=5.9000
          119.1894    ***
          119.3878    ***
          119.4940    ***
              |
              v
          119.6942    ***
          119.6990    ***
          119.7026    ***
          119.7053    ***
PHT=5.6252

SPECIES

R27= 0.0835
R28= 0.0006
R29= 0.0049
R30= 0.0510
R31= 0.4881
R32= 5.7200-06
```

PBS

(programmed by RWH)

Program Description

This program calculates the solubility of galena as ppb lead at specified temperature, pH, Cl molality, H_2S molality.

Input data are K_{PbS} (from Helgeson modified to delete $S^=$, see Chapter 9), followed by association constants and activity coefficients for Pb^I to Pb^{IV} complexes.

When a printer is attached, the program also lists the percentage of each complex present at t°C. The activity coefficients may be calculated through program ION. After each solubility calculation, the program returns to pH = ? or sequentially to m_{H_2S} =, m_{Cl} = by input 0 R/S.

Other metals

1. The percentage of each chloride species for other metal may also be calculated using this program. Where high ligand numbers do not apply insert small negative values for B_n.

2. Solubilities of sulfides other than galena may only be calculated if the dissolution equation is written in the form used in Chapter 9, e.g., ZnS.

Necessary Accessories: Optional printer

Warnings: Since the use of a printer is optional, Flag 21 is used. If changing to another program using a printer, following use of pH without this accessory, check that Flag 21 is set.

Sample Problem

Calculate the solubility of galena (PbS) in the following hydrothermal solutions.

	Salton Sea	Broadlands
t	300	260°C
m_{Cl}	2.0 *	0.033 m/kg
m_{H2S}	5×10^{-4}	4×10^{-3} m/kg
pH	4.7	6.1

* An effective value of m_{Cl} is given here as an ad hoc estimate of the free chloride ion present

If a printer is attached, the proportions of each lead chloro-complex are also obtained.

PROGRAM LISTING

PBS

```
01♦LBL "PBS"
02 FS? 55
03 GTO "GO"
04 CF 21

05♦LBL "GO"
06 FIX 3
07 ADV
08 "LOG KS="
09 PROMPT
10 XEQ 03
11 STO 01
12 "LOGB1="
13 XEQ 01
14 STO 02
15 "LOGB2="
16 XEQ 01
17 STO 03
18 "LOGB3="
19 XEQ 01
20 STO 04
21 "LOGB4="
22 XEQ 01
23 STO 05
24 "GAM PB="
25 PROMPT
26 XEQ 03
27 STO 06
28 "GAM PBCL="
29 PROMPT
30 XEQ 03
31 STO 07
32 "GAMPBCL2="
33 PROMPT
34 XEQ 03
35 STO 08
36 "GAM CL="
37 PROMPT
38 XEQ 03
39 STO 12
40 GTO "COMPLEX"

41♦LBL 01
42 PROMPT
43 XEQ 03
44 10↑X
45 RTN

46♦LBL "COMPLEX"
47 FIX 4
48 "CL=M/KG"
49 PROMPT
50 X=0?
51 GTO "PBS"
52 "CL="
53 XEQ 03
54 RCL 12
55 *
56 STO 12
57 RCL 07
58 /
59 RCL 02
60 *
61 STO 15
62 STO 20
63 RCL 03
64 RCL 12
65 X↑2
66 STO 19
67 *
68 RCL 08
69 /
70 STO 16
71 ST+ 20
72 RCL 04
73 RCL 12
74 *
75 RCL 19
76 *
77 RCL 07
78 /
79 STO 17
80 ST+ 20
81 RCL 05
82 RCL 19
83 X↑2
84 *
85 RCL 06
86 /
87 STO 18
88 ST+ 20
89 RCL 06
90 1/X
91 STO 19
92 ST+ 20
93 "PERCENT COMPLEX"
94 FS? 55
95 PRA
96 RCL 19
97 XEQ 02
98 RCL 15
99 XEQ 02
100 RCL 16
101 XEQ 02
102 RCL 17
103 XEQ 02
104 RCL 18
105 XEQ 02
106 GTO "LEAD"

107♦LBL 02
108 RCL 20
109 /
110 100
111 *
112 FS? 55
113 PRX
114 RTN

115♦LBL "LEAD"
116 "PB AS PPB"
117 FS? 55
118 PRA
119 "MH2S=M/KG"
120 PROMPT
121 X=0?
122 GTO "COMPLEX"
123 "MH2S="
124 XEQ 03
125 STO 09
```

Important Note:
Line 122, substitute GTO 05 in place of GTO "COMPLEX" and insert LBL 05 following line 023.

PROGRAM LISTING(contd) and SAMPLE PROBLEM

```
126♦LBL 04
127 "PH="
128 PROMPT
129 X=0?
130 GTO "LEAD"
131 "PH="
132 XEQ 03
133 CHS
134 2
135 *
136 RCL 01
137 +
138 10↑X
139 RCL 08
140 /
141 RCL 09
142 /
143 RCL 20
144 *
145 207 E6
146 *
147 "PB="
148 XEQ 03
149 STOP
150 GTO 04

151♦LBL 03
152 ARCL X
153 AVIEW   (struck through)
154 END
```

```
LOG KS=-2.930
LOGB1=3.180
LOGB2=4.980
LOGB3=5.030
LOGB4=4.000
GAM PB=0.140
GAM PBCL=0.590
GAMPBCL2=1.000
GAM CL=0.560
CL=0.1400
PERCENT COMPLEX
  Pb++       0.8067   ***
  PbCl+     22.7137   ***
  PbCl2     66.2912   ***
  PbCl3-     9.8837   ***
  PbCl4--    0.3048   ***

LOG KS=-2.930
LOGB1=3.180
LOGB2=4.980
LOGB3=5.030
LOGB4=4.000
GAM PB=0.036
GAM PBCL=0.430
GAMPBCL2=1.000
GAM CL=0.350
CL=2.0000
PERCENT COMPLEX
             0.0138   ***
             1.2231   ***
            23.2285   ***
            42.4279   ***
            33.1067   ***
```

```
GAM=0.015
GAM=0.330
GAM=0.251

LOG KS=-2.540
LOGB1=3.890
LOGB2=6.260
LOGB3=6.760
LOGB4=4.000
GAM PB=0.015
GAM PBCL=0.330
GAMPBCL2=1.000
GAM CL=0.251
CL=2.0000
PERCENT COMPLEX
             0.0025   ***
             0.4343   ***
            16.8671   ***
            81.1389   ***
             1.5572   ***
PB AS PPB
MH2S=0.0004
PH=4.1000
PB=25602263.98
PH=6.1000
PB=2560.2264
```

AU

(programmed by RWH)

Program Description

This program calculates Au solubility as $Au(HS)_2^-$ in mg/kg at specified temperature, log f_{O2}, pH and total sulphur (STOT in moles/kg). Data inputs are prompted in the following order;

t°C, K_{Au}, STOT, GAMMA $HSO4^-$, LOG FO2, PH

After each calculation press R/S and the program will prompt for a new pH. If the input is 0, it prompts for a new LOG FO2 and if again input is 0 it prompts for new values of STOT and then GAMMA HSO4.

The program uses regression equations from Chapter 8 to calculate the dissociation constants K_{H2O}, K_{H2S} and K_{HSO4} at the required calculation temperature.

Sample problem; See Chapter 9, Figure 9.2 for input data and sample answers.

Log K_{Au} from Seward (1973) -- see Chapter 9.

t°C	200	225	250	275	300°C
log $K_{Au(HS)_2^-}$	-1.28	-1.22	-1.19	(-1.18)	(-1.16)*

*bracketed values extrapolated.

PROGRAM LISTING

AU

```
01♦LBL "AU"
02 FIX 2
03 "T="
04 XEQ 01
05 FS? 55
06 XEQ 02
07 273.15
08 +
09 1/X
10 STO 01
11 48026.92
12 *
13 30.34
14 -
15 STO 02
16 RCL 01
17 4324.08
18 *
19 13.59
20 -
21 STO 12
22 RCL 01
23 2377.5
24 *
25 12.18

45♦LBL 03
46 "STOT="
47 XEQ 01
48 STO 10
49 XEQ 02
50 "GAMMA HSO4="
51 XEQ 01
52 FS? 55
53 XEQ 02
54 LOG
55 STO 09

56♦LBL 04
57 "LOG FO2="
58 XEQ 01
59 X=0?
60 GTO 03
61 XEQ 02
62 FS? 55
63 XEQ 02
64 STO 07

65♦LBL "SOL"
66 "PH="
67 XEQ 01

90 RCL 08
91 +
92 10↑X
93 *
94 1
95 +
96 RCL 03
97 RCL 08
98 +
99 RCL 09
100 -
101 10↑X
102 +
103 1/X
104 RCL 10
105 *
106 STO 13
107 "MH2S ="
108 FS? 55
109 XEQ 02
110 X↑2
111 LOG
112 RCL 05
113 +
114 RCL 03
```

```
26 -
27 STO 03
28 RCL 01
29 14564.13
30 *
31 CHS
32 7.6
33 +
34 STO 04
35 TONE 3
36 "LOG KAU="
37 XEQ 01
38 FS? 55
39 XEQ 02
40 RCL 04
41 2
42 /
43 -
44 STO 05
```

```
68 X=0?
69 GTO 04
70 FS? 55
71 XEQ 02
72 STO 08
73 RCL 12
74 RCL 09
75 3
76 *
77 -
78 RCL 08
79 +
80 10↑X
81 1
82 +
83 RCL 02
84 RCL 07
85 2
86 *
87 +
88 RCL 09
89 -
```

```
115 +
116 RCL 07
117 4
118 /
119 +
120 RCL 08
121 +
122 RCL 09
123 -
124 5.947
125 +
126 10↑X
127 SCI 2
128 "AU="
129 XEQ 02
130 STOP
131 GTO "SOL"

132♦LBL 01
133 PROMPT
134 RTN

135♦LBL 02
136 ARCL X
137 AVIEW
138 RTN
139 END
```

STEAM

(programmed by Jill Robinson, Geothermal Resources International, Santa Rosa, CA)

Program Description

This program consists of three parts and is designed primarily for use in chemical modelling of vapor-dominated systems. It may also be used in hot-water systems.

In vapor-dominated reservoirs in their undisturbed state, a vapor coexists with liquid. After exploitation and pressure drop, additional steam is formed by evaporation so that the extracted vapor phase is made up of original vapor plus newly formed steam; these are here designated y and (1-y) respectively - note that in this case the term (1-y) is not a water fraction.

1) Y-T, a program which uses the analysis of H_2S, H_2, CO_2 and CH_4 in steam to calculate the temperature of origin and the fraction of original reservoir vapor, y. The remaining steam, 1-y originated from vaoorization of original liquid.

2) Y-H_2S, a program which calculates y from the H_2S, CO_2 and CH_4 contents and known or assumed reservoi temperature.

3) Y-H_2, a program using the temperature and H_2, CO_2 and CH_4 contents to calculate y. The derivation of Y-H_2 is described in detail in Chapter 11.

These programs prompt the input of % gas (in dry gas), °C (in Y-H_2 and Y-H_2S), and g/s (mole). When using on hot water systems, the g/s should be replaced by the gas/total fluid ratio. The y value is then interpreted as the fraction of excess steam in the reservoir fluid.

PROGRAM LISTING

STEAM

```
 01♦LBL "STEAM"
 02♦LBL "Y-T"
"G/S?" PROMPT STO 06
"%H2S?" PROMPT 100 /
* LOG STO 09 "%H2?"
PROMPT 100 / RCL 06
* LOG STO 06 "%CH4"
PROMPT "%CO2" PROMPT
/ LOG STO 07 .25 *
CHS RCL 06 + 6.355 +
STO 08 RCL 07 .08333
* CHS RCL 09 + 2.122
- STO 10 "EST MIN T?"
PROMPT 273 + STO 03
XEQ 01

 51♦LBL 03
TONE 1 RCL 03 2 +
STO 03 GTO 01

 58♦LBL 02
BEEP RCL 04 RCL 05 +
2 / "Y=" ARCL X
AVIEW PSE CLA RCL 03
273 - "T=" ARCL X
AVIEW STOP GTO "Y-T"

 78♦LBL 01
RCL 03 .0140 *
10.0585 - 10↑X CHS
STO 11 RCL 03 LOG
-2.076 * RCL 03 1/X
951.6 * + RCL 08 +
10↑X RCL 11 + RCL 11
1 + / FIX 3 VIEW X
STO 05 RCL 03 .00981
* 6.7328 - 10↑X CHS
STO 12 RCL 03 LOG
.098 * RCL 03 1/X
2542 * + RCL 10 +
10↑X RCL 12 + RCL 12
1 + / FIX 3 VIEW X
STO 04 .005 RCL 04
RCL 05 - ABS X<=Y?
GTO 02 GTO 03

145♦LBL "Y-H2S"
"%CH4?" PROMPT "%CO2?"
PROMPT / STO 03
"T(DEG C)?" PROMPT 273
+ STO 01 273 -
.00981 * CHS 4.0547
+ 10↑X 1/X STO 02
"%H2S?" PROMPT 100 /
"G/S?" PROMPT * LOG
2.122 - RCL 01 1/X
2542 * + RCL 01 LOG
.098 * + RCL 03 LOG
12 / - 10↑X RCL 02
- RCL 02 CHS 1 + /
"Y-H2S=" ARCL X AVIEW
STOP

204♦LBL "Y-H2"
"CH4%?" PROMPT "CO2%?"
PROMPT / STO 03
"T(DEG C)?" PROMPT 273
+ STO 01 273 -
.01403 * CHS 6.2283
+ 10↑X 1/X STO 02
"%H2?" PROMPT 100 /
"G/S?" PROMPT * LOG
6.355 + 951.6 RCL 01
/ + RCL 01 LOG 2.076
* - RCL 03 LOG .25
* - 10↑X RCL 02 -
RCL 02 CHS 1 + /
"Y-H2=" ARCL X AVIEW
STOP
```

FS

(programmed by JAW)

Program Description

This program calculates the log f_{S2} from the composition of pyrrhotite using the equation of Toulmin and Barton, 1964. First, be sure you have appropriately calculated the mole fraction of FeS (N_{FeS}) in the pyrrhotite*. If this is done from a chemical analysis, you must remember that N is defined in the system Fe-S_2. When you run the program it will first ask you for the temperature in °K. After entering the value press R/S. The program will then ask for NFES which is the mole fraction discussed above. After entering this number hit R/S. The log f_{S2} will then appear. To do a second N at the same temperature, simply hit R/S. If you wish to do another temperature, enter 0 for N. To end the program enter 0 for T.

---

* N.B. The program uses the symbol N instead of X for mole fraction as specified in the text.

BI

(programmed by JAW)

Program Description

This program calculates the log f_{H2O} from the biotite composition using the equation of Wones, 1972. The program will prompt you to enter the temperature, activity of magnetite, activity of Kspar, oxygen fugacity, and mole fractions of various sites in the biotite structure. Included in the program are mole fraction of Fe in the octahedral sites, K in the A site, and OH in the hydroxyl site. After responding to each entry, hit R/S. To do a second value simply hit R/S. To end the program, enter 0 for T.

SL

(programmed by JAW)

Program Description

This program calculates the temperature in °C and the log f_{O2} from the composition of coexisting iron titanium oxides using the equations of Spencer and Lindsley, 1981. First, you must calculate the mole fraction of ulvospinel in magnetite and ilmenite in hematite-ilmenite solid solution. Stormer (1983) gives one method of doing this along with a discussion of other methods. Once these values are known, the program will prompt you to enter each of a pair of values. After entering each, hit R/S. The calculator will then display the temperature in °C. If you hit R/S again, log f_{O2} will appear. Hitting R/S again will start the cycle over again. To end the program, enter 0 for XUSP. One word of warning. This program does not check to see if the results are within the limits of reliability shown on Figure 12-3, so all resulting values should be checked to assure they lie within the limits shown by Spencer and Lindsley (1981).

PROGRAM LISTINGS

FS

```
01♦LBL "FS"
02♦LBL 10
03 "T IN K?"
04 PROMPT
05 X=0?
06 GTO 05
07 1/X
08 1000
09 *
10 1
11 -
12 STO 01
13♦LBL 15
14 "N?"
15 PROMPT
16 X=0?
17 GTO 10
18 STO 02
19 85.83
20 *
21 CHS
22 70.03
23 +
24 RCL 01
25 *
26 STO 10
27 RCL 02
28 .9981
29 *
30 CHS
31 1
32 +
33 SQRT
34 39.3
35 *
36 RCL 10
37 +
38 11.91
39 -
40 STOP
41 GTO 15
42♦LBL 05
43 STOP
44 END
```

BI

```
01♦LBL "BI"
02 ΣREG 11
03♦LBL 01
04 "T IN K?"
05 PROMPT
06 X=0?
07 GTO 05
08 1/X
09 7409
10 *
11 CLΣ
12 Σ+
13 "A KSPAR?"
14 PROMPT
15 LOG
16 CHS
17 Σ+
18 "A MAG?"
19 PROMPT
20 LOG
21 CHS
22 Σ+
23 "LOG FO2?"
24 PROMPT
25 ENTER↑
26 .5
27 *
28 Σ+
29 "XFE?"
30 PROMPT
31 LOG
32 3
33 *
34 Σ+
35 "XOH?"
36 PROMPT
37 LOG
38 2
39 *
40 Σ+
41 "XK?"
42 PROMPT
43 LOG
44 Σ+
45 RCL 11
46 4.25
47 +
48 STOP
49 GTO 01
50♦LBL 05
51 END
```

PROGRAM LISTING

SL

```
01♦LBL "SL"
02♦LBL 10
"XUSP?" PROMPT X=0?
GTO 50 STO 08 X↑2 3
* CHS RCL 08 4 * +
1 - STO 01 RCL 08
X↑2 3 * RCL 08 2 *
- STO 02 RCL 08 CHS
1 + STO 18 "XILM?"
PROMPT STO 09 X↑2 3
* CHS RCL 09 4 * +
1 - STO 03 RCL 09
X↑2 3 * RCL 09 2 *
- STO 04 RCL 09 CHS
1 + STO 19 X↑2
RCL 08 * RCL 09 X↑2
RCL 18 * / STO 05
RCL 03 102374 *
RCL 04 36818 * +
27799 + STO 06 RCL 03
71.095 * RCL 04
7.7714 * + RCL 05 LN
8.3143 * - 4.1920 +
STO 07 / 1073.15
X<=Y? GTO 20 RCL 06
RCL 01 64835 * -
RCL 02 20798 * -
RCL 07 RCL 01 60.296
* - RCL 02 19.652 *
- / GTO 30

119♦LBL 20
X<>Y

121♦LBL 30
STO 20 273.15 - STOP
CLX STO 21 STO 22
RCL 20 1073.15 X<=Y?
GTO 40 RCL 20 60.296
* CHS 64835 + STO 21
RCL 20 19.652 * CHS
20798 + STO 22

147♦LBL 40
RCL 20 7.7714 * CHS
36818 + STO 23 RCL 20
71.095 * CHS 102374
+ STO 24 RCL 08 1 -
RCL 08 X↑2 * 8 *
RCL 21 * RCL 08 2 *
CHS 1 + RCL 08 X↑2
* 4 * RCL 22 * +
RCL 09 CHS 1 +
RCL 09 X↑2 * 12 *
RCL 24 * + RCL 09 2
* CHS 1 + RCL 09
X↑2 * 6 * RCL 23 *
- RCL 20 8.3143 *
1/X * RCL 08 CHS 1
+ LN 4 * - RCL 09
CHS 1 + LN 12 * +
2.203 / 24634 RCL 20
/ - 13.96 + STOP
GTO 10

243♦LBL 50
"END" END
```

SG

(programmed by JAW)

Program Description

Program SG calculates the fugacitites of sulfurous gases from the oxygen fugacity, temperature, water fugacity, pressure, etc. for pyrrhotite-bearing assemblages after the method of Whitney, 1984. The parameters are optimized for the region from 600 to 900°C and the normal magmatic oxygen fugacities (between the Q-F-M buffer and the sulfur liquid - pyrite boundary). Calculation of the pyrrhotite-magnetite boundary with linear approximations causes a shift in the log f_{S2} value of not more than 0.06, and only 1/2 this error in log f_{SO2}, log f_{SO3}, or log f_{H2S} . The thermodynamic parameters for this calculation are derived from Barton and Skinner (1979) for the Fe-S-O system, and Robie et al. (1979) for gaseous equilibria.

To run this program, use XEQ "SG". The program will then ask for log f_{O2} (don't forget the minus sign), T (Kelvin), P (kbars), X_{mag} (mole fraction Fe_3O_4), and X_{FE} in po (mole fraction Fe in the cation site). If you are willing to accept approximate values, use 0, 1, and 1 respectively for the last three variables.

After entering the last variable, hitting R/S will produce log f_{S2}. If you hit R/S again, log f_{SO2} will appear. A third R/S will yield log f_{SO3}. Hit a fourth R/S and the program will ask for f_{H2O} (note that this is not a log). Another R/S will produce log f_{H2}. An additional R/S will produce log f_{H2S}. Another R/S starts the cycle over. To end the program, enter 0 for log f_{O2}.

PROGRAM LISTING

SG

```
01♦LBL "SG"
02♦LBL 10
03 "LOG FO2?"
04 PROMPT
05 X=0?
06 GTO 20
07 STO 01
08 "T IN K?"
09 PROMPT
10 STO 02
11 1/X
12 18413
13 *
14 19.15
15 RCL 02
16 /
17 CHS
18 1.208
19 +
20 RCL 01
21 *
22 +
23 3.495
24 -
25 STO 03
26 "P IN KB?"
27 PROMPT
28 33.736
29 RCL 02
30 /
31 0.01203
32 +
33 *
34 ST+ 03
35 "X MAG?"
36 PROMPT
37 LOG
38 2
39 *
40 3
41 /
42 ST- 03
43 "XFE IN PO?"
44 PROMPT
45 LOG
46 2
47 *
48 ST+ 03
49 RCL 03
50 STOP
51 2
52 /
53 18914.4
54 RCL 02
55 /
56 +
57 RCL 01
58 +
59 3.822
60 -
61 STOP
62 RCL 01
63 1.5
64 *
65 RCL 03
66 2
67 /
68 +
69 RCL 02
70 1/X
71 24004.4
72 *
73 +
74 8.655
75 -
76 STOP
77 "F H2O BARS?"
78 PROMPT
79 LOG
80 STO 04
81 RCL 01
82 2
83 /
84 -
85 RCL 02
86 1/X
87 12956.4
88 *
89 -
90 2.90
91 +
92 STO 05
93 STOP
94 RCL 03
95 2
96 /
97 +
98 RCL 02
99 1/X
100 4720.2
101 *
102 +
103 2.561
104 -
105 STOP
106 GTO 10
107♦LBL 20
108 "END"
109 FACT
110 END
```

APPENDIX II: ANSWERS TO SELECTED QUESTIONS AND PROBLEMS

Chapter 2.

p. 17 1. C_{Cl} = 1679 mg/kg compared with 1704 mg/kg by single stage flash to the weirbox. The steam fractions removed at each stage are 0.189, 0.053 and 0.112, respectively.

2. 1062 J/gm

p. 21 1. The plot of Cl vs Enthalpy temperature is meaningless because of the variety of recharge waters in this group of systems. The Na/K vs enthalpy-temperature plot shows a simple relationship and may be used for geothermometry--further discussion is given in Chapters 3 and 6.

2. The enthalpy temperature is 338°C compared with the downhole temperature of 305°C. This is an example of an "excess enthalpy well" whose discharge is derived from a two phase zone induced in the reservoir by fluid withdrawal. The problems of interpreting discharges like this are discussed in Chapter 11.

p. 25 CO_2 in total discharge (mg/kg)
Wairakei, 135; Cerro Prieto, 3414; Broadlands, 4466;
The Geysers, 45996. Notice how dissolved gas is a principal variable in the composition of geothermal fluids, and is as important as salinity in determining the physical and chemical properties of geothermal systems (see Chapters 4 and 5).

p. 25 1c. $t_{quartz} \simeq$ 190°C

2. Many, but since the spring occurs near the margin of the system, dilution prior to boiling in the discharge vent appears most likely.

3. The data you plotted to obtain the silica geothermometer do not extend to the silica concentrations in the springs. You may continuously extrapolate, but it is at about this temperature interval that other silica polymorphs, like chalcedony, may become stable whereas quartz solubility controls C_{SiO2} at the higher temperatures. Also you do not know the actual fraction of steam formed during discharge in the spring conduit.

4. Steaming ground occurs where steam-heated waters boil at the water table; fumaroles venting steam derived directly from the reservoir may also be present. These types of features are indicative of boiling conditions in the underlying portion of the reservoir.

5. Figure II: 2.1 illustrates the Tauhara hydrological model developed by Henley and Stewart (1983) using data from hot springs and from shallow, intermediate, and deep exploration wells. The initial features are convective upflow and boiling near the center of the system, formation of steam heated waters and dilution of chloride water near the field margin. Recharge is dominated by the Mt. Tauhara catchment area which induces near surface water flows toward Lake Taupo.

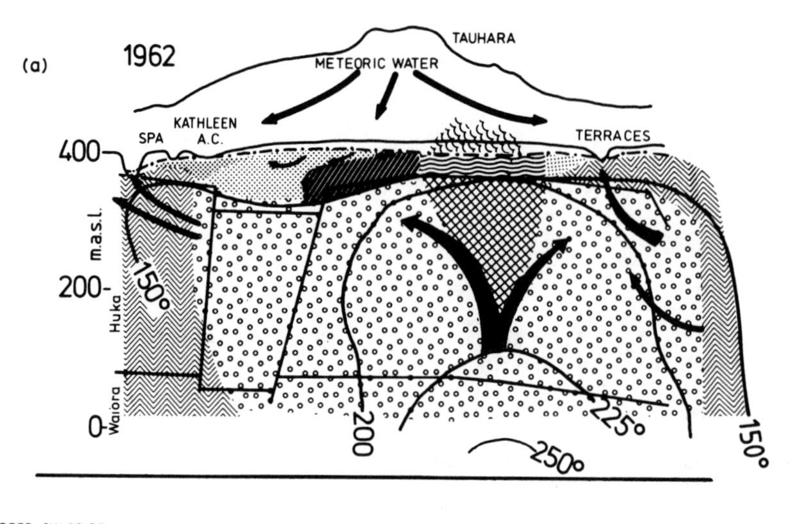

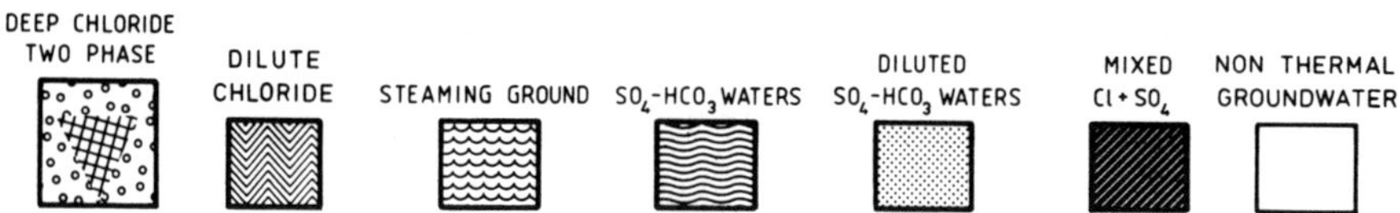

Figure II:2.1

Chapter 3.

p. 31 1. SiO_2, quartz = SiO_2,aq.

$\log SiO_2(mg/kg) = 5.085 - 1262.5(1/T)$

$$t = \frac{1262.5}{55.085 - \log SiO_2} - 273.15$$ Quartz solubility (conductive cooling)

p. 32 4. $\log SiO_2(mg/kg) = 5.721 - 1516.3(1/T)$

$$t = \frac{1516.3}{5.721 - \log SiO_2} - 273.15$$ Quartz after maximum steam loss (adiabatic cooling)

p. 37 1. Contouring is very individual, see Truesdell and Thompson (1982) for some attempts. CO_2 and HCO_3 are related by the reaction

$$CO_2 + \text{Naspar} + H_2O = HCO_3^- + Na^+ + \text{clay}$$

2, 3. The intersection of the boiling curve (maximum Cl) and the dilution curve is near 205 mg/kg Cl and 1200 J/g (= 273°C water)

4. Truesdell and Thompson (1982) drew the following cross section. Yours may be better. (see Fig. II:3.1)

6. $t_{SO_4 - H_2O} = 253°C$

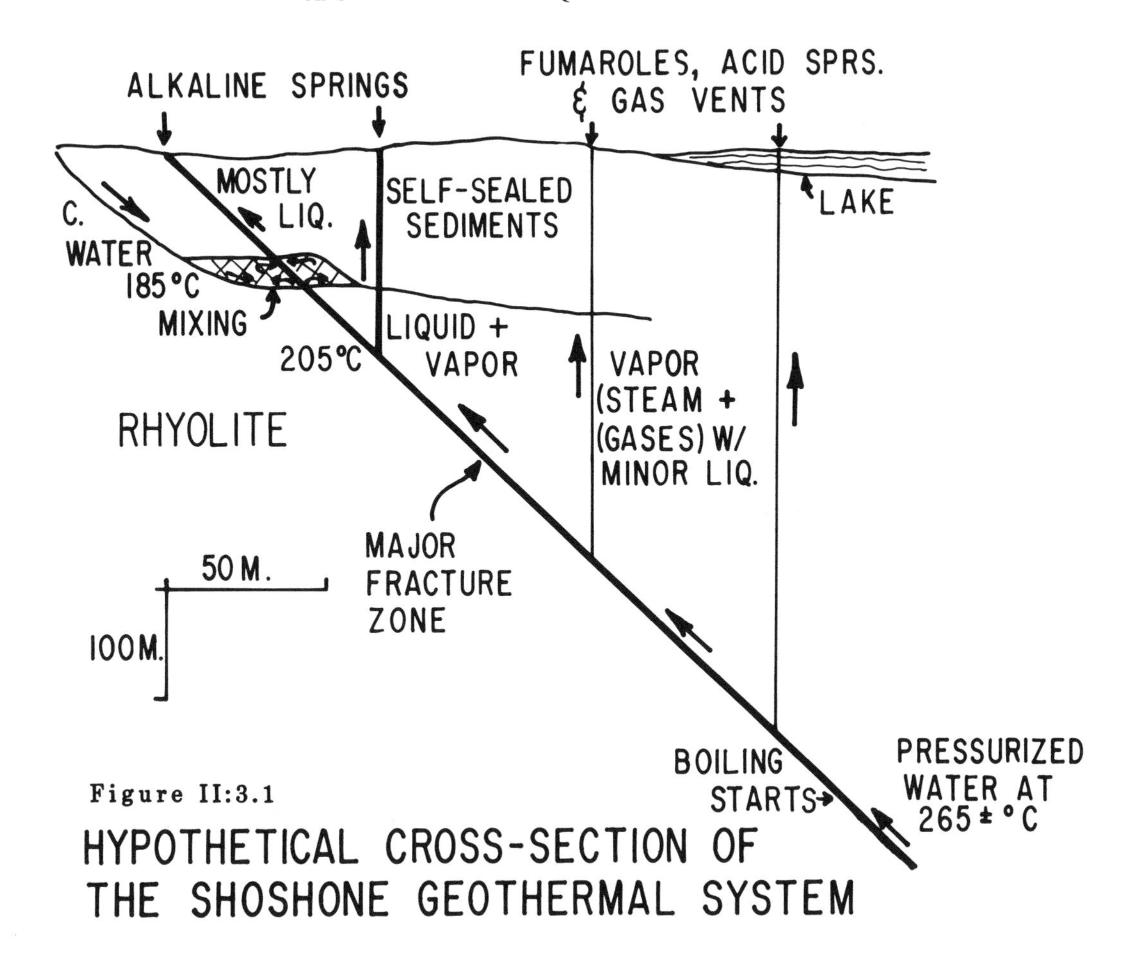

Figure II:3.1

HYPOTHETICAL CROSS-SECTION OF THE SHOSHONE GEOTHERMAL SYSTEM

Chapter 4.

p. 46 At 200°C (14.5 b.g.) for the example considered earlier, about 2% of the original CO_2 remains in the liquid phase after single step steam separation. Steam collection following separation at some lower pressure, say 5 b.g., would ensure that 99% of the CO_2 would be in the vapor phase.

p. 48 Since the gas content (Cg) and enthalpy of the total discharge may be assumed constant during sampling, we may apply the following heat and mass balances

$$H = H_{1,1} + L_1 y_1 = H_{1,2} + L_2 y_2$$

$$C \simeq C_1 y_1 \simeq C_2 y_2$$

where 1 and 2 indicate the high and low pressure sample, respectively.

Therefore setting $r = C_1/C_2 = y_2/y_1$ and $\underline{L} = L_1/L_2$, substitution leads to

$$H = \frac{(rH_1 - \underline{L}\, H_2)}{(\underline{L} - r)}$$

H_1, H_2, and $\underline{L}$ are obtained from Steam Tables for the appropriate sampling conditions.

p. 52 Provide a "bleed" valve below the master valve and keep it cracked open to vent accumulating gas.

Chapter 5.

p. 58 Table 5.2

T°K	G°R	log K	1/T
400	95088	-12.415	0.0025
500	75817	-7.919	0.0020
600	55644	-4.843	0.001667

p. 58 1. least squares linear regression R^2 = 0.9995

$$\log K = 10.278 - 9082/T$$

2.

$$\log K = \log P_{CO2} + 4 \log P_{H2} - \log P_{CH4} - 2 \log P_{H2O} = 10.278 - 9082/T$$

$$2 \log P_{H2O} = 11.02 - 4096/T$$

$$\log K' = \log P_{CO2} + 4 \log P_{H2} - \log P_{CH4} = 21.298 - 13178/T$$

p. 58 3. Converting to mole fractions $K_H = P/X$; $\log X = \log P - \log K_H$

$$\log K'' = \log K' - \log K_H(CO2) - 4 \log K_H(H2) + \log K_H(CH4)$$

This equation is then evaluated using coefficients from Table 5.3:

	+21.298 - 13178/T	
	-5.0149	+0.002515T
	-29.3336	+0.022500T
	+7.1287	-0.005425T
log K" =	-5.922 - 13178/T	+0.01959T

	CO_2	H_2	CH_4	N_2
log X	-2.996	-4.225	-4.179	-5.115

$$\log Q = \log X_{CO2} + 4 \log X_{H2} - \log X_{CH4} = -15.717$$

Equation (6) is not easily solved for T so write a program and try different values of T (or t) until log Q = log K. A simple program could be

```
LBL'GAS' 'T DEGC? PROMPT 273.15 + STO09 RCL03 x RCL02 RCL09 / + RCL01 + END
```

The equation is in the form $\log K = a + b/T + cT$. The constants a, b, and c are stored in registers 01, 02, and 03 respectively before the program is run. Make a table of results to help you interpolate

	200	250	300	350
log K	-24.50	-20.86	-17.69	-14.86

Trial and error gives log K (335°C) = -15.68 ∴ $t_{CH4-CO2}$ = 335°C

From the geothermometers equations in Chapter 3:

t_{Quartz} (adiabatic)	= 269°C	- fastest equilibration
t_{NaKCa} (B = 1/3)	= 291	
$t_{(Fournier)}$	= 306	
$t_{(Truesdell)}$	= 306	
t_{Na-Li}	= 340	
t_{K-Mg}	= 347	- slowest equilibration
$t_{CH4-CO2}$	= 335	

p. 60 log X_{NH3} = -4.425

t_{NH3} = 302°C

p. 61 t_{H2-CO2} = 293°C

$t_{Enthalpy}$ = 289°C

p. 62 $t_{D'Amore - Panichi}$ = 289°C

Chapter 6.

p. 69 pH (CO_2 - HCO_3) = 5.7, pH (calcite) = 5.9

Chapter 8.

p. 99 log m_{O2} = -39.2; 1 molecule for each $10^{15.4}$ kg; that is less than 1 molecule per thousand cubic kilometers!

p. 100 Pyrite only.

p. 101 t_{CH4} = 249°C, t_{NH3} = 246°C

p. 101 $H_2S + 2\ O_2 = SO_4^{=} + 2\ H^+$

$Fe_3O_4 + 6\ HS^- + 6\ H^+ + O_2 = 3\ FeS_2 + 6\ H_2O$

$Fe_3O_4 + 6\ SO_4^{=} + 12\ H^+ = 3\ FeS_2 + 14\ O_2 + 6\ H_2$

p. 105 $FeS + 1/2\ O_2 + H_2S = FeS_2 + H_2O$

log f_{O2} = -50.44, -43.44, and -37.6 for 200, 250, and 300°C.

f_{H2} is 11 times higher at 200° than at 300°!

The intersection of the aqueous sulfur species is angular because each species contains only one sulfur atom; the decreases in activities mutually compensate.

p. 107 N_2 at 200°C will oxidize appreciable sulfide only under very acid conditions.

At equilibrium NH_4^+ -rich alunite would be expected under the most acid and most reducing conditions under which alunite is stable. If sulfate persists metastably well down into the NH_4^+ field, it might also produce metastable NH_4^+ -rich alunite; the reduced chemistry would be expected for NH_4^+ -rich mica or feldspar, but metastable persistence in the N_2 field of NH_4^+ from the breakdown of organic matter is also possible.

p. 110 A pressure increase of 1 kbar would shift the log f_{O2} for the hematite + magnetite assemblage by +0.03.

p. 111

1. Log f_{O2} from the half-cell data = -68.19.
   Log f_{O2} from Table 8.1 = -72.30.

2. Log f_{O2} for iron + wustite at 25°C = -86.32

3. E° for wustite + magnetite = -0.199.

4. Because the E° for magnetite + iron is more negative than that for wustite + iron, the assemblage iron + magnetite must be more stable than wustite (of course, the fact that magnetite is more O-rich than wustite is also part of the argument).

p. 112

$Au_{annealed} = Au^{+++} + 3\ e^-$

$Au_{battered} = Au^{+++} + 3\ e^-$

E° = 1.485 volts

$\Delta G°$ = + 4341 J

$K°_{battered}$ = 5.77 = activity of battered gold relative to annealed gold; thus the level of saturation = 577%; and the level of supersaturation is 477%.

Chapter 9.

p. 117 $\beta_4 = K_1K_2K_3K_4$

p. 117 $Pb(HS)^+$, $Pb(HS)_2$, $Pb(HS)_3^-$ and another complex $Pb(HS)_2H_2S$ is also recognized at high temperature, see Table 9.7.

p. 120 ∿10 mg Pb/kg.

The measured concentration at Magmamax is 53 mg Pb/kg. Considering the difficulties inherent in realistically calculating the ionic strength and activity coefficients of such a concentrated solution, the apparent agreement between calculated and observed Pb contents is satisfying. Deposition from such a solution may be caused by local boiling, by which pH increases e.g., if pH increases by 1 unit, the solubility drops to less than 1 mg/kg. Galena is not abundant in the Imperial Valley systems, although it does precipitate in surface discharges.

p. 121 (2) boiling: - calcite, adularia, quartz + sulfides
dilution: - minor silicification, micas and clays in wall rocks

(3) Higher silica in solution results in slightly lower pH, hence higher PbS solubility. Wall rock buffering during boiling constrains the pH increase, so that although the solubility of galena is decreased, the effect is less marked than for isolated fluid boiling.

p. 122 The relative changes in the solubility of acanthite due to boiling, dilution, and conductive cooling are 0.015, 0.022 and 0.03 respectively.

p. 124 If f_{H2} = .33 bars, $f_{O2} = 10^{-39.5}$ bars. Gold solubility data for these conditions are shown in Figure 9.2

p. 125 1a. The principal stability boundary of interest in epithermal-geothermal studies is between $Pb(HS)_2H_2S$ and $Pb(HS)_2$. Summing the solubility reactions given in Table 9.7 you obtain

$$Pb(HS)_2 + H_2S = Pb(HS)_2H_2S \qquad \log K_{250} = 0.25$$

(Notice that you need to interpolate the solubility constant for the $Pb(HS)_2H_2S$ complex, perhaps using the linear regression program, LIN). Hence in the H_2S stability field up to pH $\simeq$ 7, for equal activities, $m_{H2S} = 1/K_{250} = 0.56$. $Pb(HS)_2$ occupies the stability field at lower total sulfur. You should also notice that at higher pH's, more sulfur is required to stabilize the $Pb(HS)_2H_2S$. Your calculation can also show that at pH 8.4 the $Pb(HS)_3^-$ complex displaces $Pb(HS)_2H_2S$.

The geothermal fluid data suggest that $Pb(HS)_2$ is the dominant thiosulfide species, but as shown above, is itself subordinate to chloride and perhaps hydroxy-carbonate species in geothermal fluids like those at Broadlands.

1b. First plot solubility curves for the two principal species, $Pb(HS)_2$ and $PbCl_2$. At pH = 6 they should intersect at about log ΣPb (mg/kg) = -5.5 and log $m_{(H2S + HS^-)} = -2.6$.

The Broadlands well plots above the solubility curves. Notice that it also plots in the field where $m_{\Sigma Pb} < m_{\Sigma S}$, which is a condition inferred to limit the metal content of ore transporting solutions. Is the Broadlands fluid really supersaturated or is there something amiss in the data? Perhaps there are other complexes to consider as is suggested for the lower temperature Mississippi Valley-type deposits considered in Chapter 14. It is less likely that the Broadlands data is in substantial error.

p. 126 2. From the reaction equations given in Table 9.7 you can derive the expression

$$\Sigma m_{Fe} = a_{Cl^-} a^2_{H^+} \{ (K_{FeCl^+} / \gamma_{FeCl^+}) + (K_{FeCl2}\, a_{Cl^-}) \} / f^{.5}_{O2}\, m^2_{H2S}$$

Substituting chemical parameters for the Broadlands fluid $\Sigma m_{Fe} = 5.6 \times 10^{-8}$ moles/kg = 0.003 mg/kg.

The analytical data of Table 9.1 suggest a much higher solubility so that, as in the previous question, you need to consider the possibility of errors in the thermodynamic data or solution analysis or that other complexes may contribute to the solubility. For example Seward (1977 in New Zealand DSIR Bulletin 218) suggests that Fe^{++} may contribute substantially to the solubility of the iron sulfides. Most important, however, is the very large temperature coefficient of FeS_2 solubility shown in Table 9.7, which is more than enough to cover the discrepancy between calculated and observed iron contents.

Chapter 10.

p. 136 1)

Temp	y	$10^3 \ln \alpha$ ^{18}O	D	water composition $\delta^{18}O$	δD
260	--	--	--	-5	40
220	0.102	2.10	0.1	-4.78	-39.99
180	0.0895	2.90	7.4	-4.52	-39.3
140	0.0812	3.91	16.3	-4.20	-38.0
100	0.0754	5.24	27.8	-3.81	-35.9

p. 138 The water is enriched with respect to the local meteoric water; $\delta^{18}O = -5.1$ and $\delta D = -34$.

p. 138 The mass flux of 110° steam (ω) is 0.81 kg/unit time so that the flux of steam heated water is (1 + 0.95 - 0.81) = 1.14 kg/unit time.

p. 138

1) $\delta D = -40$ $\delta^{18}O = -6.25$

2) $\delta D = -40$ $\delta^{18}O = -2.25$

3) no change

4) y (300 --> 200°C) = 0.253
y (200 --> 100°C) = 0.192

water temp	δD	$\delta^{18}O$
300°C	-40	-2.25
200	-39.1	-1.62
100	-33.8	-0.61

5)

steam temp	δD	$\delta^{18}O$
200°C	-42.6	-4.1
100	-61.5	-5.86

6) fraction steam = 0.107
steam heated water $\delta D = -40.3$ $\delta^{18}O = -6.02$

7) a) end member and mixture compositions before boiling

	10°C water	300°C water	2:1	1:2
H_w, J/g	42	1344	910	476
t, °C	10	300	213	113
y	—	0.41	0.22	0.025
δD	-40	-40	-40	-40
$\delta^{18}O$	-6.25	-2.25	-3.58	-4.91

b) water after boiling to 100°C

	300°C	2:1	1:2
single stage:			
δD	-28.6	-33.9	-39.3
$\delta^{18}O$	- 0.1	-2.42	-4.78
continuous:			
δD	-37.8	-37.1	-39.5
$\delta^{18}O$	-1.07	-2.76	-4.79

p.141 97% reaction equals 5 half times or 2.8 years at 250°C or 12,800 years at 25°C.

$\delta^{18}O$ (SO_4) 250° = 1.43

1) fraction equilibrated = 0.62

$\delta^{18}O$ (SO_4) equilibrium (175°C) = 5.25

Therefore $\delta^{18}O$ (SO_4) 62% equilibrated = 3.80

2) $\delta^{18}O$ (H_2O) = -3.45 $\delta^{18}O$ (SO_4) unchanged

3) SO_4 from H_2S = 14 mg/kg and original SO_4 after flashing equals 71 mg/kg

$\delta^{18}O$ (SO_4) total = 2.83

4) $\delta^{18}O$ (H_2O) before boiling = -5.0 . . apparent $t_{SO4-H2O}$ = 218°C

Chapter 11.

p. 153 A y = .364 B y = .473 C y = 0.122

The reserves are in the order C > A > B

Chapter 12.

p. 155 For a pyrrhotite with 61.5% Fe, 38.5 wt.% S, X = 0.9502, log f_{S2} = -2.72.

p. 155 For an analysis of 61 wt.% Fe, 39 wt.% S, X = 0.9380, log f_{S2} = -1.75.

p. 156 For magnetite ($Fe_{27}Ti_3O_{40}$) and ilmenite ($Fe_{11.1}Ti_{8.9}O_{30}$), T = 800°C, log f_{O2} = -13.5. Fayalite is not stable.

log f_{S2} = -2.3

p. 165 To convert the magmatic fluids in Table 12.1 to 250°C, follow the following procedure:

1) Assume SO_2 reacts using the equation given.
2) Convert concentration of total $SO_4^=$ and H_2S to molal quantities.
3) Assume H_2S and HSO_4^- dissociate according to constants given in Table 8.1, with the pH at 4.
4) Calculate a_{O2} from the reaction $H_2S + 2O_2 = HSO_4^- + H^+$ obtained by combining the first three reactions in Table 8.1.
5) Calculate a_{H2} from dissociation of water, preferably using the fugacity of water given, and constant for gaseous water from Robie et al., 1979.
6) Calculate a_{S2} from the dissociation of H_2S. This may be done using the constant from Robie et al., 1979, if the fugacity of H_2S is determined from concentration using the Henry's Law constant in Table 5.3. The results are shown below:

p. 166

Table 12.2 — Comparison of the Chemistries of Quenched Gasses from Oxidized and Reduced I-Type Magmas

	800° oxidized	800° reduced	250° oxidized	250° reduced
log f_{H_2O}	2.66	2.66	0*	0*
log P_{H_2O}	2.70	2.70	1.53	1.53
X_{H_2O}	0.74	0.98	0.65	0.98
log f_{O_2}	-11.5	-13.5	-32.6	-33.9
X_{O_2}	nil	nil	nil	nil
log f_{H_2}	-1.00	+0.22	-3.9	-3.3
X_{H_2}	$1x10^{-4}$	$3x10^{-3}$	nil	nil
log f_{S_2}	0.00	-2.3	-1.2	-3.8
X_{S_2}	$1x10^{-3}$	$1x10^{-5}$	nil	nil
log f_{H_2S}	1.10	0.92	---	---
log P_{H_2S}	1.14	0.96	-0.46	-1.18
X_{H_2S}	0.02	0.02	0.11	0.02
m_{H_2S}	---	---	8.9 (2.7M)	1.1
log f_{SO_2}	2.3	-0.9	---	---
log P_{SO_2}	2.2	-1.0	---	---
X_{SO_2}	0.24	$2x10^{-4}$	---	---
log $f_{SO_3^=}$	-3.54	-7.69	---	---
$X_{SO_3^=}$	nil	nil	---	---
$m_{Total\ SO_4}$	---	---	20.0 (6.2M)	$8.5x10^{-3}$
P_{Total}	673 bars	511 bars	34.2 bars	33.9 bars

* Note that we have changed standard states for H_2O, from 1 bar vapor at 800°C to liquid water at 250°C. Remember also that m is molal concentration, or moles per kilogram of water.

p. 167

If sulfur is dominantly dissolved in the melt as HS^-, its concentration will not vary strongly since f_{H2S} does not vary strongly.

The melt will never "run out of sulfur." About 50% will be lost, however, with the first 1 wt.% water evolved. If we recalculated the distribution constant for the oxidized case in the previous exercise on magmatic gases, the distribution constant would be closer to 500 and 50% of the sulfur would be lost with the first 0.2% of exsolved water.

p. 168 For 4 wt.% water on Figure 12.8, vapor separation will begin at 1.7 kb for 750° and 1.1 kb for 900°C.

p. 168 Upon intrusion to 1 kb pressure, a melt containing 8% water would exsolve approximately 3.9 to 4 wt% water immediately, and up to 0.2 wt.% more upon crystallization at 900°C. The rest would remain in the melt until crystallization at lower temperatures occurred.

p. 168 Since a large amount of sulfur is removed with the first few percent water, the largest portion of the sulfur would be evolved during initial vapor saturation in the case described.

Chapter 13.

p. 178 Input data required are

1) Water and steam sample analyses for each well, discharge enthalpies and flash plant separation pressures.

2) Output data (tonnes/hr) for each well. Designating the flow from each well as M, the total fluid flow to the flash plant is

$$\Sigma M = M_A + M_B + M_C$$

The chemistry of the combined discharges is given by

$$C_\Sigma = C_A M_A/\Sigma M + C_B M_B/\Sigma M + C_C M_C/\Sigma M$$

where C_Σ is the concentration of a component in the input to the flash plant and C_A to C_C are the total discharge concentrations from each contributing well. Similarly Δ_Σ (or HTOT if you prefer) is obtained from Δ_A to Δ_C and the mass fraction of each contributing well.

p. 179 167°C. The BR22 discharge appears to be supersaturated under these conditions. At 4.5 b.g. $C_{SiO2} \simeq 745$ mg/kg.

p. 179 Amorphous silica solubilities at pH = 3, 6.5 and 9 are 650, 652 and 1263 mg_{SiO2}/kg, respectively. At the pH of the separated water the solubility is 727 mg/kg so that in fact the water is much closer to saturation than we calculated earlier on the simple basis of H_4SiO_4.

p. 180 Fig. 13.1 Caption. More effective acid gas (CO_2 + H_2S) removal results from multistage flash than for single stage separation.

p. 188 Condensate pH $\simeq$ 5.3. Notice how the relatively low pK of NH_4^+ and the concentration of NH_3 dominate the ion balance under these particular conditions.

Chapter 14.

p. 194 607 J/mol for gold in 0.1 micrometer cubes.

a_{Au} = 1.28; supersaturation = 28%.

p. 194 Composition of the gold, esp. silver content; also mechanical deformation.

p. 197 1. $Au° + 2H_2S + 1.5$ kaolin $+ K^+ + 2\ HSO_4^- =$

$Au(HS)^- + 0.5\ H_2 +$ alunite $+ 3\ SiO_2 + 1.5H_2O$

2. d log c/d log a_{SiO2} = -3; thus chalcedonic silica suggests a <u>lower</u> solubility than the equilibrium calculation would suggest.

3. The K-feldspar + Kmica + silica buffer also gives d log c/d log a_{SiO2} of -3.

p. 197 The discussion on gold applies only to the H_2S-dominant field.

p. 198 In the context of Figure 14.3, pyrite might be expected to replace galena in the region where galena increases and pyrite decreases in solubility: any trend within the pyrite field having a bearing between N45E and N63.5E (North = up in Figure 14.3).

The solubilities of Cu and Fe vary in the same direction for solutions buffered by pyrite + Pyrrhotite, although log Fe changes twice as fast as that for Cu. Thus Figure 14.3 is not an appropriate format within which to interpret simultaneous replacement of pyrite and pyrrhotite by chalcopyrite. Perhaps a plot of Cu/Fe in solution vs t or pH would be more useful.

APPENDIX III: STEAM TABLES — THERMODYNAMIC DATA FOR WATER AT SATURATED VAPOR PRESSURES AND TEMPERATURES

0 - 374.136°C

Derivation: Equation of State of Keenan et al. (1969); and calculated by P. Delaney (U.S. Geological Survey)

Reference:

Keenan, J.H., Keyes, F.G., Hill, P.G. and Moore, J.G., 1969, Steam Tables - thermodynamic properties of water including vapor, liquid, and solid phases (International Edition - metric units): Wiley, New York, 162 p.

Temp.	Press.	Specific Volume		Enthalpy		
(°C)	(bars)	(cc/gm)		(J/gm)		
		vap.	liq.	vap.	liq.	evap.
0.01	0.01	206136	1.000	2501	0.01	2501
1	0.01	192577	1.000	2503	4.16	2499
2	0.01	179889	1.000	2505	8.37	2497
3	0.01	168132	1.000	2507	12.57	2494
4	0.01	157232	1.000	2509	16.78	2492
5	0.01	147120	1.000	2511	20.98	2490
6	0.01	137734	1.000	2512	25.20	2487
7	0.01	129017	1.000	2514	29.39	2485
8	0.01	120917	1.000	2516	33.60	2482
9	0.01	113386	1.000	2518	37.80	2480
10	0.01	106379	1.000	2520	42.01	2478
11	0.01	99857	1.000	2522	46.20	2475
12	0.01	93784	1.001	2523	50.41	2473
13	0.01	88124	1.001	2525	54.60	2471
14	0.02	82848	1.001	2527	58.80	2468
15	0.02	77926	1.001	2529	62.99	2466
16	0.02	73333	1.001	2531	67.19	2464
17	0.02	69044	1.001	2533	71.38	2461
18	0.02	65038	1.001	2534	75.58	2459
19	0.02	61293	1.002	2536	79.77	2456
20	0.02	57791	1.002	2538	83.96	2454
21	0.02	54514	1.002	2540	88.14	2452
22	0.03	51447	1.002	2542	92.33	2449
23	0.03	48574	1.002	2544	96.52	2447
24	0.03	45883	1.003	2545	100.7	2445
25	0.03	43360	1.003	2547	104.9	2442
26	0.03	40994	1.003	2549	109.1	2440
27	0.04	38774	1.003	2551	113.2	2438
28	0.04	36690	1.004	2553	117.4	2435
29	0.04	34733	1.004	2554	121.6	2433

Temp.	Press.	Specific Volume		Enthalpy		
(°C)	(bars)	(cc/gm)		(J/gm)		
		vap.	liq.	vap.	liq.	evap.
30	0.04	32894	1.004	2556	125.8	2430
31	0.04	31165	1.005	2558	130.0	2428
32	0.05	29540	1.005	2560	134.1	2426
33	0.05	28011	1.005	2562	138.3	2423
34	0.05	26571	1.006	2563	142.5	2421
35	0.06	25216	1.006	2565	146.7	2419
36	0.06	23940	1.006	2567	150.9	2416
37	0.06	22737	1.007	2569	155.0	2414
38	0.07	21602	1.007	2571	159.2	2411
39	0.07	20533	1.007	2572	163.4	2409
40	0.07	19523	1.008	2574	167.6	2407
41	0.08	18570	1.008	2576	171.7	2404
42	0.08	17671	1.009	2578	175.9	2402
43	0.09	16821	1.009	2580	180.1	2400
44	0.09	16018	1.009	2581	184.3	2397
45	0.10	15258	1.010	2583	188.4	2395
46	0.10	14540	1.010	2585	192.6	2392
47	0.11	13861	1.011	2587	196.8	2390
48	0.11	13218	1.011	2589	201.0	2388
49	0.12	12609	1.012	2590	205.1	2385
50	0.12	12032	1.012	2592	209.3	2383
51	0.13	11486	1.013	2594	213.5	2380
52	0.14	10968	1.013	2596	217.7	2378
53	0.14	10476	1.014	2597	221.9	2376
54	0.15	10011	1.014	2599	226.0	2373
55	0.16	9569	1.015	2601	230.2	2371
56	0.17	9149	1.015	2603	234.4	2368
57	0.17	8751	1.016	2604	238.6	2366
58	0.18	8372	1.016	2606	242.8	2363
59	0.19	8013	1.017	2608	246.9	2361
60	0.20	7671	1.017	2610	251.1	2358
61	0.21	7346	1.018	2611	255.3	2356
62	0.22	7037	1.018	2613	259.5	2354
63	0.23	6743	1.019	2615	263.7	2351
64	0.24	6463	1.019	2617	267.9	2349
65	0.25	6197	1.020	2618	272.0	2346
66	0.26	5943	1.020	2620	276.2	2344
67	0.27	5701	1.021	2622	280.4	2341
68	0.29	5471	1.022	2623	284.6	2339
69	0.30	5252	1.022	2625	288.8	2336
70	0.31	5042	1.023	2627	293.0	2334
71	0.33	4843	1.023	2629	297.2	2331
72	0.34	4652	1.024	2630	301.4	2329
73	0.35	4470	1.025	2632	305.5	2326
74	0.37	4297	1.025	2634	309.7	2324
75	0.39	4131	1.026	2635	313.9	2321
76	0.40	3973	1.027	2637	318.1	2319
77	0.42	3822	1.027	2639	322.3	2316
78	0.44	3677	1.028	2640	326.5	2314
79	0.46	3539	1.028	2642	330.7	2311
80	0.47	3407	1.029	2644	334.9	2309
81	0.49	3281	1.030	2645	339.1	2306
82	0.51	3160	1.030	2647	343.3	2304
83	0.53	3044	1.031	2649	347.5	2301
84	0.56	2934	1.032	2650	351.7	2299

Temp. (°C)	Press. (bars)	Specific Volume (cc/gm)		Enthalpy (J/gm)		
		vap.	liq.	vap.	liq.	evap.
85	0.58	2828	1.032	2652	355.9	2296
86	0.60	2726	1.033	2654	360.1	2293
87	0.63	2629	1.034	2655	364.3	2291
88	0.65	2536	1.035	2657	368.5	2288
89	0.68	2446	1.035	2658	372.7	2286
90	0.70	2361	1.036	2660	376.9	2283
91	0.73	2278	1.037	2662	381.1	2281
92	0.76	2200	1.037	2663	385.3	2278
93	0.79	2124	1.038	2665	389.5	2275
94	0.81	2052	1.039	2667	393.7	2273
95	0.85	1982	1.040	2668	398.0	2270
96	0.88	1915	1.040	2670	402.2	2268
97	0.91	1851	1.041	2671	406.4	2265
98	0.94	1789	1.042	2673	410.6	2262
99	0.98	1730	1.043	2674	414.8	2260
100	1.01	1673	1.043	2676	419.0	2257
101	1.05	1618	1.044	2678	423.3	2254
102	1.09	1566	1.045	2679	427.5	2252
103	1.13	1515	1.046	2681	431.7	2249
104	1.17	1466	1.047	2682	435.9	2246
105	1.21	1419	1.047	2684	440.1	2244
106	1.25	1374	1.048	2685	444.4	2241
107	1.29	1331	1.049	2687	448.6	2238
108	1.34	1289	1.050	2688	452.8	2236
109	1.39	1249	1.051	2690	457.1	2233
110	1.43	1210	1.052	2691	461.3	2230
111	1.48	1173	1.052	2693	465.5	2227
112	1.53	1137	1.053	2695	469.8	2225
113	1.58	1102	1.054	2696	474.0	2222
114	1.64	1069	1.055	2697	478.2	2219
115	1.69	1037	1.056	2699	482.5	2217
116	1.75	1006	1.057	2700	486.7	2214
117	1.80	975.6	1.058	2702	491.0	2211
118	1.86	946.7	1.059	2703	495.2	2208
119	1.92	918.8	1.059	2705	499.5	2205
120	1.99	891.9	1.060	2706	503.7	2203
121	2.05	865.9	1.061	2708	508.0	2200
122	2.11	840.8	1.062	2709	512.2	2197
123	2.18	816.6	1.063	2711	516.5	2194
124	2.25	793.2	1.064	2712	520.7	2191
125	2.32	770.6	1.065	2713	525.0	2189
126	2.39	748.8	1.066	2715	529.2	2186
127	2.47	727.7	1.067	2716	533.5	2183
128	2.54	707.3	1.068	2718	537.8	2180
129	2.62	687.6	1.069	2719	542.0	2177
130	2.70	668.5	1.070	2720	546.3	2174
131	2.78	650.1	1.071	2722	550.6	2171
132	2.87	632.3	1.072	2723	554.9	2168
133	2.95	615.0	1.073	2725	559.1	2165
134	3.04	598.3	1.074	2726	563.4	2163
135	3.13	582.2	1.075	2727	567.7	2160
136	3.22	566.6	1.076	2729	572.0	2157
137	3.32	551.4	1.077	2730	576.3	2154
138	3.41	536.8	1.078	2731	580.5	2151
139	3.51	522.6	1.079	2733	584.8	2148

Temp. (°C)	Press. (bars)	Specific Volume (cc/gm)		Enthalpy (J/gm)		
		vap.	liq.	vap.	liq.	evap.
140	3.61	508.9	1.080	2734	589.1	2145
141	3.72	495.6	1.081	2735	593.4	2142
142	3.82	482.7	1.082	2736	597.7	2139
143	3.93	470.2	1.083	2738	602.0	2136
144	4.04	458.1	1.084	2739	606.3	2133
145	4.15	446.3	1.085	2740	610.6	2130
146	4.27	435.0	1.086	2742	614.9	2127
147	4.39	423.9	1.087	2743	619.2	2124
148	4.51	413.2	1.088	2744	623.6	2120
149	4.63	402.9	1.089	2745	627.9	2117
150	4.76	392.8	1.090	2746	632.2	2114
151	4.89	383.0	1.092	2748	636.5	2111
152	5.02	373.5	1.093	2749	640.8	2108
153	5.15	364.4	1.094	2750	645.2	2105
154	5.29	355.4	1.095	2751	649.5	2102
155	5.43	346.8	1.096	2752	653.8	2099
156	5.57	338.4	1.097	2754	658.2	2095
157	5.72	330.2	1.098	2755	662.5	2092
158	5.87	322.3	1.100	2756	666.9	2089
159	6.02	314.5	1.101	2757	671.2	2086
160	6.18	307.1	1.102	2758	675.5	2083
161	6.34	299.8	1.103	2759	679.9	2079
162	6.50	292.7	1.104	2760	684.3	2076
163	6.66	285.9	1.106	2761	688.6	2073
164	6.83	279.2	1.107	2762	693.0	2070
165	7.00	272.7	1.108	2764	697.3	2066
166	7.18	266.4	1.109	2765	701.7	2063
167	7.36	260.2	1.110	2766	706.1	2060
168	7.54	254.3	1.112	2767	710.5	2056
169	7.73	248.5	1.113	2768	714.8	2053
170	7.92	242.8	1.114	2769	719.2	2050
171	8.11	237.3	1.116	2770	723.6	2046
172	8.31	232.0	1.117	2771	728.0	2043
173	8.51	226.8	1.118	2772	732.4	2039
174	8.71	221.7	1.119	2773	736.8	2036
175	8.92	216.8	1.121	2774	741.2	2032
176	9.13	212.0	1.122	2775	745.6	2029
177	9.35	207.3	1.123	2775	750.0	2025
178	9.57	202.8	1.125	2776	754.4	2022
179	9.79	198.4	1.126	2777	758.8	2018
180	10.02	194.0	1.127	2778	763.2	2015
181	10.25	189.8	1.129	2779	767.6	2011
182	10.49	185.8	1.130	2780	772.1	2008
183	10.73	181.8	1.132	2781	776.5	2004
184	10.98	177.9	1.133	2782	780.9	2001
185	11.23	174.1	1.134	2782	785.4	1997
186	11.48	170.4	1.136	2783	789.8	1993
187	11.74	166.8	1.137	2784	794.3	1990
188	12.00	163.3	1.139	2785	798.7	1986
189	12.27	159.9	1.140	2786	803.2	1982
190	12.54	156.5	1.141	2786	807.6	1979
191	12.82	153.3	1.143	2787	812.1	1975
192	13.10	150.1	1.144	2788	816.5	1971
193	13.39	147.0	1.146	2789	821.0	1968
194	13.68	144.0	1.147	2789	825.5	1964

Temp.	Press.	Specific Volume		Enthalpy		
(°C)	(bars)	(cc/gm)		(J/gm)		
		vap.	liq.	vap.	liq.	evap.
195	13.98	141.1	1.149	2790	830.0	1960
196	14.28	138.2	1.150	2791	834.5	1956
197	14.59	135.4	1.152	2791	839.0	1952
198	14.90	132.6	1.153	2792	843.4	1949
199	15.22	130.0	1.155	2793	847.9	1945
200	15.54	127.4	1.156	2793	852.4	1941
201	15.87	124.8	1.158	2794	857.0	1937
202	16.20	122.3	1.160	2794	861.5	1933
203	16.54	119.9	1.161	2795	866.0	1929
204	16.88	117.5	1.163	2796	870.5	1925
205	17.23	115.2	1.164	2796	875.0	1921
206	17.59	113.0	1.166	2797	879.6	1917
207	17.95	110.7	1.168	2797	884.1	1913
208	18.31	108.6	1.169	2798	888.7	1909
209	18.68	106.5	1.171	2798	893.2	1905
210	19.06	104.4	1.173	2798	897.8	1901
211	19.45	102.4	1.174	2799	902.3	1897
212	19.84	100.4	1.176	2799	906.9	1892
213	20.23	98.51	1.178	2800	911.5	1888
214	20.63	96.63	1.179	2800	916.0	1884
215	21.04	94.79	1.181	2801	920.6	1880
216	21.46	92.99	1.183	2801	925.2	1876
217	21.88	91.23	1.185	2801	929.8	1871
218	22.30	89.52	1.186	2802	934.4	1867
219	22.74	87.84	1.188	2802	939.0	1863
220	23.18	86.19	1.190	2802	943.6	1859
221	23.62	84.58	1.192	2802	948.2	1854
222	24.08	83.01	1.194	2803	952.9	1850
223	24.54	81.47	1.195	2803	957.5	1845
224	25.00	79.96	1.197	2803	962.1	1841
225	25.48	78.49	1.199	2803	966.8	1836
226	25.96	77.05	1.201	2803	971.4	1832
227	26.44	75.64	1.203	2804	976.1	1828
228	26.94	74.26	1.205	2804	980.8	1823
229	27.44	72.90	1.207	2804	985.4	1818
230	27.95	71.58	1.209	2804	990.1	1814
231	28.46	70.29	1.211	2804	994.8	1809
232	28.99	69.02	1.213	2804	999.5	1805
233	29.52	67.77	1.215	2804	1004	1800
234	30.06	66.56	1.217	2804	1009	1795
235	30.60	65.37	1.219	2804	1014	1791
236	31.15	64.20	1.221	2804	1018	1786
237	31.71	63.06	1.223	2804	1023	1781
238	32.28	61.94	1.225	2804	1028	1776
239	32.86	60.84	1.227	2804	1033	1771
240	33.44	59.76	1.229	2804	1037	1767
241	34.03	58.71	1.231	2804	1042	1762
242	34.63	57.68	1.233	2804	1047	1757
243	35.24	56.67	1.236	2803	1052	1752
244	35.86	55.68	1.238	2803	1056	1747
245	36.48	54.71	1.240	2803	1061	1742
246	37.11	53.75	1.242	2803	1066	1737
247	37.76	52.82	1.244	2802	1071	1732
248	38.40	51.90	1.247	2802	1076	1727
249	39.06	51.01	1.249	2802	1081	1721

Temp. (°C)	Press. (bars)	Specific Volume (cc/gm)		Enthalpy (J/gm)		
		vap.	liq.	vap.	liq.	evap.
250	39.73	50.13	1.251	2802	1085	1716
251	40.40	49.26	1.254	2801	1090	1711
252	41.09	48.42	1.256	2801	1095	1706
253	41.78	47.59	1.258	2800	1100	1700
254	42.48	46.77	1.261	2800	1105	1695
255	43.19	45.98	1.263	2800	1110	1690
256	43.91	45.19	1.266	2799	1115	1684
257	44.64	44.42	1.268	2799	1120	1679
258	45.38	43.67	1.270	2798	1124	1674
259	46.13	42.93	1.273	2797	1129	1668
260	46.89	42.21	1.276	2797	1134	1663
261	47.65	41.49	1.278	2796	1139	1657
262	48.43	40.79	1.281	2796	1144	1651
263	49.21	40.11	1.283	2795	1149	1646
264	50.01	39.43	1.286	2794	1154	1640
265	50.81	38.77	1.289	2794	1159	1634
266	51.63	38.12	1.291	2793	1164	1629
267	52.45	37.49	1.294	2792	1169	1623
268	53.29	36.86	1.297	2791	1174	1617
269	54.13	36.25	1.300	2791	1179	1611
270	54.99	35.64	1.302	2790	1185	1605
271	55.85	35.05	1.305	2789	1190	1599
272	56.73	34.47	1.308	2788	1195	1593
273	57.61	33.90	1.311	2787	1200	1587
274	58.51	33.34	1.314	2786	1205	1581
275	59.42	32.79	1.317	2785	1210	1575
276	60.34	32.24	1.320	2784	1215	1569
277	61.26	31.71	1.323	2783	1220	1563
278	62.20	31.19	1.326	2782	1226	1556
279	63.15	30.68	1.329	2781	1231	1550
280	64.12	30.17	1.332	2780	1236	1544
281	65.09	29.67	1.335	2778	1241	1537
282	66.07	29.19	1.338	2777	1246	1531
283	67.07	28.71	1.342	2776	1252	1524
284	68.07	28.24	1.345	2775	1257	1518
285	69.09	27.77	1.348	2773	1262	1511
286	70.12	27.32	1.352	2772	1268	1504
287	71.16	26.87	1.355	2771	1273	1498
288	72.22	26.43	1.359	2769	1278	1491
289	73.28	26.00	1.362	2768	1284	1484
290	74.36	25.57	1.366	2766	1289	1477
291	75.45	25.15	1.369	2765	1294	1470
292	76.55	24.74	1.373	2763	1300	1463
293	77.66	24.33	1.376	2761	1305	1456
294	78.79	23.94	1.380	2760	1311	1449
295	79.93	23.54	1.384	2758	1316	1442
296	81.08	23.16	1.388	2756	1322	1435
297	82.24	22.78	1.392	2755	1327	1427
298	83.42	22.40	1.396	2753	1333	1420
299	84.61	22.04	1.400	2751	1338	1412
300	85.81	21.67	1.404	2749	1344	1405
301	87.02	21.32	1.408	2747	1350	1397
302	88.25	20.97	1.412	2745	1355	1390
303	89.49	20.62	1.416	2743	1361	1382
304	90.75	20.28	1.420	2741	1367	1374
305	92.02	19.95	1.425	2739	1372	1366

Temp. (°C)	Press. (bars)	Specific Volume (cc/gm)		Enthalpy (J/gm)		
		vap.	liq.	vap.	liq.	evap.
306	93.30	19.62	1.429	2737	1378	1358
307	94.59	19.29	1.434	2734	1384	1350
308	95.90	18.97	1.438	2732	1390	1342
309	97.23	18.66	1.443	2730	1395	1334
310	98.56	18.35	1.447	2727	1401	1326
311	99.92	18.04	1.452	2725	1407	1318
312	101.2	17.74	1.457	2722	1413	1309
313	102.6	17.45	1.462	2720	1419	1301
314	104.0	17.16	1.467	2717	1425	1292
315	105.4	16.87	1.472	2714	1431	1283
316	106.8	16.58	1.477	2712	1437	1275
317	108.3	16.30	1.482	2709	1443	1266
318	109.7	16.03	1.488	2706	1449	1257
319	111.2	15.76	1.493	2703	1455	1248
320	112.7	15.49	1.499	2700	1461	1239
321	114.2	15.22	1.504	2697	1468	1229
322	115.7	14.96	1.510	2694	1474	1220
323	117.2	14.71	1.516	2691	1480	1210
324	118.8	14.45	1.522	2687	1486	1201
325	120.3	14.20	1.528	2684	1493	1191
326	121.9	13.95	1.534	2681	1499	1181
327	123.5	13.71	1.541	2677	1506	1171
328	125.1	13.47	1.547	2673	1512	1161
329	126.8	13.23	1.554	2670	1519	1151
330	128.4	13.00	1.561	2666	1525	1141
331	130.1	12.76	1.568	2662	1532	1130
332	131.7	12.54	1.575	2658	1539	1119
333	133.4	12.31	1.582	2654	1545	1109
334	135.2	12.09	1.589	2650	1552	1098
335	136.9	11.87	1.597	2645	1559	1086
336	138.6	11.65	1.605	2641	1566	1075
337	140.4	11.43	1.613	2636	1573	1064
338	142.2	11.22	1.621	2632	1580	1052
339	144.0	11.01	1.629	2627	1587	1040
340	145.8	10.80	1.638	2622	1594	1028
341	147.7	10.59	1.647	2617	1601	1016
342	149.5	10.39	1.656	2612	1609	1003
343	151.4	10.18	1.665	2606	1616	990
344	153.3	9.983	1.675	2601	1624	977
345	155.2	9.784	1.685	2595	1631	964
346	157.1	9.587	1.695	2589	1639	951
347	159.1	9.391	1.706	2583	1647	937
348	161.1	9.197	1.717	2577	1654	923
349	163.1	9.005	1.728	2571	1662	908
350	165.1	8.813	1.740	2564	1671	893
351	167.1	8.623	1.753	2557	1679	878
352	169.2	8.435	1.765	2550	1687	863
353	171.3	8.247	1.779	2542	1696	847
354	173.4	8.060	1.793	2535	1704	830
355	175.5	7.873	1.807	2527	1713	814
356	177.6	7.688	1.822	2518	1722	796
357	179.8	7.502	1.839	2510	1731	778
358	182.0	7.317	1.855	2501	1741	760
359	184.2	7.131	1.873	2491	1751	741

Temp. (^{o}C)	Press. (bars)	Specific Volume (cc/gm)		Enthalpy (J/gm)		
		vap.	liq.	vap.	liq.	evap.
360	186.5	6.945	1.892	2481	1761	721
361	188.7	6.759	1.913	2471	1771	700
362	191.0	6.571	1.934	2459	1781	678
363	193.3	6.381	1.958	2448	1792	655
364	195.7	6.190	1.983	2435	1804	631
365	198.0	5.995	2.011	2421	1816	606
366	200.4	5.797	2.041	2407	1829	578
367	202.8	5.593	2.076	2391	1842	549
368	205.3	5.380	2.114	2374	1857	517
369	207.7	5.162	2.159	2354	1873	482
370	210.2	4.925	2.213	2332	1890	442
371	212.8	4.671	2.280	2306	1911	395
372	215.3	4.380	2.369	2274	1936	338
373	217.9	4.019	2.509	2229	1971	258
374	220.5	3.404	2.880	2140	2049	91
374.136	220.9	3.155	3.155	2099	2099	0

GLOSSARY OF SYMBOLS

We have attempted to use symbols most commonly employed in the literature even though in some cases this involves some overlap--for example in the use of α as a fugacity coefficient in gas calculations and as an isotope fractionation factor. Note that in expressing analyses and for clarity in some equations valence symbols may be omitted. Text format has required raising some subscripts, and for clarity, omission of some valence symbols e.g., HCO_3^- may sometimes be written HCO_3 or HCO3 and H_2 as H2. Units used in this text are specified in parentheses and for some symbols the chapter where they are first used is indicated. A table of useful constants and common conversion factors follows this Glossary.

Subscripts

aq component in aqueous solution

g gaseous component

i one of a set of i species or gaseous components

j one of a set of j reactions

l liquid phase

o initial concentration of specified component

r a reference state

v vapor phase

w water

English characters

a_i activity of component i (Chapter 1)

$\mathring{a}_i$ ion size parameter (10^{-8}) of the Debye-Hückel equation (Chapter 1, Table 1.1)

b Deviation function of the extended Debye-Hückel equation (Chapter 1)

c_p specific heat of an element or compound at constant pressure (joules/gm)

f_i fugacity of a gas component (i) or specified component, eg., bars CO_2

g acceleration due to gravity at the earth's surface (9.78 m/sec^2)

h depth (meters)

m molality (moles/kg solution) (Chapter 1)

$\underline{m}$ mass

$\underline{n}$ number of steps in a multistep process (Chapter 4)

n_e number of electrons exchanged in a redox half cell reaction (Chapter 8)

pE -log electron (Chapter 8)

pH pH = -log a_{H^+} (Chapter 1). The temperature of measurement may be given as a subscript.

pK pK = -log K. A convenient method for the expression of equilibrium constants

q heat transfer between a system and its surroundings (Chapter 2)

r Ratio of the concentrations of a specified gas in two samples (Chapter 4)

t temperature in degrees Centigrade (°C)

t_{AM} amorphous silica geothermometer temperature - no steam loss (Chapter 3)

t_{CH} chalcedony geothermometer temperature - no steam loss (Chapter 3

t_{NAK} Na to K ratio geothermometer temperature, with author of calibration equation identified by initial (see Table 3.2, Chapter 3)

t_{NKC} or t_{NaKCa} NaKCa geothermometer temperature (Chapter 3)

t_{QA} quartz geothermometer temperature with allowance for adiabatic boiling (Chapter 3)

t_{QC} quartz geothermometer temperature - no steam loss (Chapter 3)

x water fraction in a two phase mixture (Chapter 2)

xg or x_g gas/steam molal ratio (Chapters 4 and 5)

y steam fraction in a two phase mixture (Chapters 1 and 2)

z_i charge on the ith ionic species

A Debye-Hückel parameter (Chapter 1, equation (7), and Table 1.2)

$\underline{A}$ Alkalinity (Chapter 7, equation (18))

B Debye-Hückel parameter (Chapter 1, equation (7), and Table 1.2)

B gas distribution coefficient for liquid-vapor mixtures (Chapter 4)

B' an average value for the gas distribution coefficient B over a specified temperature interval (Chapter 4)

C_i concentration of component i in either molal or specific units (e.g., mg/kg)

E cell potential (volts) (Chapter 8)

E° standard half cell potential (volts) (Chapter 8)

Eh oxidation potential relative to hydrogen electrode (volts) (Chapter 8)

F Faraday constant (96487 joules/volt) (Chapter 8)

FE fraction equilibrated (defined in Chapter 10)

GFW gram formula weight

H specific enthalpy (joules/gm). In this text H usually refers to water as liquid or vapor or a two phase mixture and values are given in Appendix III.

HTOT molal sum of ionisable protons (Chapter 7, equation (19))

$\underline{I}$ integrated isotopic enrichment between a reference and final state (Chapter 10, equation (23))

$\underline{I}_n$ integral related to calculation of water fraction during continuous steam loss (Chapter 10, equation (27))

I Ionic strength (moles kg^{-1}) $I = 1/2 \quad \Sigma\, m_i z_i^2$ (Chapter 1)

J Joule

K" Analytical concentration quotients for the specified gas reactions at T and specified reservoir steam fraction (Chapter 11)

K_H Henry's Law constant (bars/mole fraction)

K_j Equilibrium (or mass action) constant for specified reaction (Chapter 4); (subscript refers to principal reactant e.g., K_{H4SiO4})

P pressure (bar)

R gas constant (8.3144 joules/deg mole)

$\underline{R}$ ratio of concentrations of two isotopes, e.g., $^{18}O/^{16}O$ (Chapter 10)

T temperature in degrees Kelvin (°K)

$\underline{V}$ molal volume (cc/GFW)

V specific volume (cc/gram)

W/R water to rock mole ratio (Chapter 10)

Xi or X_i mole fraction for component, i (Chapter 1)

Z compressibility factor for specified component or phase (cc/bar)

Greek characters

α_i fugacity coefficient for component i

α fractionation constant for an isotope exchange reaction (Chapter 10)

β_n cumulative formation constant for a metal complex; n is the number of ligands (Chapter 9)

β a constant in the NaKCa geothermometer equation (β =1/3 or 4/3, Chapter 3)

γ_i activity coefficient of component i (Chapter 1)

δ del notation used to express isotope concentrations relative to a reference standard. (Chapter 10, equation (2))

Δ sum of charged weak acid dissociation products (Chapter 7, equation (16))

$\Delta^{18}O$ oxygen isotope fractionation geothermometer (Table 3.2)

ΔG_R Gibbs Free Energy change of a reaction at specified temperature and pressure (Joules) (Chapters 1 and 6)

ΔG°_R Standard Gibbs Free Energy change of a reaction (Joules)

ΔG°_f Standard Gibbs Free Energy change for the formation of a component from its elements (Joules/mole) (For discussion of standard state see Chapters 1 and 6)

ρ density (gm/cc)

ψ flux of high temperature steam (kg/unit time)(Chapter 10)

ω flux of low temperature steam (kg/unit time)(Chapter 10)

Useful constants and conversion factors

The following are conversions of metric units into other units commonly used in geothermal studies (m = meters, kg = kilogram, s = second, t = tonnes h = hours, J = Joules)

	Typical Unit	Equivalent
length	m km	= 3.28 feet = 100 cm = 0.62 miles
volume	liter	= 10^{-3} m^3 = 0.264 U.S. gal = 0.22 Imperial gal
volume flow	liter/s	= 10^{-3} m^3/s = 15.8 U.S. gal/min = 13.2 Imperial gal/min
mass	kg t (tonne)	= 2.2 pounds = 32.15 troy ounces (oz) = 1.1 U.S. ton = 10^3 kg
mass flow	kg/s	= 3.6 t/h = 7.92 x 10^3 lb/h
density	kg/m^3	= 6.24 x 10^{-3} lb/ft^3 = 1/V
pressure	kPa	= 10^{-2} bar = .00987 atm = 0.010197 kg/cm^2 = 0.145 psi (pounds per square inch) = 7.5006 mm Hg
temperature	°C	= 0.556(°F + 32) = T -273.15
energy	kJ	= 0.984 x 10^{-3} Btu = 1/4.184 kilocalories(thermochemical) = 1/4.1864 kilocalories (international) -- used in the Steam Tables in Appendix III
specific enthalpy	kJ/kg	= 0.43 Btu/lb = 1/4.186 kcal/kg = 1 Btu/lb = 1.8 kcal/kg
permeability	m^2	= 10^{12} darcy
power	J/sec	= 1 watt
Avogadro's number	mol/GFW	= 6.022094 x 10^{23} mole/GFW
Faraday	J/volt.electron	= 96487 J/volt.electron
Gas constant, R	J/(mole °K)	= 8.3144 J/mole K = 1.987 cal/mole K = 8.309 x $10^{-5}m^3$ bar/mole
Time	sec	= 3.2 x 10^7 sec/year

FIGURE, TABLE, AND PLATE CREDITS

Figure	1.1	Reprinted with permission from Economic Geology Publishing Co.
	2.1a	Reprinted by permission of John Wiley & Sons, Ltd.
	2.1b	Reprinted with permission from Academic Press, Inc.
	2.2b	Reprinted with permission from Economic Geology Publishing Co.
	2.3	Reprinted with permission from Elsevier Science Publishers B.V.
	2.4	Reprinted with permission from Elsevier Science Publishers B.V.
	2.5a	Reprinted with permission from New Zealand Journal of Science
	2.5b	Reprinted with permission from Economic Geology Publishing Co.
	2.8	Reprinted with permission from Elsevier Science Publishers B.V.
	3.1	Reprinted by permission of John Wiley & Sons, Ltd.
	3.2	Reprinted by permission of John Wiley & Sons, Ltd.
	3.3	Reprinted with permission from Wyoming Geological Association
	3.4a-c	Reprinted with permission from Wyoming Geological Association
	6.1	Reprinted with permission from Elsevier Science Publishers B.V.
	6.5	Reprinted with permission from Economic Geology Publishing Co.
	7.2a-d	Reprinted with permission from Pergamon Press, Ltd.
	8.2	Reprinted by permission of John Wiley & Sons, Inc.
	8.3	Reprinted with permission from Economic Geology Publishing Co.
	9.1	Reprinted with permission from Pergamon Press, Ltd.
	10.1	Reprinted with permission from Elsevier Science Publishers B.V.
	10.4	Reprinted with permission from the International Atomic Energy Agency
	11.1-11.3	Reprinted with permission from Pergamon Press, Ltd.
	11.5	Reprinted with permission from American Journal of Science
	11.6	Reprinted with permission from Pergamon Press, Ltd.
	11.7	Reprinted with permission from Pergamon Press, Ltd.
	12.3	Reprinted with permission from American Mineralogist
	12.8	Reprinted with permission from Economic Geology Publishing Co.
	12.9	Reprinted with permission from Journal of Geology
	13.1	Reprinted with permission from Pergamon Press, Ltd.
	13.2	Reprinted with permission from R. Hurtado, Instituto de Investigaciones Electricas, Centro Cerro Prieto
	13.3	Reprinted with permission from American Journal of Science
	13.4	Reprinted with permission from Springer-Verlag, Heidelberg
	13.5	Reprinted with permission from the American Society for Testing and Materials, 1916 Race Street, Philadelphia, PA, 19103
	14.2	Reprinted by permission of John Wiley & Sons, Inc.
	14.3	Reprinted with permission from Economic Geology Publishing Co.
	14.4	Reprinted with permission from Pergamon Press, Ltd.
	II 2.1	Reprinted with permission from Elsevier Science Publishers B.V.
	II 3.1	Reprinted with permission from Wyoming Geological Association
Table	6.2	Reprinted with permission from Pergamon Press, Ltd.
Plate	2	Reproduced with permission from the Rotorua Museum
	5	Reproduced with permission from Pan Eimon

INDEX

Three categories of entries are distinguished--text pages, figures, and tables -- each under its own column; appendix entries are included. Entries listed by chapter number without specific page(s) are discussed throughout that chapter.

A separate Locality Index follows this general Subject Index.

Locality Index